Biological Beekeeping

The Way Back

by Ed & Dee Lusby

Biological Beekeeping—The Way Back

ISBN: 978-161476-104-4

X-Star Publishing Company
Nehawka, NE 68413, USA

648 pages
113 illustrations
12 pt typeface

Dedication

I am dedicating this book to my husband Edward, since in passing on, while being held in my arms, he asked me to bring it home for teaching what we'd been doing, for the betterment of all beekeeping worldwide, to put honey bees back into harmony/alignment with Nature again, and her ways, for clean evolution going forward for all life. So this is my dedication for that is what I do, and wanting this book now finally out to as many beekeepers as possible for reading, while I do another with more pictures and information Edward wanted taught to others for helping feed a hungry world, just like my forefathers and his did and wanted, for life is working in harmony and people helping people.

Foreword

Well, this book and my doings are for helping all beekeepers worldwide for how to keep clean and sustainable honey bees without chemical treatments of any kind, including acids and essential oils, and then only feeding real honey and pollen, while breeding in accordance with Nature locally/regionally with smaller cell size as breaks out by latitude and altitude, with local outmating doings...while discussing with them daily, how to do basic field workups of individual hives throughout the active beekeeping year, along with how to balance whole beeyards into working as a whole collective working unit, as honey bees are social insects/animals.

About The Book

This book is about how to move from conventional beekeeping methods to a system of natural beekeeping. The content of this book has been available for many years now on Beesource.com. It is offered here in the form of a book. These are Dee Lusby's methods unedited, in her own words. It is organized into several sections. The first is Dee & Ed's method of getting from mainstream beekeeping methods, back to biological methods. The appendices are sorted into research and articles in which Dee and Ed are credited, articles about Dee and Ed, a section of historic documents on cell size and the origins of the idea of enlarging the bees, and a section on other supporting documents on Varroa and treatments and anything that didn't fit in one of the other sections.

Table of Contents

Italian honey bees were originally brought to this country in the late 1800s along with several other European strains to replace the common black bees brought by our forefathers in the settling of our Way Back to Biological Beekeeping.

Italian honey bees were originally brought to this country. This is what the beekeepers in the United States have been told for generations. But depending upon whose heritage you are describing, so changes the story ever so slightly. However, it wasn't that the other black strains were bad also, that caused Italian honey bees to be favored, but the fact that common everyday beekeepers couldn't tell them apart from other blacks, as it was much more difficult to detect a mixture of Caucasians or Carniolans, while, with the Italians, if the yellow bands are not present one can immediately know that they have been mixed or hybridized.

The original real Italian bees, the bee today that has almost lost its identity among beekeepers in the United States was originally three-banded, leather-colored uniformly in color, bred true to form, color and habits. How unfortunately today often we see ads for queens and honey bees with no reference to strain, traits, etc., except to say that "boy are these good bees, and at a price you can't pass up."

Our present day bee breeders are not so much to blame for placing emphasis on breeding in good looking yellow characteristics into the Italian honey bees,..."the industry demanded it". Nearly all queen breeders went with Italian breeding because the old German Black Bees were difficult to handle and more importantly because a hybridized yellow strain would make more honey than one that wasn't. All this means is that, good business is the ability to give your customers what they think they want and don't mention the shortcomings.

One must always remember that the common beekeeper can be accommodated with any race or strain he may desire, only the real trick is to know what you are looking for. As the pendulum swung during the late 1800s and early 1900s in this country and around the world to a bee different in color from what was commonly kept so that any amateur beekeeper could distinguish from their own then prominent common black bees, bear in mind then, that yellow strains only became popular because they were different in color, not that they were all around better. Now at the end of our century, the late 1900s, the pendulum again is swinging. But, instead of common black bees beekeepers wish to get rid of, it is now the dreaded yellow colored Killer Africanized Honey bees in the Americas. *How true the old saying really is: "History really does repeat itself."*

Beekeeping Today in Southern Arizona

Today during our present period of economic liquidations and retrenchments we are beginning to realize that in an increase in the quantity and quality of our production per hive lies much of our hope for a profitable return on investment from our bees. Food Safety is much again in the public eye, both here in the United States and around the world, having been buried for a few years behind the need to find remedies for maladies from parasitic mites, scavengers, and secondary infectious diseases. This has been compounded as we beekeepers try to reduce the cost of producing a pound of honey or generate more income by migratory pollination of food crops, in order to meet our future new costs associated with externally and eternally having to confront the so-called Africanized Honey Bees which some have said have gained a foot hold on our American continents.

It is my firm belief that in the selection and adaptation of the various *naturally sized* races and strains of honey bees to our own individual localities lies our best chance for our industries continued survival. So it is a matter worthy of new particular commendation to note and reestablish large interest in the keeping of biologically kept natural races and strains of honey bees...even heated arguments. Nothing but good can come out of what I now write, for if even one beekeeper can hold the line and continue to succeed no matter what natural strain or race he may desire to keep without the use of chemicals, essential oils, and antibiotics, etc., we have all won.

In order to avoid misunderstandings, I shall write from the standpoint of Southern Arizona area conditions and confine my thoughts to a discussion of necessary data and observations gleaned in a semi-arid, but still temperate location. As I write, I hope certain self-learned principles will become obvious which I hope may become helpful in applying the results we have achieved in retrogressing our own honey bees back onto a natural system of biologically sound beekeeping without the use of chemicals, essential oils, and antibiotics or foreign feed/food supplement, so other beekeepers in other localities throughout our nations temperate zones can succeed in replicating also.

The Southern Arizona desert areas where our apiaries are located is a much varied topography ranging from the flats of the Sonoran Desert to the fast rising mountains that encircle our local valleys and plains. Our elevation ranges from about 2500-4600, elevation. The average growing season varies from a low of 262 days in the Tucson area to 318 days in the Phoenix area with fall frosts rarely beginning before Thanksgiving. These flats are an arid belt, surrounded for the most part by mountains. The honey plants are largely dependent upon sporadic rains and a brief monsoon season normally occurring during July and August (sometimes lasting until the middle of September). Farmers must irrigate their crops in order to raise crops. When honey flows do come, especially outside of the farming areas, they rarely last more than six weeks and many only last two to three.

Consequently, beekeepers here must have stock that can anticipate a flow and pack stores away fast. The summer Spring/Summer season is hot with temperatures often over 100° F (38° C). The Fall/Winter are relatively mild and cool. The mean maximum temperature for the warmest month runs about 95° F (35° C). and the mean minimum temperature for the coolest month runs about 50° F (10° C)., happening during July and January respectively. Keep these means in thought, for they are very important to beekeeping and we will be going over their full meaning and impact towards honey bees later!

Although quite arid here, normally fall rains are sufficient enough to allow for spring growth with the main honey flows coming on during the end of April, continuing through May and ending the first part of June. Unfortunately for us, our main rainy season hits during July and August with the summer monsoons after the main flow is over, but in time to help cool our hives off enough and grow enough vegetation so the bees can make a secondary short honey flow for surplus, allow for requeening, and the accumulation of sufficient stores for winter. The major honey flow source comes from acacia, mesquite and catsclaw with a varied season blend in the later Fall months of greasewood, rabbit and burrow weed, wild Indian buckwheat, pigweed, poppy, golden eye, daisy etc.

The Way Back to Biological Beekeeping...The Saga Continues

Articles have appeared from time to time advocating various races and strains of honey bees as superior to any other. Yet, at the same time beekeepers complain about the undesirable qualities found while trying other races or strains. It is interesting to notice the wide discrepancies between their various statements comparing one rational to the other. Why should beekeepers differ so radically? Is it because the various honey bee races and strains vary in productiveness about as much as in disposition or is there some underlying causative effect taking place? I believe the latter, *that there is an underlying causative effect taking place.*

Some beekeepers feel that bees must necessarily be vicious in order to be productive. This old way of thinking going back to the turn of the century is, however, a gross mistake showing absolutely no correlation between viciousness and productiveness. It is not necessary to have vicious honey bees to make large amounts of honey and pollen. Besides, one certainly can't work more violent bees than one can work gentler ones. Besides, as our historical pendulum again is swinging and beekeepers are under pressure to not keep today's vicious so-called "Killer Africanized Honey bees in the Americas" beekeepers are again turning to the black races and strains they abandoned so long ago. Yet, is this fair to Italian honey bee bee breeders? Could this be another bum rap going full circle? We have found no evidence that smaller naturally sized honey bees are more vicious than their artificially oversized domesticated siblings. In fact, the opposite was found. We have found smaller naturally sized honey bees, no matter what the color variation, to be gentler. Therefore, one should ask and we do: "Is there an underlying causative effect taking place?"

One of the safest and best ways to obtain good honey bees is from one's own best honey and pollen gathers from one's own apiaries or to purchase them from a good queen breeder. We have several good queen breeders here in the United States and I am sure most countries overseas have several good queen breeders too. However, beekeepers sending away for new stock gamble on most any queen or package that has traveled a long way to be composed of bees that will produce well. They also gamble upon the bees safe arrival without transportation problems that could affect the performance of the bees they have bought. Further complicating, many beekeepers then have no follow-up plan keeping records to show that their apiary was much improved temper-wise, production-wise, etc., with feed-back to the breeder the bees or queens were purchased from.

There is one thing to remember when you read claims for different kinds of races and strains of honey bees: *That much of the difference is in the purchasing beekeeper and in his management style, and not in the purchased bees or it could also be an underlying causative effect not taken into consideration.* Again I now state: I have found nothing wrong with the honey bees and queens being produced by our nation's bee breeders. They are some of the best to be found around our world. Yet complaints abound today about inferior stock and inferior queens. Again, could there be an underlying causative effect not being taken into account in the keeping of today's modern domesticated honey bees? Probably, but what is it?

Perhaps, investigation of our past for clues will help with answers. How often has this occurred to help those in distress learn from past mistakes so as not to repeat them in a continuous recycling mode? We are now at the end of our century and the pendulum is swinging. It is time for beekeepers to relearn their traditions and history, and to undo and retool that which is wrong for our honey bees survival. We must follow the long hard way back to biological beekeeping to overcome our problems. It's not hard, but it must be done. Chemicals, essential oils, drugs, artificial feed, size, and to some degree inbreeding, are killing our nations industry and it must be reversed.

It is important for beekeepers today to keep in mind that there are basically only three major races of honey bees in Europe: 1) Carniolans 2) Italian and 3) Mellifera (the dark bee). Caucasian honey bees are from Eurasia. These are the main races and strains of honey bees to be found in our United States. All of these races are excellently adapted to the original climate and nectar plants of their respected areas, but unfortunately if exported to another area or climate, let's say a different continent, these major races very soon break down by natural selection and are hybridized, if left by themselves without man's interference, into new strains and then races, fully adapted to the new climate and nectar plants of their adopted new geographical areas. All three of the major races of bees from Europe have played key roles in beekeeping in the Americas along with the lone Caucasian bees from Eurasia (Asia Minor).

If one reads back over the years of acquired beekeeping knowledge, one learns through Mackensen and Roberts in the 1940s, that every major breed of animal and fowl that modern man knows of today was originally introduced to the European mainland during the 1,000 year plus reign of the Roman Empire. Bearing this in mind, one would also assume that the Romans introduced yellow hot-weather (tropical zone) honey bees to the European mainland too. If so, then probably other strains of bees throughout the Mediterranean area were imported also as early civilizations traded wares back and forth.

Through reading, beekeepers can learn that Egyptians kept honey bees up and down the Nile river prior to 2600B.C., and early prehistoric man through his drawings, left behind in caves in Africa, show that man kept bees long before this. It is also a known fact that the color of gold was a worshiped color in ancient times in temples and at numerous ceremonies. To keep animals the same color was a blessing to be sought. That man kept bees up and down the upper and lower Nile river, and followed the planting seasons of many crops via Nile barges, just as we migrate bees today from crop to crop, only instead using modern day vehicles is also a well-known fact.

Easy it was for early civilization to bring the yellower hot-weather (tropical zone) honey bees out of Africa and scatter them throughout the Mediterranean states, only to have them very soon break down by natural selection and be hybridized into strains and then races to create the various yellow strains of honey bees to be found throughout the Mediterranean area. Not to say that the honey bees could not have come north on their own, just like the purported so-called Africanized honey bees that came north through the Americas in today's age, but it's probable that man did not slow down their progress any over the centuries, especially if one looks at how today beekeepers spread bees around in migratory vehicles. While the various strains and races of European, North African, and Eurasia (Asia Minor) honey bees took several centuries for Nature to perfect through natural selection, this is not true for our American continents.

First there is the issue as to whether or not honey bees are native to the Americas. This fact has never been qualitatively proved either pro or con, but popular belief, the politically correct answer is, that the American tropics, void of true honey bees, were artificially colonized by European races. These European races and strains were able to survive because they had little competition. Today in modern times it is claimed that a true tropical honey bee has recently been introduced and adapted, from South Africa, and the earlier introduced European races had no chance in tropical conditions of the Americas to survive if exposed to natural selection. But is this a true assessment of the situation relative to natural selection in Nature? I think not. It does not parallel history in Europe and other parts of the world as bees were spread by man. But note, it is politically correct and modern popular thinking! I can see it being accurate in a Tropical Zone this assessment, but not in a Temperate Zone. History dictates otherwise.

This would be true in the tropics because the purported Africanized honey bees would have truer hot-weather (tropical zone) characteristics and traits, while bees kept artificially of European descent would be either races and strains of cold weather (temperate zone) characteristics and traits or a hybridization of hot/cold weather traits which would be at a breeding disadvantage in a pure Tropical Zone.

Africanization of honey bees is a political issue that must be addressed by beekeepers on the way back to biological beekeeping because it is a force that is shifting the pendulum back from a centuries tradition in the United States of selecting for yellow Italian honey bees away from the Old Black Bees of Europe. It (Africanization) has now caused our industry to shift away from breeding yellow strains to the darker black strains of the true Temperate Zones: i.e. Carniolan and Caucasian, etc. Because of this pendulum swing, compounded with a problem of parasitic mites and accompanying secondary diseases, necessitating the use of chemicals, essential oils, and antibiotics, and an underlying causative affect not apparent to the common everyday beekeeper, we have major problems today in keeping honey bees alive.

Beekeepers must understand that names on honey bees technically mean nothing to Nature in the real world. They are notions given by man. Neither color is of real use in discriminating races and strains of honey bees because Nature adapts over a given period of time. Having the same color does not mean that bees are related. The black color of bees in Madagascar has nothing to do with the black bees in Norway, and the yellow of Saharians has nothing with that of Italians. Other characteristics have to be used to discriminate many of the world's bee races and strains.

While all hot-weather (Tropical Zone) type bees are various shades of yellow, and all cold-weather (Temperate Zone) type bees are various shades of brown to black, and hybridized honey bees exhibit traits of each other where the cuff Temperate Zones are located (where yellow and black bee areas overlap), each are different in characteristics and traits because of the climates and nectar plants of each of their respected areas, that dictated that this had to be so in order to survive, by natural selection of the fittest. In keeping this in mind, beekeepers must remember that latitude equates with altitude in defining Tropical and Temperate habitat for honey bees.

However, since popular belief is that our so-called Africanized honey bees are yellow, then it is an excellent idea to work exclusively with darker (black/brown) cold-weather bees in ones bee yards here in the United States in order to make any hybridization instantly evident. Thus to win against the purported Killer honey bees, a theoretical truer yellow hot-weather bee of the Tropics, beekeepers must use a bee of a dramatically different color, which is a truer cold-weather honey bee of Northern latitudes to win a breeding battle in a Temperate Zone. This would be because not only would a bee of a dramatically different color give us instant recognition of hybridization evidence in our beehives, but would also give beekeepers concerned, dramatically different enough racial characteristics and traits to be able to either throw-off the so-called Africanized honey bees through out-of-season cold weather stress mating; or a better chance to hybridize the Africanized honey bees hot-weather traits with the milder cold-weather traits or northern bees, for a newer more gentler strain similar to the way racial honey bee strains and races developed in the Mediterranean area thousands of years ago with the help of the Roman Empire and early Egyptians thus giving rise to our Modern Italian honey bees so popular for so long.

Now, since the so-called Africanization we hear about is almost like a doomsday honey bee going through our American continents and in the real world this does not actually happen, then there must be another explanation for what is going on with the racial spread of these aggressive honey bees. *Is there an underlying causative effect taking place to artificially overturn nature's natural selection of new strains and races with adaptation to new climates and nectar plants of their adopted new geographical areas?* We shall have to investigate this aspect in our writings here, for in the real world in our cuff zones (where tropical meets temperate), Nature never slams shut the breeding door of the seasons putting one race or strain of honey bee more dominate over another or one could not have evolution going forward or into reverse to correct problems of survival of the fittest. Instead, in Nature there is a gradual seasonal change-over of climate, a transition zone of breeding overlap that is always there taking place year-round allowing for interracial bee breeding pressure as all elements dictate that follows centuries of Earth, Fire, Wind, etc., beliefs of many cultures worldwide.

If we believe that there is an underlying causative effect not being taken into account in the keeping of todays' modern domesticated honey bees, causing the pendulum to swing back to black strains and races of honey bees for popular keeping again, then will this swing alone correct the problem? We *do not believe so if there is an underlying causative effect story not being told that does not follow reality.* We must therefore investigate the past for clues that will give us answers. After all, how often has this occurred to help those in distress learn from past mistakes so as not to repeat them in a continuous warp loop?

It is therefore time for beekeepers to relearn of their traditions and history and to undo and retool that which is defective and wrong to overcome our problems. We must follow the long hard way back to biological beekeeping to overcome our problems without the over use of artificial inbreeding (nothing in Nature inbreeds for long without evolution collapse), chemicals, drugs, essential oils, and artificial feed. Everything must be done naturally or as close to natural as we can if we are to alleviate and correct our problems of parasites, secondary diseases and so-called Killer Africanized Honey bees to put Nature back into synchronization.

We have talked a bit about movement of animals by the Roman Empire now let's go over a little bit about some needed bee traditional history that is relevant and necessary to solving today's problem and could be a major underlying causative effect bringing all these situations about, *namely comb size and its ramifications.*

Honey Bee Comb: Brief History, Size and Ramifications Part 1

From very early times, the comb built by honey bees has been studied and admired as a solution, to the problem of combining light weight and great strength, to be duplicated in the building of structures. The first known research on the structure on honeycomb dealt with the hexagonal form of the cells by Zenodorus of Sicily. This was done in the 2nd century B.C., right after the time of Archimedes. Zenodorus proved back then that, of the three regular figures that will completely fill a plane surface (namely, the equilateral triangle, the square, and the regular hexagon), the hexagon has the greatest content for a given circumference.

Pappus of Alexandria later, around A.D. 500 copying from Zenodorus, also found that bees wisely choose the hexagon form for the cell-mouth which they suspect will contain and hold the most honey for the same expenditure of wax in its construction. He was the first one to put forth the suggestion that honey bees economize wax, a notion believed for many years, though in today's world now known to be far removed from the realities of the matter. After Pappus there was no known study of honeycomb construction until a person by the name of Kepler, an astronomer in 1611, published a very good cell description. He was credited with being the first to notice the rhombs at the base of individual cell construction.

In the 1700s there was a terrible misunderstanding / calculation concerning the measurement of the angles of the rhombuses in the pyramidal bottoms of comb cells. It started in 1712 and continued for several years during the century.

To wit: An Italian, Maraldi, an astronomer, studied bee cells and measured the bottoms of the cells and found them to be approximately equal to one another. He then calculated that, if these angles really were equal, they must each be of about 109° 28′. Anne D. Betts in an article in July 1921 related the story further this way, Maraldi is an "awful warning" to us all to express ourselves quite clearly, so as to avoid all danger of being misunderstood. By using somewhat involved phraseology, he succeeded in conveying to the French naturalist Reaumur (some years later) the idea that he had found this value of 109° 28′ by measurement! A feat which, as several writers have since remarked, was impossible with the instruments then in existence, even if the cells were regular, which they are not.

Reaumur suspected that the bees economized wax, so asked a mathematical friend, Koenig, to work out the "problem of the bee's cell" above referred to. Koenig did so and gave the larger angle of the rhombs as 109° 26′. Later investigations showed that 109° 28′ was the correct answer (to the nearest minute and that Koenig had made a slip in his arithmetic).

Kent L. Pellett in an article in June 1929 added more information to the story when he wrote, "Reaumur marveled that their economy came so close to perfection." But other scientists were troubled that the bees came so close to the correct angles and yet missed them. They took the trouble to solve the problem for themselves, with the same result as that of Koenig. But the bees refused to correct their error, slight as it was, and continued to construct the cell bottoms with the old angles.

In later years there was a shipwreck. Investigation of the accident showed that the captain had steered off his course due to calculation from logarithm tables that were defective. The tables were corrected to prevent further error. Then it was discovered that these were the same tables from which Koenig had made his calculations. The cell bottom problem was again solved, with the corrected tables, and this time the same angles were obtained that the bees had always used. The end effect was that the bees were right and the mathematician wrong.

The first artificial comb foundation was made in Germany in 1842 by Gottlieb Kretchmer. It was made by a pair of engraved rollers, and starch was used to prevent the wax from adhering to the rollers. The device consisted of a strip of tracing linen, coated with a composition of white wax and starch, and upon which the comb-foundation or base of the cells were impressed, by passing it through a pair of engraved rollers.

From there others followed, namely Jean Mehring (Dutch). In 1857 he used pure wax cast between metal molds, and A.I. Root (USA) in 1876 first used a metal roller press. Otto Schenk in 1872 produced and showed foundation with projecting starters for the side walls and John Long (USA) in 1874 produced a similar product. D.S. Given (USA) about 1879-1881, produced wired foundation made in a press, but it was not until 1892 that E.B. Weed (USA) produced sheet wax in long lengths for use between rollers. All this advancement in the making of artificial comb foundation set the stage for our present century's achievements in technology relative to modern beekeeping, as well as, all of today's pressing problems of parasitic mites and their associated secondary diseases.

I would say our present era of problems began around 1891 in Belgium with the introduction of artificial comb foundation with 920 cells to the square decimeter which would equate to about midway between 4.6 cm and 4.7cm for 10 worker cells. The beekeepers there all adopted this size of cell. The experts of that time believed that it was advantageous to produce as many bees as possible on the least possible surface of comb. Thus there was said to be a premature narrowing of the cells throughout Belgium, and at the end of a few years the bees were miserable specimens. (We could say that this was then the opposite of today's problem of bigger is better.)

It was then that to combat so harmful a tendency that an idea was born with a proposed magnificent final (which we are still playing out today). A Prof U. Baudoux of Belgium published an article in Progress Apicole in June, 1893, advocating the use of larger comb cells, as a result of experiments duly described. It seems Prof Baudoux wanted to rear bees of extraordinary vigor, able to forage over a more extended flight-radius and to visit a multitude of flowers the nectar of which was, then (and probably still is), out-of-reach of their tongues.

He experimented with cells up to the limit of 750 cells per square decimeter, the sizes of cells which he obtained by stretching wax comb foundation. Then encouraged by his experiences, he wished to do still better—*"to go to the bounds of possibility"*. (Here is where our current modern-day problems begin with parasitic mites and their secondary diseases.)

Prof Baudoux experimented with various sizes of foundation per the square decimeter, namely: 750, 740, 730, 710 and down to 675. He also experimented with various ways of measuring cells and devised his own measurement system for it. (Unfortunately, there was no corresponding correlation chart made for his devised measurement system, vs. the traditional way of measuring comb foundation that had been in use for over 2,000 years back to before the time of Christ, so beekeepers could go back and forth between the two measurement systems.)

Prof Baudoux was so successful with his writing and his experiments, and so convincing, that manufacturing houses all started selling foundation with enlarged cells and claiming good results for the use of the same. Most of this work was done around the late 1920s through the 1930s and 1940s. (The result has been that this process of bigger is better with its resultant selling has never stopped, and continues up to modern day to the detriment now, that only enlarged oversized foundations (well beyond the bounds of possibility for bigger honey bees as envisioned by Professor Baudoux) are now only sold and standardized large at that, i.e. 5.7cm for 10 worker cells being about the largest).

Could this continuing trend towards bigger is better be an underlying causative effect creating todays problems of parasitic mites and their accompanying secondary diseases? Probably. But on what evidence would one place such a thought?

Unfortunately for us all, it was not the large cell in itself for which Professor Baudoux was working, but rather instead for the selection of a better bee. *It is important for today's beekeepers to remember the time frame here, and its place in history, and also the fact that the measurement system devised by Professor Baudoux was not correlated with traditional measurement methods. Why? Because, an underlying causative effect for today's parasitic mite problems and their accompanying secondary diseases was born here!*

Equally unfortunate for us all is the fact that Prof Baudoux was a follower of what is called Lamarckian Theory, and believed that it was possible to improve the honey bee permanently, by giving her the chance to grow larger in each succeeding generation. However, to a follower of Darwinian or Mendelian theory, this is indeed a very uncertain doctrine; and the bees themselves appear to confirm this criticism (even today any beekeeper can compare the feral to those bees confined domestically), since even back in the early 1900s it was common knowledge that they tend to retrogress in the size of the worker comb cells they build when let alone, back to what is *natural* as put forth by the laws of Nature and not by man's artificially created and politicized rules.

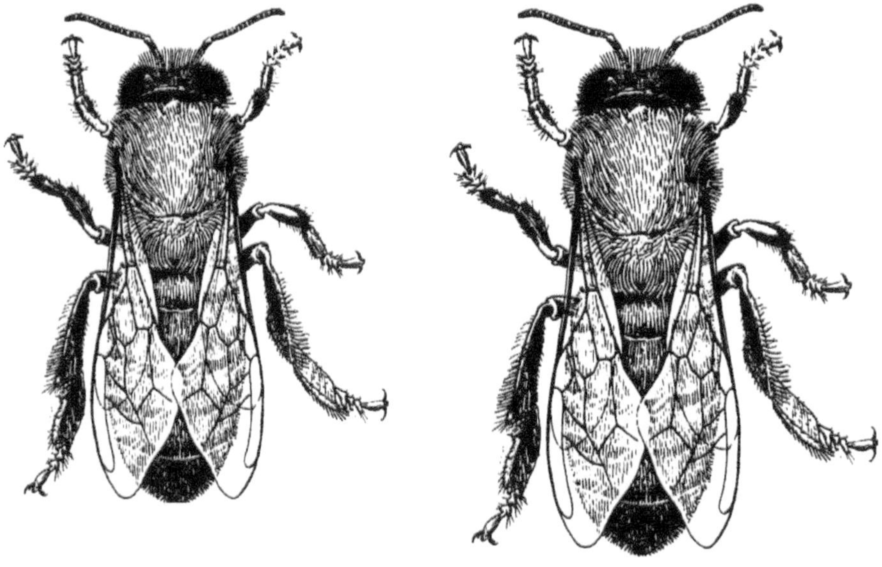

Proportion of workers from 4.9mm vs 5.4mm foundation.

We have found through reading, that mills have been made in many sizes over the years, all the way up to 3-1/2 cells to the inch. *This is where our industry has gotten into so much trouble. cell sizes, their size, and how to measure them*. Most beekeepers universally agree that five cells to the inch is worker size and four cells to the inch is drone size in the feral population. But, the domestic sizes our bees today have ended up on, are quite different and have wrought havoc, as artificial combs have gotten bigger. The stress upon our honey bees caused by being out-of-balance with natural flora has opened Pandora's box to foulbrood diseases, chalk, and viral infections. By being too big, our domesticated honey bees have taken on parasitic mite infestations as our now pseudo-drones, aka: worker bees, are perceived as a new food source by both Varroa and Tracheal mites. With all this damage being done, we find no teaching anywhere on the history of use of artificial comb foundation sizes in the USA, so our beekeepers can make rational decisions concerning proper usage.

The various comb cell sizes (706, 711, standard, drone, etc.) were originally designed with meaning, which is now forgotten in today's modern world. No one person or company is to blame in the USA or elsewhere in our world for artificially increasing honey bees so big as to cause disease and parasitic mite problems that now encompass over 135 nations at catastrophic level, placing them in a situation out-of-tune with the reality of Nature. Nothing was hidden and everything was written and published out in the open. The only thing that has happened is that artificial supposition and an artificial domesticated system of beekeeping has so far prevailed over that found naturally occurring in Nature and has been taught. It began with a simple idea everyone wanted to participate in, i.e. scientists, manufacturing houses. Now through the passing of each successive generation (figure 20 years to each new generation), the growing disparity of size between feral and domestic hives dictates this bigger-is-better trend must be corrected and its underlying causative effects must be brought to light and taught, if beekeeping as we know it is to continue.

Remember the time frame we went over previously for a proposed magnificent final: "The need to rear bees of extraordinary vigor, able to forage over a more extended flight-radius and to visit a multitude of flowers the nectar of which was then and still is today, out-of-reach of their tongues? In actuality, the need for selection to breed a better bee. *Have we really succeeded?* Hindsight would say no. But one must admit, everyone tried hard to do so. We had a great "Roaring 20s", even fought a War. We learned a lot this century, but at what price? We all strived in every nation to breed bigger livestock, bigger and better plants for food, a better human race, we vanquished diseases with various treatments, we have practically redesigned the world, and lastly, we have now gone to the heavens. But as everything goes full circle in the end in evolution, if it is not right, the real question is: Will it last and stand the test of time eternal? I do not think so.

Much is already collapsing or has already collapsed. Antibiotics are playing out, and inbred plants and animals are naturally retrogressing in spite of man's attempts to prevent otherwise. We have learned the hard way this century that there is no superior race of man. Why should beekeeping be any different? It's not except for one thing: most beekeepers in the field are not schooled on the history of what transpired concerning the artificial mutation of making our honey bees bigger. Why? Was it not explained to them through various journals and publications? Explained it was, but not all at once, nor in the same journal, so someone could learn through continuous reading. The information seems to have been scattered, a trait common even today in the publishing of scientific information or any information technically, for that matter.

The question now seems to be, how does a worldwide industry put honey bees that are artificially oversized back in tune, locality by locality, and geographic region, by geographic region so that domestic and feral honey bees alike can intertwine in natural evolution? Further, the question must now be asked, when science becomes based upon an artificial system, out-of-tune with the natural environment, where does science stop and supposition/politics begin?

Here one must remember, that if experimentation with various ways of measuring cells brought forth a new measurement system, and it was not correlated to go back and forth between the systems, and everyone wrote about comb cell size or referenced it from time to time during purported scientific study; and each study and generation began building upon one another; and because science is supposed to be exact, then what kind of science do we have concerning our USA and more importantly worldwide beekeeping community? How does one reference traditional history back to before the time of Pappus and Archimedes? You cannot take the math from one era and apply it to another to lend credibility if they both measure radically different, because you end up with different outcomes relative to the size of the honey bee and a major underlying causative effect bringing forth plagues of parasitic mites, secondary diseases, and field breeding variances in today's world.

It would seem we as an industry have forgotten basic traditional history and stopped teaching it, instead preferring to rely upon quick fix contaminating chemicals and negotiated politics, that are becoming an ever deeper and deeper chemical treadmill and political nightmare, to eventually send us all in the end into massive hive collapses, both locally and regionally. We cannot rely upon these quick fix gimmicks to solve our problems, for in truth, they have created them. The information is there to solve our disease and parasitic mite problems if we are willing to study and learn from the past so we can protect our joint futures as a beekeeping community. Isn't it about time tradition was gone over again for our new generation of beekeepers to review and use; to make constructive decisions with, so they can manage honey bees to overcome today's problems?

The basic principles are simple. Basically it's believing in a natural biologically controlled system to correct the situation without the use of essential oils, antibiotics, chemicals, artificial feed, and overuse of insemination by placing our domesticated honey bees back onto a beekeeping system compatible with that of Nature and her feral populations. The only catch is, going back to naturalist beekeeping, with a clean sustainable is hard labor-intensive field work and will take years to accomplish, but then it took years to create our current situation.

Honey bee Comb: Size and Ramifications – Part 3

Honey bee comb cells are measured parallel wall to parallel wall in three directions. They are not measured point to point, nor measured with a mixture of both, one way in each direction differently, i.e. parallel wall to parallel wall across, then straight down through the points.

Metal mill rollers were originally made by making the bottom of the cells out of three chip-out little lozenge shaped plates, that when put together formed the bottom of the cell. This was done so that the bees could beautifully build what is called a "Rhombic Dodecahedron". Beekeepers know this figure as a common bee cell. When beekeepers measure comb foundation, they should measure the combs using the dimensions inside that of a rhombus, because in doing so they measure parallel wall to parallel wall and can arrive at an accurate figure that corresponds to that used by the mill maker in creating the mold that duplicates Nature.

This is also the way traditional comb cells were measured back to our earliest times (Zenodorus, Pappus, Maraldi, etc.). When beekeepers measure comb foundation today, many make the mistake of measuring parallel wall to parallel wall across the first row and then down straight to make a cell count determination. This is not a traditional way of measuring combs, only becoming popular just after the turn of this century. *Combs are measured in what is called a "square decimeter", but a square decimeter can be measured one of two ways*. One way is traditional going back over two thousand years of recorded history and is what our industry was founded upon, along with much of our culture relative to building, transportation, and medicine. The other way again, only became popular just after the turn of this century.

A square decimeter can be measured either with a perfect square or by a rhombus method. By changing to a perfect square measurement, we have gotten into deep trouble because...the numbers arrived at in the totals are vastly different. It is this vast difference that has wrought down upon us our parasitic mite problems as many of us try to use what we think is the proper size foundation our honey bees should be using, but in actuality it is not. By trying to approximate the old USA standard of 856 and the old world (European Mainland) standard of 800 cell sizes to the square decimeter many beekeepers have used foundation bases geared to a square decimeter using square measurements, rather than a square decimeter using rhombus measurements. The error is proving fatal to say the least.

Example: The cubic content of a cell is of interest here. Why? The comparison of worker comb with the values for drone comb, if one were to do it, would probably approximately confirm Mullenhoff's results announced some 115 years ago, that the drone cell has a volume double that of a worker cell. Why is this important today? Basic survival necessity, equating to a "*Food source attraction for parasitic mites!*

Here it should be convenient to remember the figures 3, 4, and 5 as traditionally applying to the size of cells, that are their diameters in a common feral beehive as given at the beginning of this century. Three queen cells placed side by side, would then traditionally measure an inch, 4 drone cells placed side by side would then traditionally measure an inch, along with 5 worker cells placed side by side, measured traditionally, would measure an inch; or, the size of drone cells built by bees of a given size bears a constant ratio to the size of the worker cells, as does the size of the queen. In addition, the size of the bee is correlated with the capacity of the cells inside diameter, which regulates the size of the thorax, which then regulates the size of all body parts. So when you inadvertently change the size of the worker bee by changing the size of the cell, you change everything in the hive.

Question: If in Nature with Apis cerana, only the drones are attacked by parasitic Varroa mites, why would bigger be better, if it triggers a pseudo-attack by parasitic mites perceiving oversized worker larva as just another food source like drones?

Question: If one believes that Tracheal mites are historically external parasitic mites on honey bees that were for centuries non-threatening, and the only place these mites can get into honey bees to do damage is through the first thoracic spiracle on the thorax of a honey bee, and changing the size of a comb cell bigger would therefore change the size of the thorax bigger and consequently enlarge the entrance of the first thoracic spiracle on a honey bee, then why would bigger be better, if it triggers an attack by parasitic external mites to today turn them into internal parasitic mites, and is there evidence to back this conclusion up? I believe that there is ample evidence to back this conclusion up, especially when it can be shown that retrogression back onto smaller natural comb size of 5.0mm stabilizes the death curve and further retrogression back onto 4.9mm comb size foundation further eliminates accompanying secondary diseases.

Question: If retrogressing honey bees back onto smaller sized worker comb foundation, traditionally used for centuries, can be shown to eliminate parasitic mite attacks of both Tracheal and Varroa mites and also their accompanying secondary diseases, then what does this say about modern genetics breeding concerning honey bees, if say, retrogression to smaller traditional sizing for worker cells, eliminates parasitic mite problems, secondary disease problems, and lastly, even inbreeding problems?

Question: This would then necessitate the question as to what is modern day genetic breeding for diseases and parasitic mite problems, really based upon 1) a traditional beekeeping historical background or 2) an artificially derived system developed just after the turn of the century on the supposition that bigger is better 3) a hodgepodge of both poorly correlated? *What are examples of normal sizing?* In any geographical area there is a range in Nature that takes into account that which is acclimatized to an area (pure) and that which is in the process of adapting (hybrid). A pure honey bee will always be smaller in size than a hybrid honey bee of the same race in a given geographical area.

In Belgium, Prof Baudoux measured workers in the range of 5mm to 5.17mm and 5.35mm per cell. (Ordinary drone cells he placed at 5.5mm next to 5mm for worker cells.) Even he found 5.5mm drone cells next to 5mm worker cells to be double in volume of cell contents for food confirming Mullenhoff's results. When we then consider having today, foundation on the market as large as 5.7mm, it's not hard to see how it would not be disastrous for beekeepers to use, because of parasitic mite attraction with a pseudo-effect on worker bees perceived as a food source just like that of drones. Yet, look how much foundation on today's market is 5.44mm in size (close to the 5.5mm drone size measured by Baudoux). Wouldn't this too, also be dangerous for attraction for parasitic mites, causing a pseudo-effect, with mites mistaking worker bees for drones? *Just how big then, is too big*? I would say too big is when a colony of bees starts to become distressed, which leads to stress, which leads to disease, which lets a beekeeper know through visual perception that something is wrong within the hive.

How were worker combs measured around the turn of the century about a hundred years ago, just before a newer measurement system was devised and presented at a newly formed "Apimondia of world renowned beekeepers/scientists in the 1930s?" Let's look at a few.

From A.I. and E.R. Root in their 1913 edition of *ABC and XYZ of Bee Culture* we learn: "If the worker-cells were exact hexagons measuring five to the inch, there would be exactly 28-13/15 cells to the square inch on one side of a comb. But there is not this exactness, as will be shown by careful measurement, although the eye may detect no variation. Count the number of cells in a given length in a horizontal row of cells, and then make the same count in one of the diagonal rows, and you will find they are not precisely the same. That shows that the cells are not exact hexagons. Measure the cells in a number of combs built by different colonies, or even by the same colony, and it will be found that they are by no means all of them five to the inch.

This, of course, refers to natural comb built by the bees without any comb foundation being supplied to them. Comb foundation is generally made with cells of such size that worker comb built upon it contains about 27 cells to the square inch."

Knowing that combs are measured in what is called a "Square Decimeter" then if we take the measurements of 27 cells to the square inch, times 16 for one side of a comb and multiply it times 2, for the two sides of a comb it would refer to, we would find out the foundation the Roots are referring to would have 864 cells for the square decimeter, which would equate to 4.8mm sizing, and the size of the foundation our industry in the USA was founded upon, if one were to read old editions *of ABC and XYZ of Bee Culture.*

Going further, if we look at the natural measurements given via the rhombus method, explained by the Roots, by noting the measurements given on the diagonal, we might say: 28 x 16 x 2= 896 cells or 29 x 16 x 2 = 928 cells, or depending upon your interpretation, say approximately 900 cells give or take for cell size. Does 900 cells for a size sound familiar? But herein is the problem. How does 900 cells for a size equate with 800 cells for a size relative to history and actual measurements at the beginning of our 1900s? Now measure straight across the parallel walls and then straight down the rows for a square decimeter measurement. The numbers would be vastly different. Like an old slide of hand trick exchanging one measurement for another. However, Prof Baudoux and others back then knew about the measurement disparity. It was common knowledge. Somehow, because it was not written down, it became forgotten. But today it is paramount to understanding our parasitic mite and secondary disease problem as a major underlying causative effect and the error needs to be corrected. *Now take any piece of foundation and do the same thing, look at the numbers count and the area involved, and compare the differences. interesting, wouldn't one say!*

By the way, the 900 cell size range would place foundation in the range of 4.7mm, which would be the feral size for the sea coast of the USA Gulf port states. We calculate 4.7mm as the beginning of the range, of comb cell sizes in the USA, with 4.9mm average in a major part of the USA below 3500 feet above sea-level, and 5.0mm to 5.1mm the top of the range by our own latitudinal plotting of feral comb cell sizes by latitude and altitude. However, only on 4.9mm size comb foundation could we drop off secondary diseases in our own bee colonies. On 5.0mm we stabilized with tracheal mites and Varroa mites with our hives not dying due to mites, but secondary diseases would finish the job during drought years triggered by high stress. Since Nature is *harmonious* and secondary diseases abated by going one step smaller to 4.9mm worker cell size, we stopped our retrogression to smaller cell sizes here, because extraction of honey is more difficult the smaller the cells; and mites and secondary diseases are no longer a problem at our latitudes and altitudes in Southern Arizona between Tucson and Nogales.

Now let's look at measurements for honey bee cells around the turn of the century for England and see what we get. We will reference E. B. Wedmore and his book "A Manual of Beekeeping". In the third edition on page 78 we find some interesting information.

Namely, Wedmore like the Roots, talks about the size of hexagon cells being generally measured across the flats, which would be the parallel walls. In fact to quote E.B. Wedmore for the range of cell sizes to be found in England we note: "Foundation for worker brood is generally made to give comb with about 4 3/4 to 5 cells per inch run (measured across the flats), the latter more commonly, and 4 cells per inch for drone brood. The area of the hexagon of a cell then becomes such as to give 26 to 29 worker cells per square inch and 18.5 drone cells, in each case counted on one side of the comb, and not allowing for stretching". E.B. Wedmore then goes further saying: "In nature the *inside* dimensions of worker cells across the flats may vary as much as from, say, 0.195 to 0.235 inch with the same lot of bees, and will vary still more as between races having the smallest and largest bees. Similarly, in one lot of bees drone comb may run from 0.22 to 0.26 inch".

Again, knowing that combs are measured in what is called a "Square Decimeter" then if we take Wedmore's measurements of 26 cells to the square inch, times 16 for one side of a comb and multiply it times 2, for the two sides of a comb it would refer to, we would find out the comb would have 832 cells for the square decimeter, which would equate to 4.9mm sizing, for the large cell-range size. If we then take Wedmore's measurements of 29 cells to the square inch, times 16 for one side of a comb and multiply it times 2, for the two sides of a comb it would refer to, we would find out the comb would have 928 cells for the square decimeter, which would equate to 4.66mm approximately for small cell-range. Lastly, 27 worker cells per square inch would equate to 864 cells for square decimeter, and 28 worker cells per square inch would equate to 896 cells for square decimeter.

Notice that if one is referring to a small black bee of Europe at sea level then the 4.66mm would be the very smallest sizing reported by Wedmore and if the same were found in Belgium at sea level would show why some beekeepers believed the small black bees of Europe were quite small indeed. If you look at the 864 cell sizing, it is close to the 4.8mm sizing then used in the USA and quoted by the Roots. If you look at the 896 cell sizing you are looking at the old quoted 900 sizing or 4.7mm for honey bees found on several coastal plains from Belgium

South through France and Spain to Italy around the Mediterranean Sea including the North of Africa. (Also 4.66mm cell size at its smallest). If one looks at measurements given by Cheshire in England in 1888 one would see that he gives 28 13/15 worker cells per square inch for an answer for cell size in England. Here, just like with the Roots in the USA, we might say: 28 x16 x 2 = 896 cells or 29 x 16 x 2 = 928 cells, or depending upon your interpretation, say approximately 900 cells give or take for cell size.

So what does all this mean? It means that for old small black bees found in England the range of sizing went from 4.66mm to 4.9mm in worker cell diameter. Cheshire's measurement for drone cell size was 18 178/375 per square inch relative to Wedmore's. If one references Anne D. Betts of England, she gives the number of cells at 830 worker cells in "Bee World" January 1934, which would also equate to 4.9mm worker cell sizing. One might ask the question here: In England, like in the USA, could the variance from traditional worker cell sizings quoted by respectable noted early beekeepers, be a major underlying causative factor in today's world for secondary diseases, and parasitic mites brought upon by ever bigger and bigger cell sizes?

My husband Ed and I have told beekeepers calling and emailing, that if man should ever seek to change honey bees so that they no longer relate to Nature's and *God's* law, they would likely intervene in such a way as to preserve the necessary balance originally created. For there is some reason to believe that in the plan of Nature, the honey bee was not only created to conform to the necessity of its mission as a pollinating agent, but that the plants and their bloom may have been fashioned to conform to the convenience of the bee also in one large masterful plan.

There is a barrier we have crossed as an industry both locally and worldwide we need to retreat from, that seems to have been deliberately placed there by *God* and Nature to prevent any wide deviation of the honey bee in size and action from what they designed that it should be, this being accomplished by limiting the size of the bee to that of the cell in which it is developed, as set down in the feral bee, beyond which it cannot go far off size without being forced back for fear of extinction. Diseases and parasitic mites are forcing us back now into balance with native regional floras. Shouldn't beekeepers heed their traditional past, to learn from it to protect their future as an industry?

Time is getting short now as our politics catch up with us. Time is short now for our industry as our breeding mistakes, breeding for superior races catch up with us over that chosen naturally. Time is short now for our industry as our chemical treatment mistakes catch up with us. We are deepening the chemical treadmill we have put our bees upon, which can only end with total brood nest contaminations and hive collapses of the very hives we love. Time is short now for our industry as our oversized now pseudo worker bees, now perceived as drones are eaten for food. We as an industry want to run fast and cheap and it won't work. For those who want to go back to naturalist beekeeping with a clean sustainable system, it is not hard, but it does take a lot of field management and work, and *time*. Isn't it about time we as an industry got in tune with traditional history again and Nature's system?

Cheshire in 1888 "Bees and Beekeeping" noted that: "Flowers and bees have been constantly interacting. The build of every floret is adapted to that of its fertilizer, and, could we suddenly increase the dimensions of our hive bees, we should throw them out of harmony with the floral world around them, decrease their utility, by reducing the number of plants they could fertilize, and diminish equally their value as honey gathers. Mechanics, physiology, economics, and botany alike, show any craving after mere size to be an ill-considered and unscientific fancy, for which it would be even difficult to find an excuse."

E.B. Wedmore's "A Manual of Beekeeping" 3rd Edition adds by saying "Too large an increase in cell diameter involves increased size of brood chamber and some loss of economy in wintering, the cluster being less compact. Undoubtedly the beekeeper needs to study foundation in relation to the size of his bees. Although larger cells produce larger bees, there is no evidence that they are better bees. They are of lighter build."

These two paragraphs mean a lot to today's beekeepers since both plants and honey bees have evolved together, they are linked together in the evolutionary chain, each dependent upon each other for survival. Change the size of the honey bee by artificially making her bigger and you decrease the variety of floral sources available in her diet which leads to nutritional induced stress, just like other undernourished animals, which can lead to a lowered internal immune system for increased risk of affliction by disease, or parasitic attack by foreign organisms. Also breached is overall pollination of the natural range of plants pollinated, and if the unpollinated plants cannot be pollinated by other species of insects, then reproduction is compromised leading to disappearance of native flora. Could this be a contributing factor in today's world with the hybridization of artificial crops for bigger yields, pollinated by bigger and bigger artificially created honey bees, which has simultaneously led to diminished natural vegetation/plant species in a number of areas around the world?

Compounding, if you increase the size of the brood chamber and create loss of economy in overwintering by the cluster being less compact, you have added even more stress upon the honey bee colony. You have also added more cost of upkeep to the beekeeper in the way of extra equipment. When you have bigger artificial worker cells and bigger honey bees, you have less brood per comb raised, which means you need more frames of brood to equal a normal size brood chamber containing the same number of bees to carry out proper colony functioning with division of labor. To have bigger artificial worker cells and not add additional frames of comb for brood to equal a normal size brood chamber containing the same number of bees would necessitate a restructuring of division of labor within the colonies for continued smooth functioning.

If restructuring cannot be accomplished to cover all required tasks, then stress appears and compounds as more and more tasks cannot be accomplished, for the successful continuance of the colony (we will talk more on this later). You have loss of economy in overwintering of the cluster, with the cluster being less compact, in many ways with the artificial use of bigger honey bees. First, the cluster in winter is only capable of physically covering a certain number of worker cells in the brood nest area and yet move, for stored feed of pollen and honey for daily maintenance. If even a 100 cells difference, say hypothetically, in square decimeter cell count were to occur (in actuality there is much more cell count difference), from traditional counting, to an artificially oversized count, as would occur with today's more modern square square decimeter counting, this would equate to over the space of 8 square decimeters, on just one brood frame, as being an 800 cell difference.

Hypothetically here, that would be 800 less worker bees or physical bodies available for work within the colony on just one frame for just one brood turn. How many worker bodies can be lost, before economies of scale becomes an item, of much merit to be watched, so spring buildup is not compromised or winter carry over (shivering for warmth) is not compromised? In addition, as the bees bodies themselves increase in size with their various parts as cells are artificially made larger, all changes are not in the same proportion. One would think that the bees flight muscles must increase in proportion to the wing length, but it does not. When looked at internally, it's like mass/muscle has been expanded over a larger surface and gone from dense muscle to less dense muscle. It's as if pockets of nothingness are created, making for perfect cavities for parasitic mites to nest in, as the first thoracic spiracle on the thorax is artificially enlarged on bigger combs, to allow for external parasitic mites to freely walk in and dine!...and stay!...and reproduce!

Might not then, this be considered a contributing factor for an underlying causative effect taking place helping to set the stage for today's problems of disease and parasitic mites? When looked at externally, it's like body mass has been expanded over a larger surface, but more loosely. Equate this to a suit-of-armor on insects i.e. the exoskeleton. On small insects it is very tight and close fitting. Even with small hot-blooded animals i.e. armadillos this is true. Now look at bigger insects and bigger similar animals. As size increases the plates are not so tight. With bigger honey bees the exoskeleton is looser than the exoskeleton of naturally sized honey bees. The bigger the honey bees get, the looser and less close fitting, the exoskeleton and various body parts become. What this means is that the tergites on the honey bees body, as one example here, are artificially enlarged enough to allow for parasitic mites to crawl under and suck bees blood. Regulate the bees body back to normal size and naturally tighten the tergites and this practice stops!

Let's talk about *retrogression back to normal* for a while now.

Just how does a beekeeper and an industry at the upper limits of bigger is better sizing, retrogress domesticated honey bees back to normal sizing? After all, the industry has been on this path for a good 100 years now, and it should be obvious that we as an industry do not have a 100 years' time to get back to traditional comb sizing and stability for our bee stocks. We must remember that beekeepers are not all scientists with laboratories to back themselves up, and the most that many beekeepers can only afford to hire for workers are other common folks like themselves. Whatever is done must be kept simplistically simple where possible. It may in truth be labor intensive, but it must be simplistically simple (KISS principle i.e. Keep It Simple Stupid) to follow and do.

To achieve a successful outcome in the field, beekeepers must be able to breed freely between the feral population and domestically kept honey bees. They must be willing to accept what honey bee races/strains will live in their regional areas acclimatized to their own local areas. Here it is best said that live bees make honey and can be traded, while dead bees are, well just dead!

To achieve a successful outcome in the field, beekeepers must be willing to make field management changes, keeping what modern mechanized advances will work, and get rid of those that will not.

To achieve a successful outcome in the field, it will also take looking at honey bees with a new perspective to breeding and field management, as much of today's modern methods evidently are just not working, having been developed on an artificial oversized domesticated system not relative to the true feral population.

Nothing is hard to do or understand, to retrogress back to natural feral comb sizing, but it will take some time. While it won't take 100 years to accomplish, for some beekeepers it will take upwards of 10-15 years, depending upon how many colonies are currently managed and their degree and willingness to participate. Many will not make it through the change-over back to traditional comb sizing, but many must for our industry to survive the long way back to biological beekeeping without the use of chemicals, drugs, and essential oils.

The way back to biological beekeeping through retrogression is a multi-stepped process. Just as it took several steps upward in sizing through-out the last 100 years, it will take more than one to down-size back to natural sizing parameters for domesticated honey bees so they can racially mingle again with the feral. It cannot be accomplished in one retrogression step smaller, due to the extent that we as an industry went bigger, in search of a better honey bee! Basic tools for every beekeeper to have should include: 1) a supply of ready-made swarm-catching frames, 2) a supply of queen excluders, 3) a supply of 4.9mm comb foundation(without pronounced sidewalls), 4) basic grafting supplies for queen rearing.

Retrogression Back to Normal – Part 1

Picture in your mind the world as a basic map of honey bee thermal/cell size zones based on composites of hot and cold land area masses. Now picture this as lateral belts going around the world with the hottest zones and smallest sizes near the Equator and volcanic hot thermal zones, and the largest sizes near 60, latitude in belts going around the world and the tops of colder mountain ranges. Now picture in your mind the world surrounded in layers with the smallest sizes being on the bottom where compression is greatest, and the largest sizes being on top where compression is least. Now picture in your mind that every honey bee thermal/cell size zone has a natural range of small, medium, and large cell sizes to allow for bees to transition into and out of habitat areas as vegetation and rain occur throughout a yearly cycle. With smaller cell sizes a beekeeper would gain variability and with larger cell sizes a beekeeper would gain less. As you go from the Equator, both North and South, picture smaller yellow hot-weather bees getting bigger to about the 30 latitude belts around the world.

Picture a transition zone here, where yellow hot-weather honey bees and dark (black or brown) cold-weather honey bees come together and overlap co-mingling. Yellow hot-weather honey bees will naturally be at their largest sizing and dark cold-weather honey bees will naturally be at their smallest. As you go from the 30 latitudes towards the poles, picture the dark cold-weather honey bees getting larger to about the 60 latitude belts, where their natural migration north transitions to an end. Now picture in your mind, in the tropics, only yellow hot-weather honey bees at sea-level and when you reach about 3500 – 4000 elevation, a transition to dark cold-weather honey bees again. Now picture in your mind, in the temperate zones, only dark cold-weather honey bees at sea-level, except where volcanic areas create hot thermal-zones allowing for yellow hot-weather bees to survive naturally. Again within these limited yellow bee zones, transition would occur towards dark honey bees, as elevation is increased.

Now picture in your mind this only occurring with pure strains/races of honey bees, except in transition zones where hybridization would occur. Now picture in your mind hybrid mixes of different honey bees as always being bigger in size than the pure strains/races they originate from. Now picture in your mind simple hybrids being naturally occurring in Nature and complex hybrids only occurring in an artificial environment.

Now, let's ask some questions. (We will re-state the above prior to discussing bee breeding, applicable to field mechanics/management only). *Question:* If honey bees, when they swarm out of oversized domesticated hives, retrogress downward in size, then how stable is a hybridized mixture in this scenario, especially if bigger flys slower? Same scenario relative to Africanization? If honey bees for evolution security, acclimatize to natural sizing, then should not we as an industry be following them to solve our problems of diseases and parasitic mites? If domesticated honey bees in a given area swarm and retrogress downward, should not this be a sign that the artificial comb size in use is wrong for that particular area? If retrogression back to natural sizing, in any given area/region cannot be accomplished in one retrogression-step smaller, due to the extent that we as an industry went bigger artificially, then what is the fastest way back to natural biological beekeeping without the use of chemicals, antibiotics, and essential oils; and it taking a large amount of time, which we seem to be running out of?

Retrogression: The first step in retrogression is to survey colonies in each area/region that are both domestic and feral (Note – Colonies on oversized artificial brood foundations do not fully correlate with naturally occurring breeding cycles, necessitating that differences be taken into account or excluded from survey. Further, colonies in the feral, whether established or un-established, need to be separated visually by subcaste matings, to establish their true degree of being natural, feral, i.e. all small subcastes, all large subcastes, or mixed-size subcastes noting the degree.)

While this is taking place, a historical survey needs to be accomplished to ascertain what traditional comb cell sizes where, prior to use of artificially enlarged foundation, by each area/region. Comb cell sizing here needs to be split into two categories: 1) That what was, prior to importation and subsequent hybridization; 2) That what was, after importation and hybridization; but before the use of artificially enlarged foundation.

Where historical data is insufficient for verification of traditional sizings, physical observation of retrogression by physically removing domesticated honey bees from artificially enlarged brood combs will be required. Note: This may require several steps, to finally arrive at a standard range of smaller sizing by use of "V-cut " top bars. Beekeepers can gauge very quickly what size brood foundation their local honey bees prefer. All they need is to makeup a few brood supers with top bars in them only, with 5 degree angle cuts for feral swarms to build comb upon, precoated with beeswax to facilitate comb building during the swarming season. Then all that has to be done, is measure the inside diameter of the cell wall to ascertain the common worker cell size for the geographic area. Beekeepers doing this should find that their feral bees (large caste/mixed-size castes) and even their own domesticated bees will prefer a smaller size brood comb than what they are probably now using. Here, first rule-of-thumb to clear up parasitic mite problems should be...do what the bees prefer!

Keep in mind, the idea is to figure out how many retrogressions it will take to bring the honey bees in a given area back to traditional natural sizing. Repeat the above mentioned process, with V-cut top bars on several domesticated hives and several well established feral colonies (not less than 10 each). We found the average here in Southern Arizona to be about 4 for our domesticated hives and 2 for established feral colonies. We found it impossible to make the retrogression back to natural sizing in one jump (This was our retrogression attempt to 5.0mm sizing). Two retrogression would be required, but it became apparent that the time-frame would be a limiting factor, because we didn't have the time to retrogress our whole beekeeping outfit for each retrogression required (figured 10 years for each retrogression for changing 1,000 hives with foundation) for our domestic hives. We decided to see if we could speed the process up to match what we observed on the feral side. Consequently, we decided beekeepers cannot do an old-fashioned comb shake-down, from today's domesticated hives and restart on new undrawn 4.9mm foundation, to match the top-of-the-sizing-spectrum for traditional sizing before artificial hybridization by man, without modifying the technique to fit today's needs relative to stress by parasitic mites/secondary diseases, limited time-frame within which to work, and the different requirements for field management, between domestic and feral.

Process For Speeding-Up Retrogression:

This is a definite multi-year application to accomplish. Depending upon the size of the beekeeping operation, it can take anywhere from 3 to 15 years average to accomplish. The first year is a preparation year for creating "seed-frames" for what will be pot-progressive work, the years following. Work is begun by the preparation of "removable swarm-catching frames" to act as stimuli for producing seed-frames of drawn-out 4.9mm brood-comb foundation. It also fills a two-fold purpose of speeding-up retrogression, while supplying a renewable source of clean uncontaminated beeswax for foundation making (We will be going over making foundation by hand later). Depending upon the size of the operation to be converted back to natural comb sizing, figure making about 50 supers of swarm-catching frames for every 1000 hives, or 5 supers of swarm-catching frames per 100 average.

Begin by catching feral swarms. When hiving unestablished swarms, separate by worker caste sizing, keeping small-cast worker bee swarms for production of seed-frames, by immediately placing upon 4.9mm foundation. Hive the swarm into a super of undrawn foundation setting on a queen excluder, which is setting upon a bottom board. Use a tight top cover to close. Transport to desired location. D*o not remove queen excluder from between bottom board and super until foundation is drawn and queen is laying on a minimum of 2-3 frames*.

When hiving an established feral colony, cut out feral combs and mount into swarm-catching frames. Take care to keep brood together, filling each frame as much as possible. When hiving during a good nectar flow, discard pollen and honey stores (bring an empty bucket to put in to take home). Most established swarms cut-out, will fill 2-5 frames with mounted brood when transferred into swarm-catching frames. *Note*: If 3 mounted brood frames or less, place in super on top of queen excluder, on top of bottom board, and complete filling the super with frames of undrawn 4.9mm foundation. If 4 to 5 frames of mounted brood when transferred into swarm-catching frames, place 1 (*no more*) frame of undrawn 4.9mm foundation in the center of the cut feral comb mounted into swarm-catching frames. Then fill out the rest of the super with frames on undrawn 4.9mm foundation. Make sure super again is setting upon a queen excluder, setting upon a bottom board. Use a tight top cover to close. Transport to desired location.

Do not remove queen excluder from between bottom board and super until foundation, on frames in center of swarm-catching frames and at sides of swarm-catching frames, are drawn and you have had a chance to reposition them together in the center of the swarm-catching frames, for the queen to lay in as a consolidated working unit.

Note: By removing pollen and honey while hiving the bees, you speed up comb drawing, because by creating no place for the bees to put stores, you trigger wax production in honey bees. Continue cycling of drawn 4.9mm foundation from the sides of the swarm-catching frames to the established consolidated working unit. When the swarm-catching frames are adjacent to the sides of the super, remove and replace with more 4.9mm foundation. Next melt-down the feral comb and recycle wax into foundation as a clean renewable resource. Clean and recycle swarm-catching frames with another colony. Upon completion of first super with drawn-out foundation and stores of brood, pollen, and honey, super a second box and continue, repeating supering as desired.

Retrogressing domesticated colonies, established on oversized foundation, requires a different approach. First, beekeepers must separate the comb sizes within their colonies to be retrogressed. This is best done going into winter, leaving the broodnest to settle into the smallest drawn comb available to overwinter upon. Then when the honey bees are at their smallest body sizing going into Spring for the year, just before brood-rearing begins, an old-fashioned hive shake-down should be accomplished. This is done by physically shaking the bees off of the combs and restarting like a shook-swarm, into a super filled with new undrawn frames of 4.9mm foundation, sitting upon a queen excluder, sitting upon a bottom board. Honey-syrup and a pollen patty (made with honey and pollen only) may need to be supplied to induce bees to draw wax foundation. Use a tight top cover to close. *Do not remove queen excluder from between bottom board and super until foundation is drawn and queen is laying on a minimum of 2-3 frames.*

Note: For those colonies that will not draw comb out properly, stop, and remove all undrawn foundation. Continue drawing comb with hives that will draw out 4.9mm foundation properly. When excess drawn comb is available in hives that will draw out comb properly (not necessarily filled with honey or pollen, just drawn out enough to hold pattern (1/8″ to 1/4″on cell walls), remove and add to those colonies that will not draw comb properly. Again shake down bees to restart queen laying on correct pattern. Once queen is laying on correct pattern of 4.9mm comb foundation and brood is sealed, again add frames of undrawn 4.9mm foundation and finish filling out super. Beekeepers will find that once, bees reluctant to draw comb are given properly drawn frames of foundation from another colony, and go through a full brood-cycle to size down newly emerging bees, the colony will straighten out. Upon completion of first super with drawn-out foundation and stores of brood, pollen, and honey, super a second box and continue, repeating supering as desired.

This is the process to be followed the first year for catching feral bees or retrogressing domesticated colonies from oversized brood foundation. The objective is to create as much correctly drawn-out 4.9mm comb foundation as possible, to act as seed-frames for the second year's work, and to stabilize as many colonies as possible with stores of pollen and honey. Foundation not correctly drawn is to be culled and melted down, cycling back into undrawn foundation for reuse.

Note: Only by careful culling of misdrawn comb foundation will beekeepers bring parasitic mites and their accompanying secondary diseases under control so no chemicals, essential oils, and antibiotics are necessary for field maintenance. R*epeat: The objective is to create correctly drawn-out comb to act as seed-frames for the second year's work. nothing else will work in the end. bees that will not correctly draw out foundation over the course of the year will succumb to disease, die and/or not overwinter properly. do not try to save them or you will perpetuate your mite and disease problem. treat this as survival of fittest only, and extinction for that which will not retrogress to solve the problem biologically back to traditional beekeeping.*

Removable Swarm Catching Frame

Frames can be made from old supers. Cut wood strips of 1X pine into 1/2" strips. Once you place recovered feral honey bee combs in the removable "swarm catching" frames, simply take a few frame nails and nail hinged frame together thus making permanent the combs placed therein. Simply place in super, reconstructing brood nest to allow bees to settle down.

Making a removable swarm catching frame is quite easy to accomplish.

1. Each half frame is 3/4" wide by 1/2" thick.
2. You can drill 4 to 5 holes to lace frame wire to hold comb that you cut out of feral colony.
3. Mount honey filled comb pieces together in a frame.
4. Mount pollen filled comb pieces together also.
5. Mount brood laid-up pieces together in frames.
6. Place brood frames together in center of super.
7. Place pollen frames on each side of brood frames.
8. Place honey filled frames on outside of pollen and brood frames to complete makeup of supered colony.
9. Remove to calm place for bees to settle down.

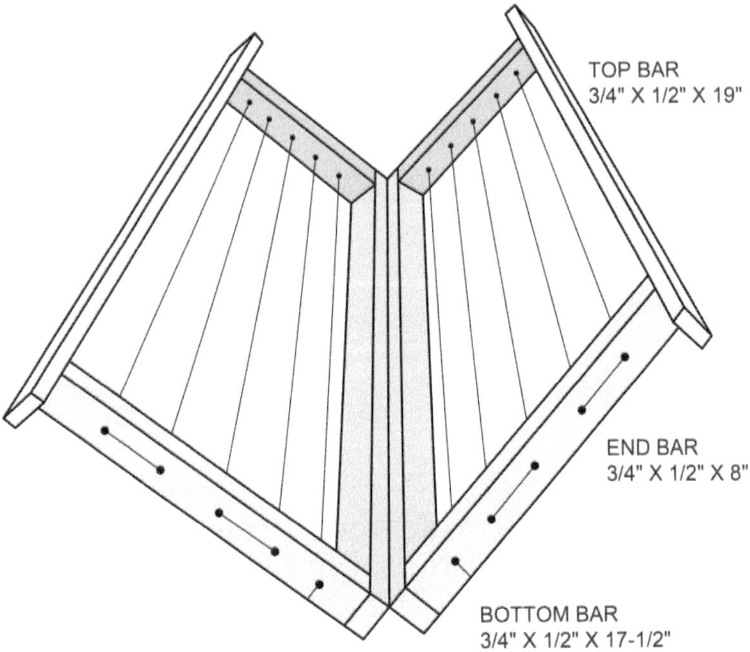
TOP BAR
3/4" X 1/2" X 19"
END BAR
3/4" X 1/2" X 8"
BOTTOM BAR
3/4" X 1/2" X 17-1/2"

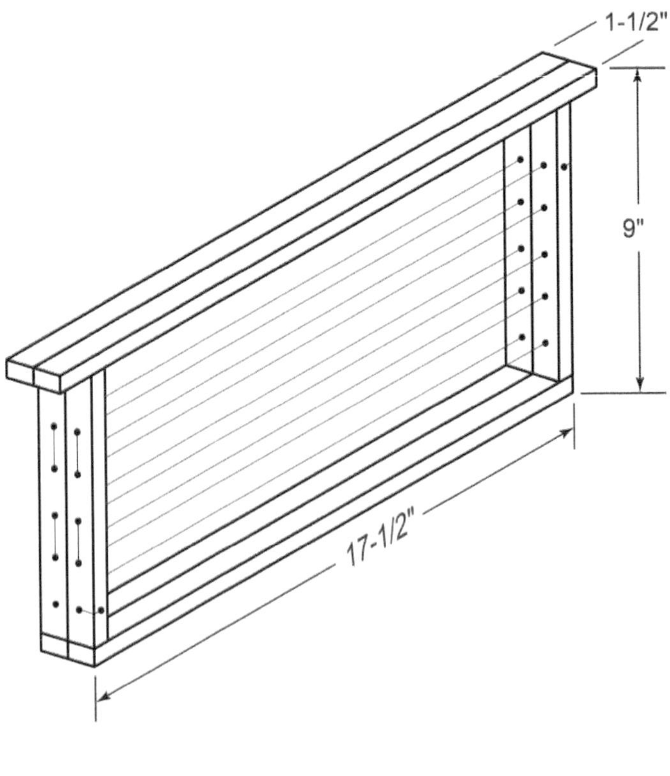
1-1/2"
9"
17-1/2"

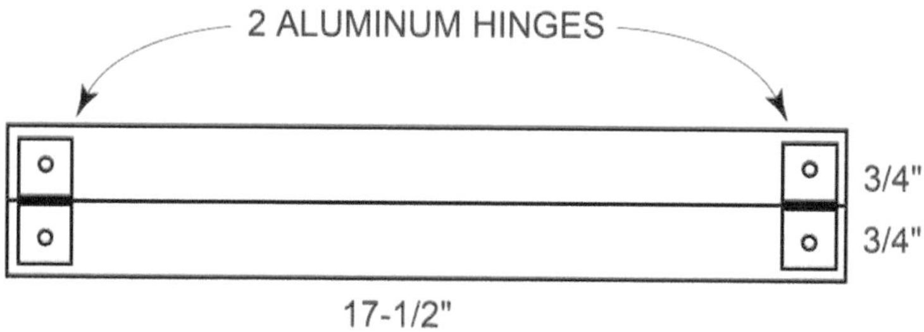

This is a typical swarm trap (these swarm removal pictures are all taken at night in complete darkness) ready for cutting bees out, that someone dropped off at the Lusby's front industrial yard gate, taped shut. (Since the Lusby's (family) have been in Tucson since

1927, often they find swarms left off at their gate, be it swarm traps, chest of draws, old storage boxes, cans, misc. containers, etc., to deal with on a daily basis throughout the active year by well meaning neighbors.)

Here Dee has opened up the swarm trap and is beginning to remove the bees and comb for transferring into standard deep super bee equipment. All of this being done by flashlight.

Dee merely flips the swarm trap cover, turning the bees and combs upside down, and starts cutting off comb and bees for mounting into ready made swarm catching frames using a Swedish hive tool.

Each piece of brood comb is laid on top of the wires and fit like a puzzle till the frame is full, at which time, the hinged frame is closed and wired shut, trapping the comb between the wires

After closing the swarm catching frame together like a book, Dee simply twists a piece of wire around the top bar ends to hold in place.

Brood comb in swarm catching frame.

All different size combs in swarm catching frame. The bees build comb to tie them all together quite nicely.

Dee is placing the swarm frame with comb into a standard deep hive body.

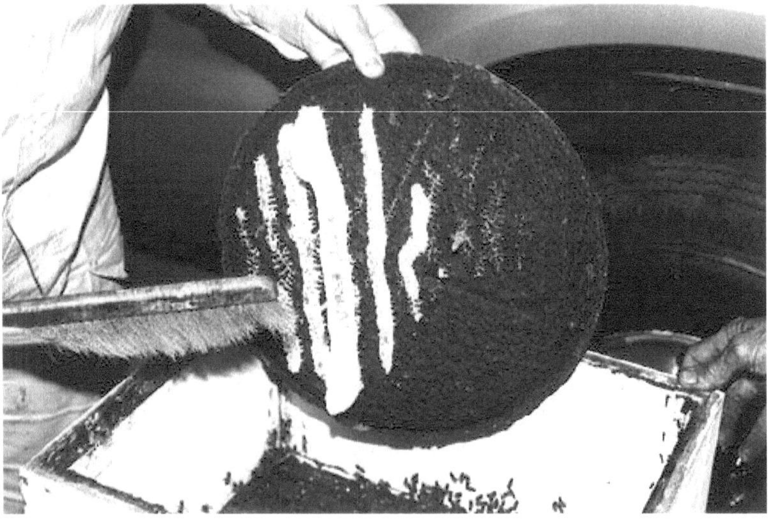

The last remaining bees are brushed off into the new hive and the swarm trap lid will be placed back on the base, leaving the small amount of comb as an attractant, and given back to the owner to be used again. Notice the small vial (center right) that holds the pheromone.

Retrogression Back to Normal – Part 2

We have talked about the process followed the first year for catching feral bees or retrogressing domesticated colonies from oversized brood foundation. Writing from the standpoint of Southern Arizona area conditions and confining thoughts to a discussion of necessary data and observations gleaned in a semi-arid, but still temperate location, note to those following this Saga, that this is basically a non-migratory area and we are working with colonies in apiary sites that have been permanently set (some for decades, generation to generation), making these sites perfect for determining field feasibility for natural biological controls for parasitic mites and secondary diseases without the use of chemicals, essential oils, and antibiotics. To assess, understand, and retool colonies to fit a natural biological system, beekeepers must work with stable apiary sites from which to observe and learn, by process of trial and error, that which works, figuring out why; and that which does not, also understanding why not. The objective then being, to implement.

Transitory colonies of bees, caught-up within a perpetual migratory yearlong loop are incapable of completely adapting due to constant area/regional movement, not allowing for natural acclimatization necessary for evolution. Though associated inherent problems of parasitic mites and secondary diseases for transitory colonies of bees, may be *controlled* with proper field manipulative management, relative to use of small cell foundation matching the top-of-the-range natural traditional sizing for a given area for acclimatized honey bees, this does not apply to the breeding aspect of migratory colonies which has no relativity to real-world natural selection.

Once the process of catching feral bees or retrogressing domesticated colonies from oversized brood foundation has been accomplished the first year, with the creation of as much correctly drawn-out 4.9mm foundation as possible with stores of pollen and honey, the colonies need to be assessed going into Winter to make sure they contain enough stores to overwinter. If reserves are not adequate they need to be supplemented. Preferably feed honey and pollen only. Do not feed substitutes. When feeding honey, feed it as granulated honey, in manila paper packets with 1/4 punched feeding holes. If feeding pollen, feed it as a mixture of pollen/granulated honey patties. Place both as needed right next to or directly above the brood nest or cluster. Replace as necessary. (Note: granulated honey, fed in packets this way, is assimilated within the hive like regular stores of gathered honey). To stimulate brood rearing in the Spring, switch out the granulated honey in manila paper packets and change to LIQUID honey in manila paper packets, with pin-holes (4-6 each side, just enough to let honey slowly drip, but not enough to run/stream). With the feeding of liquid honey, pollen/honey patties must be fed, if internal stores of pollen are inadequate to initiate brood rearing.

Prior to initiation of brood rearing in the Spring, while the bees are still clustered, pull surplus drawn-out foundation where stores have been consumed. Leave hives as singles with drawn-out combs. Where supplemental feed is still necessary, or you are feeding to stimulate brood rearing for Spring build-up, use undrawn frames of foundation in a second story super around the feeder packets positioned over the brood/cluster. As brood rearing increases, and drawing- out of new wax commences, this restarts the process of drawing new frames of foundation for the coming year.

Store pulled surplus drawn-out foundation temporarily in supers standing on end, in a cool place until used. Store butting, facing bottom to bottom, with top-bars facing out (placement this way will keep out rodents).

As soon as enough fresh nectar/pollen is being collected to sustain bees and initiate natural brood rearing in the field, begin shake-down of selected colonies from oversized combs onto your seed-frames of surplus drawn-out foundation. Use 3-5 frames, depending upon your supply of seed-frames made the previous year. Remember, like the previous year, this is done by physically shaking the bees off of the combs and restarting like a shook-swarm, only this time instead of using all undrawn frames of foundation, you are now using 3-5 frames of drawn-out foundation (Note: Place drawn seed-frames together in center of super with undrawn foundation on either side to establish a compact broodnest/cluster site). This will speed up the process of retrogression, by giving the queens an immediate place in which to lay brood, instead of having to wait for comb to be drawn-out to lay in. Like the previous year, place the shaken down colonies upon a queen excluder, upon a bottom board, to prevent swarming until the queens are laying upon a minimum of 2-3 frames. You should notice that this year, once the queens have finished the first brood cycle and new bees are emerging, the bees will be sized down to proper sizing and should be drawing small natural comb foundation more easily, necessitating less culling of misdrawn foundation.

Having accomplished the field shake-down of selected colonies from oversized combs, continue the objective of creating correctly drawn-out combs. *Cull any combs with more than 10% drone cells drawn on any one side. make this a mandatory field management technique. there is valid reason for doing this*. It has been previously demonstrated that Varroa mites prefer drone brood to worker brood for reproduction in the feral population of honey bees. Generally, about 40% of drone cells are infested, while for workers, the average is close to 10% (Note: For Tracheal mites the feral average is also about 10% for workers for infestation levels). It has been demonstrated that the larvae food is the stimulant in oversized bigger cells for attracting Varroa infestation. For many years it was traditionally taught to cull drone combs as much as possible.

Since the advent of Varroa, this practice has been reversed to the detriment of our industry. Beekeepers should go back to the old traditional way of culling drone combs, as there will always be plenty of drones reared in corners of the frames or in cells that become enlarged by accident. It should always be remembered that while the drones do no work physically in the hive, *they do act as the best attractant to pull disease and parasites to themselves as first targets, so workers can survive throughout the active season. Then, when the honey is in and new queens are mated, their jobs done, they are cast out to initiate cleansing the hive of its disease and parasite problems.*

On a natural biological system, the few phoretic mites that remain are quickly filtered out through the brood nest by the workers chewing-out and/or removing mites from infected larvae cells (cleansing). By culling brood frames which have excessive (more than 10%) drone cells, beekeepers limit colony infestations and reduce Varroa infestation down, using the 40% vs 10% infestation level difference to their own manipulative hive management advantage. Additionally, by changing out and shaking-down colonies from oversized brood combs, they further reduce the attraction for Varroa to enter pseudo-drone cells, (aka: artificially oversized worker comb acting as an attractant with more larvae food for mites) and reproduce at a higher than natural 10% infestation level. By beekeepers actively culling drone comb to less than 10% drone cells per drawn frames, beekeepers can reduce the naturally occurring 10% infestation level in Nature to below 5% with field manipulative management.

In the second year of retrogression back onto a natural biological system of keeping bees, beekeepers must learn to look for visual signs that their colonies are cleansing themselves of parasitic mite and disease problems. *Question*: Just what are the signs to look for, to see if worker bees are chewing out and/or removing Varroa mites (cleansing) from infested larvae cells?

In Southern Arizona, the chewing-out of Varroa mites (cleansing) happens on the downside of the honey flow. It starts slowly as the queens stop laying drones, picks-up speed as the drones are expelled from the hive, then tapers-off just prior to the end of broodnest cleansing time. By the time the broodnest is resituated and cleaned by the workers, with the pulling-out of old larvae cocoons and reshellacked, beekeepers will find Varroa mite infestation reduced to a non-detectable level in most cases; and under control by the worker bees. In Southern Arizona, this happens approximately twice a year with the primary cleansing season occurring in the fall. Other times beekeepers will see it occurring, will be right after a colony requeens, when the hive workers are throwing out drones and getting ready to roll again.

Beekeepers will see cleansing mostly around the periphery of the broodnest of sealed worker cells, although it can occur as a buckshot brood pattern in weaker hives, or in a strong hive where large numbers of mites are transferring from drone to workers. In some hives beekeepers will see a combination of both patterns, starting from the sides of the supers and progressing towards the center of the broodnest. When this happens, the buckshot brood pattern is normally nearest the pollen and honey frames, changing to the periphery around the edge of the broodnest of sealed worker cells, as frames are looked at nearer the center of the brood nest.

Beekeepers should look for signs, like uncapped worker brood with the pupae exposed and in many cases cannibalized. If there was only one Varroa and it was located on the head between the eyes, many times the pupae will be unharmed, as the worker bees have only to remove the mite to rectify the situation. If the Varroa is situated on the back of the head between the thorax, the worker bees will eat the head off to get to the Varroa. If the Varroa and/or another is situated on the thorax, worker bees will eat down to that also. If those Varroa and/or more are located on the abdomen, lodged between the tergites, the bees will continue eating down. Beekeepers should notice that when the worker bees are doing this and working only with removing Varroa mites from healthy bees, the pupae will be a *healthy white color*, which shows that the worker bees are not removing diseased or infected larvae/pupae. (Note: What they are doing is cleaning-out infestation, like our human livers clean out impurities from our bodies. They are also recycling protein as a food source, being labor efficient.)

When Varroa is removed from the top of the head and the pupae left unharmed, beekeepers will usually notice that the pupae, are at-a-state of purple, darkening eyes.

Beekeepers observing colonies, will find that the bees seem to chew-out Varroa when other chores of the hive are not pressing, i.e. honey gathering and major brood rearing, showing that they have a way of prioritizing work. Until worker bees are ready to cleanse the broodnest on the downswing of honey flows, Varroa mainly infest drone larvae and pupae.

Thus the drones, although they do no work physically in the colonies, *do act as the best attractant by body mass and therefore a better basic food target, to pull disease and parasites to themselves as an inherent defense mechanism within the colony caste structure's division of labor*, so workers can survive throughout the active season raising vital brood and gathering stores of honey and pollen. Then as the season winds down, the drones are thrown-out, and the *worker brood acts as a living liver in the hive purging the overpopulation of Varroa mites, by chewing out (cleansing)*, to bring the colony back into a balanced parasitic mite host relationship (similar to that of Apis cerana).

Each new brood rearing season, the cycle repeats. Uncapping of sealed worker brood (manually by hand, for field observation), *not uncapped by workers during this time*, have revealed non-infested pupae by Varroa. When beekeepers see these signs, they can know that their bees are doing what they should, to handle the problem naturally without the use of chemicals, essential oils, and drugs. *Caution*: Do not confuse this phenomena with starving bees that need pollen and/or honey, and are driven to eat their own for nourishment to survive hard-times. Beekeepers must learn to see with their eyes and understand the difference.

As progress continues in retrogression, beekeepers will notice more colonies stabilizing. As parasitic mites become more controllable, by the end of the second year, beekeepers should notice the worker bees beginning to chew-out drone cells, cleaning them as they do worker brood cells. (Note: While all this is going on throughout the active year, beekeepers should be actively culling combs for irregularities, i.e. incorrectly drawn size, disease, etc.) Beekeepers should notice, that just as culling for incorrectly drawn foundation subsides, so will the culling of diseased combs. This signals that the colonies are coming into balance with a natural system of beekeeping, working with Nature effectively instead of to the detriment - side, by being out-of-tune with natural flora and climatic conditions, etc..

During the second year, beekeepers should continue catching feral swarms as a renewable resource for bees and clean, uncontaminated wax for recycling into foundation. Beekeepers also need to start planning how many hives to get ready for production for the third year (End year with colonies, a minimum of 3 deep supers of bees, pollen and honey, with all drawn 4.9mm foundation), and how many hives to keep using for the production of seed-frames to continuing shake-downs each Spring until all colonies maintained are retrogressed back onto a natural biological system of beekeeping. Here it is suggested that newly caught feral swarms be allowed to do the balance of the drawing-out of foundation work, along with forced-splits (swarm cells in colonies, necessitating divides rather than having bees go to the bushes).

During the second year beekeepers need to also plan to start recycling empty shook-down boxes of frames of oversized combs. If chemical treatments were used in management, beekeepers also need to plan decontamination wax procedures prior to recycling wax into new undrawn foundation.

Recycling Beeswax: Part 1

Background – Decontamination. Is it necessary?

Looking at today's happenings relative to beekeeping with artificial supplemental treatments for parasitic mites and secondary diseases one could wonder, just how do our present-day honey bees continue to survive? Can all this wondrous stuff (chemicals, essential oils, and artificial drugs) really keep being placed and dumped into our hives, without ill-effects happening, that will take years for both ourselves and future generations to reverse the effects of and clean up?

From time to time short blurbs about the seriousness of the problem are published, but is anyone paying heed? Taken from the Australian Bee Journal and published by Bee Culture, in the USA in June 1996, we learned of: *"Organic problems – Contaminated wax in Europe* – Scientists at the Specialists meeting of the 34th Apimondia Congress were in general agreement that the use of Apistan over the past 10+ years for the control of Varroasis has brought about a situation in which every kilogram of beeswax in Europe is contaminated with fluvalinate (the active ingredient).

It is most likely that recycling contaminated wax for the manufacture of foundations is largely responsible for the increase in residue levels. Scientists who spoke informally to the session, however, drew back from asserting that if the present usage rates continue the time must come when all beeswax has no value due to unacceptable high residues. However, they did note that importing foundation from countries with no fluvalinate usage could help stabilize the situation and may possibly reduce the problem and if the use of Apistan was also curtailed. A return to non-residue wax was possible after the next 50+ years! Many beekeepers only recycle their combs through a 10 year period and if wax recycling for foundation continues it will take 50+ years to get down to unreadable levels and this still assumes an almost complete lack of chemical usage commencing immediately.*Reading this, it is realistic to say.......it's not happening, this stoppage of deadly use, and the problem is becoming worse! Beekeepers reading this, look hard now at the USA and say the same scenario is not happening here, after all, we started using the same chemical approximately 2 years later!*

If Europe is already contaminated, and the situation is becoming comparable in the USA, and many countries in other parts of the world are treating also, *just where is all the uncontaminated beeswax going to come from to decontaminate foundation, to allow beekeepers the chance to retrogress back onto a clean sustainable system of keeping honey bees? Just who is watching the store? Question:* How can beekeepers put something into their colonies without knowing *how to get it out*? Further, How can governments approve the use of chemicals for active treatments and overlook environmental concerns relative to *health concerns of contamination and decontamination*, without asking the question during the approval process – How do you get it out, especially with food products concerned? For what started out to be short-term treatments, until a long-term biological solution can be found for our mite and disease problems, it would appear that not very serious business has been going on, due to the actual detrimental long-term results acquired and verified so far, that will now require years to clean-up.

Look at the question, just where is all the uncontaminated beeswax going to come from to decontaminate wax during recycling, for foundation to get down to unreadable levels? Just what makes beekeepers believe there will be much clean wax available to purchase for decontamination recycling within our beekeeping industry? Ours is an *economically driven world*. When the time comes, what makes beekeepers believe that the pharmaceutical dollar for uncontaminated wax in cosmetics and medicine won't prevail, and beekeepers will not be left to work the problem out, or go out of business because they won't legally be allowed to market contaminated hive products? As to what these same beekeepers might do for a living if they can no longer sell relatively-clean hive products?

They might end up becoming mandatory honey bee pollinators for food crops. No longer will they walk the fence reaping rewards from both hive products and crop *pollinating.* When their hives and combs become dirty with *contamination residues*, their hive products will eventually not be sold, and without sales, can their beekeeping outfits survive? Then, when their *migratory pollination* hives and combs become so *contaminated* brood will not live and the *sterility* of the hives, breached, *with sales of hive products having been previously lost due to excessive contamination and previous demonstrated inability to decontaminate combs, necessitating their migratory treadmill; and decontamination now cannot take place again due to lack of availability for non-contaminated foundation, either actual or economically, the final becomes irreversible. when this happens and it will, then mass food production will be irreparably harmed both here in the USA and around the world!*

Beekeepers should read the article in the December 1995, American Bee Journal by Klaus Wallner, Titled "*The Use of Varroacides and Their Influence on the Quality of*

Bee Products." In the article, Wallner goes over the fact that the more frequently pesticides are put into the beehive, the higher becomes the risk that residues can be detected in bee products such as honey, wax, and propolis. Further, that it is obvious that the wrong use of preparations creates residues. In temperate zones, contaminated winter food, is a possible reason for residues. This is because normally pesticides are used during the feeding period, with the consequences that winter food is sometimes contaminated with treatment chemicals. (We all know that the bees restore their food in Springtime to create more cell space for breeding. Thus, the winter food with its residues may penetrate the Spring honey thru bee relocation of honey stores). Also gone over in the article, was the fact that beeswax has an influence on honey quality depending upon the uncapping procedure used during honey extraction. Thus it is always important to skim honey as carefully as possible before it is transferred to end users. With a filter all coarse wax particles can be eliminated. During the honey extraction process, a great many fine wax particles float to the surface and can be skimmed. The remaining floating particles represent a source for residues in honey, if they contain pesticides themselves (Note: from contact strips as an example).

It was noted by him that this residue transfer really happens and can be easily detected. the higher the concentration in the wax, the more residues can be detected in the honey. He explained that a thin layer of contaminated wax was put into petri-dishes and a layer with approximately 2.5mm of honey free from residues was put onto the surface. The closed petri-dishes were stored in an incubator with 30,C for 30 days(about 86,F). Then Aqua dest. was added into the honey layer and after 24 hours the honey solution was poured off and analyzed. Among the chemicals very easily detected were coumaphos and fluvalinate. Wallner asked the following question concerning this matter: *Why should pesticides (wax-soluble) leave their favorite fat milieu and go to a watery medium, the honey?* He then seemed to give the answer, in that pesticides with a low tendency to migrate, accumulate very easily in the beeswax and particles may lead to residues in honey, and went on the state, In practice this means: The wax quality of the beekeeper also influences the honey quality.

From combs with high concentrations you cannot get unpolluted (uncontaminated*) honey*. Thus, the beekeeper has to prevent the concentration of residues in the comb during the years he uses pesticides. (*Question*: Knowing many European beekeepers renew their combs, that is, melt down old combs and start new ones with foundation every 3-4 years; few American beekeepers do this, it being common to find combs 50+ years old in everyday use, then how is the industry going to avoid the problem of contaminated honey on the open market?) With the simple test Wallner did, *it is now possible for the first time to fix a limit for allowed quantity of residues in wax without having dangerous effects on the honey quality. for some pesticides this limit is a 1mg/kg (1ppm), based on laboratory tests* (fluvalinate). Tests done by Wallner showed that in all countries where Varroacides are used, uncontaminated beeswax can hardly be found (Note: Australian Bee Journal article in Bee Culture), and often various pesticides can be detected in one sample. Further, that all fat-soluble pesticides are being preserved in wax for many years. Wallner then noted that this phenomenon does not only appear with Varroacides, but has *also been proved with paradichlorobenzene,* which is used against wax moths in some countries.

The following from Wallner, *"can residues be washed out of wax?"*, I am going to quote for beekeepers to read and heed, but it also may point to a solution to our problem which I shall write on later in Part #2, Recycling Beeswax. "Fat-soluble pesticides are stored in wax and are conserved. A decrease or a degradation does not occur, as far as we know. Therefore, one important question is whether or not there exists a chemical cleaning process that could solve this problem. Laboratory tests showed that heating wax to over 100'C and high pressure did not solve the problem. The residues in wax were destroyed with this procedure, but also the wax itself had been destroyed. The addition of bleach showed similar effects. The exposure to uv-light showed that only the residues on the wax surface were destroyed by this aggressive radiation. *With the solar melter the uv-part of the sunlight is not able to penetrate the sheet of glass.* In practice these attempts brought no solution. Many beekeepers have been using steam heated wax melters, but they did not bring remarkable effects either on the residue concentration in beeswax. Experience has shown that an effective residue cleansing of the beeswax is not possible with our state stage of knowledge. *These pesticides are movable in wax. They can move just like in Apistan or Bayvarol which permanently supply the plastic strip surface with residues from the inside of the plastic strip. If the pesticides that stick on the bees' legs fall down when the bees move in the hive, new pesticides are added to the wax layers. In the course of time, the chemical concentration in the comb wax of a colony thus is elevated and also the foundation in the new frames and the uncapping wax are contaminated. a real separation between the breeding place, where pesticides are used, and the honeycomb does not exist. The bees distribute these pesticides even through queen excluders. Naturally, the walls of the hive and the frames are not excluded. Together with the honeycomb wax, they provide a long-lasting source for residues that also influence the propolis quality. Propolis analysis showed that even there residues of Varroacides are to be expected, possibly in higher concentrations that those in the beeswax."*

Just as pesticides have created problems for beekeepers and necessitate caution, so should beekeepers be careful when using essential oils. While generally less toxic than conventional pesticides, caution and care should always be a consideration. Essential oils are also commonly known as "volatile oils," especially since they can be heated and distilled with little decomposition. Because of this, once in honey it's almost impossible to get out, especially as relates to taste and smell. That's why many individual components of essential oils, including thymol and menthol are produced synthetically in the laboratory for use in the perfume, pharmaceutical, and flavor industries. Concerning safety? Just because a compound is a "Natural Product" does not mean it is entirely safe. While the large majority of essential oils are reasonably safe in small amounts, some contain compounds that are not particularly safe. For instance, the compound thujone, a component of many essential oils, is quite toxic. Likewise, methyl salicylate, a component of wintergreen oil and teaberry oil can be dangerous. The Merck Index states that "ingestion of relatively small amounts of methyl salicylate may cause severe poisoning and death with average lethal dose 10 ml in children and 30 ml in adults"

Some industry beekeepers see essential oils as a "magic bullet" because of their reputation for controlling mites and their relatively low toxicity to the user (humans). But this does not mean it's technically good for the bees, nor natural to a beehive. If we as an industry are to overcome our problems, we must go back to a system of beekeeping paralleling the natural environment of feral honey bees as much as possible. Further, it must be simple (though labor intensive) and capable of being applied universally. While certain plants may be used by honey bees in localized areas, that used as essential oils may give limited control, if they were intended for use across the full spectrum of honey bees throughout our world, they would already be in place naturally controlling our parasitic mite problems, which they are not. If beekeepers are to travel the long way back to biological beekeeping without the use of chemicals, antibiotics, and essential oils, to a clean sustainable system not unnatural to Nature, i.e. technically organic beekeeping, then all that is not paralleling the natural environment of feral honey bees needs to be eliminated as management control tools. For over 2000 years going back to before the time of Hippocrates, beekeepers have kept honey bees without overwhelming problems of parasitic mites and secondary diseases. Chemicals, antibiotics, and essential oils have not been necessary all that time until recent history. The only change with recent history is the fact that man decided he could change what was natural and breed everything bigger and better. With the conception of this idea began many of today's woes as the eternal environment tries to correct itself and cleanse itself of that which does not belong naturally in Nature. Only wax, pollen, honey, and propolis come from a hive and are used by the hive workers to build their fabulous city naturally free of disease, and harmonious with our environment. Beekeepers should traditionally go back and keep it that way to solve today's problems.

Concerning antibiotics, of which we have only one approved in the USA, it is known in our industry to store Terramycin Soluble Powder (oxytetracycline) in a dry and cool place and to protect it from sunlight, for to do otherwise would break it down chemically, but does our industry really know how long it would take. It is universally used and assimilated in many hives throughout the off-seasons of beekeeping for medication purposes for diseases of foulbroods. Yet, because of contamination residues terramycin too, must have usage stopped a minimum of 30 days prior to a honey flow. But how many beekeepers would use it so lavishly, if they knew that it's been documented and scientifically proven by Villeneuve in 1980 concerning treatment for American Foulbrood, that when terramycin is used, the life expectancy of the bees is reduced by 50%. Further, terramycin kills its beneficial bacterial flora, enhancing the growth of yeasts and molds, particularly Ascosphaera apis which causes chalk brood. Because of this beekeepers should ask, besides the possibility of contaminating hive products, how safe to honey bees are the use of antibiotics?

Basically, all that this boils down to is that chemicals, antibiotics, and essential oils all have detrimental effects. Beekeepers should be asking what they are before they are used inside of a colony of bees because they bear the burden of correcting any detrimental effects, be it within the hives treated or the market the hives' products have been sold to. The hardest detrimental effects to correct being within the hive, because field management must change to correct the underlying causative effects, once they have been identified, that compromise the hives' well-being. The easiest being correcting the marketplace, for you merely stop selling contaminated hive products. In actuality, better said than done, for to correct todays' problems of parasitic mites and secondary brood diseases, many beekeepers will not survive the long way back to biological beekeeping without the use of chemicals, drugs, and essential oils, without the back-up of a stable marketplace in which to support their families on their long-road back to traditional beekeeping.

Decontamination: What will it involve to clean up my combs?

We have gone over some of the background as to why decontamination is necessary for beehives, if beekeepers today are going to retrogress their colonies back onto a natural system of beekeeping, without the use of chemicals, essential oils, and antibiotics in their field management. We know from articles published that decontamination is not easy and will take several years to accomplish. Further, that it has been noted by Scientists at the Specialists meeting of the 34th Apimondia Congress that a return to non-residue wax is possible after the next 50+ years, assuming that an almost complete lack of chemical usage commences immediately (this was written in 1996). Of course we know that this stoppage across-the-board will not happen in countries throughout our world. But it can happen one beekeeper at a time, willing to say enough is enough, and willing to make the long journey back to biological beekeeping.

Decontamination cannot take place without cessation from the use of chemical, antibiotic, and essential oil treatments. Herein lies the problem: How to best get off of the *"chemical treadmill"*. There is no way to help individual beekeepers as a uniformly handled industry to proceed, as each has varying needs and personal economic thresholds, relative to liquidity of their beekeeping operations, that must be considered. Further, since it is possible to fix a limit for the allowed quantity of residues in wax without having dangerous effects on honey quality, some beekeepers may not want to decontaminate all the way back to non-residue detection parameters.

Note: While this would allow the sale of filtered clean honey, it *would not* have bearing on other *still contaminated* products of the hive, namely: beeswax and propolis. Still further, some beekeepers may want to decontaminate all the way back to non-residue detection parameters, but due to the size of the beekeeping outfit they maintain, they may not be able to make it in one decontamination stage due to 1) either the type or amounts of chemicals, essential oils, or antibiotics they have used, and 2) the amount of time (number of years) they have been treating, or 3) whether or not their bees in their hives have *hit-the-wall* by crashing in numbers or becoming totally resistant. These beekeepers will have to decontaminate in stages, by reducing the level of contamination concentrations by addition of uncontaminated wax, until they reach the level of non-detectable residue levels (Maybe 3+ recyclings of their hives, combs), while in the interim selling filtered clean honey, until the level of selling natural biologically clean honey can be achieved along with that of clean beeswax and propolis.

It should be obvious that the smaller beekeeper here has the advantage in that 1) there are less combs to replace making it easier to acquire uncontaminated beeswax on the open market or decontaminate the hives he already has, and 2) his livelihood is not all at stake as his sole source of family income (or a major part thereof).

Commercial beekeepers can probably be broken-down into four distinct groups, each which will have differing decontamination capabilities and needs. 1) Those that are small family beekeeping farms, self-contained in their workday needs, and 2) Those that are large beekeeping farms, employing outside help in their workday needs. 3) Those that are part-time migratory sometime during their beekeeping year, and 4) Those that are full-time migratory throughout the beekeeping year.

Small family beekeeping farms

Most mom and pop beekeeping farms run anywhere on average from 300 - 1200 colonies, and like the name implies, mom and pop do the work. These beekeeping farms should estimate retooling approximately the equivalent of 5 deeps for every hive operated, not counting extracting supers used to rotate the yearly crop of honey from the field. In an operation like this you could easily expect to retool (decontaminate) over 60,000 frames. If retooling involves melting-down and changing-out frames to new processed uncontaminated foundation bases, estimate now half this much again to draw-out foundation right, with decent culling of misdrawn combs. Then estimate how much your bees can successfully draw-out each year conservatively. Then figure out how to separate extracting clean honey from decontaminated frames, from extracting honey from non-decontaminated frames, as in the future, it is sure to have great bearing on honey marketing with the new food safety laws coming into effect. If possible, figure two separate extracting lines, in lieu of washing down extracting equipment to keep end products separate and distinct.

High-grade selling hive products where you can, while you salvage with a packer what you cannot that needs decontaminating filtering (think about sending honey from non-decontaminated frames to a packer for purchase and/or filtering (contract work), unless you have the necessary equipment to do the job yourself (with the new food safety laws being written, figure filtering out any floating contaminated wax, pollen, and propolis particles). Figure retooling combs one entire beeyard at a time. Do not try to hodgepodge or you will soon lose your place and have to start over due to co-mingling. Figure depending upon the number of colonies owned taking an average of 10 years to complete (less for closer to 300 colonies maintained and up to 15 years for closer to 1200 colonies maintained). If possible to reduce costs, suggest the purchase of a hand-crank foundation mill or even a motorized foundation mill to self-contain, along with a Kelly wax-press to have fuller control of decontamination of wax, and the processing of recycled wax into foundation, so you know exactly the nature of the wax you are working with product-wise, and where you are at, at all times.

Large beekeeping farms

Most larger, beekeeping farms run anywhere on average from 2,000-6000 colonies, figuring approximately 1,000 colonies maintained for each worker hired. Figure approximately the same number of frames (60,000 + 30,000) to be retooled per 1,000 colonies over the course of 10 -15+ years. Now figure hiring 1-2 extra workers to do nothing but retooling of frames (especially the closer to 6000 colonies you maintain). If you wish to self-contain to maintain control of decontamination of wax and the processing of recycled wax into foundation, suggest the purchase of a motorized foundation mill or even a small foundation assembly-line. As most larger beekeeping farms probably already own their own wax melting/processing equipment, suggest now an extra Kelly wax-press to have fuller control of decontamination of wax.

(The other two beekeeping groups I am not going to go over, as information needed can be retrieved from either of the two groups already gone over).

Preparation for the wall: There will come a time when all beekeepers who are treating with essential oils, chemicals, and antibiotics will be forced to make a decision, for the treatment of parasitic mites and secondary diseases, due to their hives being on oversized bigger-is-better combs. At that point many will throw-up their hands and without family-support, go out of the farming business! You will know you are getting close to the *wall* when your broodnests collapse from oversaturation (absorption) of treatments, built one upon each over, during the course of several years of treatments, breaching the sterility of the broodnest. When your broodcombs will no longer support life, you have a choice: *throw your cards on the table and walk away, or pick up the pieces and retool correcting the problem by walking away from the oversized combs, chemicals, antibiotics, and essential oils with your head-up.*

At this point, since retooling is mandatory, and decontamination with salvage of your combs is the only way out, consider now taking the long way back to biological beekeeping without all the quick-fix gimmickry to get yourself off the treatment treadmill. If you are serious about staying in the beekeeping farming business you might consider striving to get your outfit 25% converted to biological beekeeping before the wall hits, because the drawing of new foundation combs is the hardest process to accomplish after decontaminating and recycling your wax. With 25% of your outfit preconverted, and combs drawn-out to a minimum of 3-4 deep supers with accompanying bees, you have a chance to split to take-up numbers when your outfit hits the wall, because not all of your other colonies hopefully will not all collapse at once. Here, separation of extracting lines and equipment in the field will pay big dividends as to whether or not your beekeeping outfit goes another generation or not. Fore-warned is fore-armed as to what to expect and anticipate! (Third Option:- You could stay on the treadmill, rotating oversized combs and treating and treating as if in a time-warp until you give up, bankrupt, or pass the treadmill onto your own next generation, to either continue or make the decision to get-off and go back to traditional biological beekeeping methods.)

Recommended Decontamination Procedures

Having gone over some of the background as to why decontamination is necessary for beehives and some of what it will involve to clean-up contaminated comb, let's rationalize a decontamination plan.

There is no way presently known (published) to totally decontaminate a beehive and it's combs once contaminated with foreign substances to the point, that the hive and it's combs will no longer support life (the rearing of brood). Beekeepers must keep in mind that the application of foreign substances to a colony of bees, be it chemical, essential oil, antibiotic drug, or feed substitutes, always leaves residues. There are no exceptions! This means it is a long way back to natural biological control without the use of chemicals, antibiotics, essential oils and even feed substitutes, because unless beekeepers can immediately stop *total use of them all*, to retrogress back onto a natural system of beekeeping that equates parallel, totally to the feral, including comb size, they can never make the leap back to traditional beekeeping food safety – organically sound practices.

The ideal situation for decontamination would be to be able to purchase new clean woodenware and clean foundation (either residue clean or residue significantly reduced) and shake your bees onto it and restart. This would be costly for many, because they would still have to draw out new combs, providing the right natural size is available for purchase in large quantity. Of course we know that this is not the scenario at hand presently.

Where woodenware is to be saved and reused, traditional field practice since the turn of this century has been to separate the wax combs from frames, scraping clean, thus separating wax from woodenware. They are both then processed differently for salvage. One is cleaned (woodenware) and one is melted and recycled into wax blocks for sale or made into comb foundation or candles for own use or sale. Today with the advent of treatments for parasitic mites and secondary diseases, beekeepers have to deal with learning how to bring down the level of residue contamination for both woodenware and wax, whether it be for reuse or sale. Concerning reuse, if available, the purchase of bulk, clean decontaminated wax, or residue significantly-reduced wax must be a consideration in the future for all, whether bought in wax blocks or pre-embossed into natural sized wax foundation.

New plastic combs could also be another alternative (Note: used even though plastic could still contain residues). However, if purchasing plastic combs keep in mind, while honey and pollen and propolis can be gathered for marketing later, rule out the production of comb honey and possibly organic honey due to the plastic base. If purchasing new pre-embossed foundation, beekeepers should make sure whether or not it is clean of all residues (uncontaminated), clean decontaminated wax (residues processed out), or residue significantly reduced wax (most residues removed but not all). This will have bearing on their honey and wax sales later under food safety laws as they gain momentum in the new millennium. Beekeepers must keep in mind they will still have to dispose of contaminated wax if they opt not to reprocess their own for personal use or sale, probably at substantially lower prices that what they are used to, whether to a wax buyer or contract to have someone else clean-up their wax for them.

Woodenware and metal hive parts can be easily cleaned with boiling water by either pouring over or dipping within, once scraped clean prior to cleaning. Residues are not easy to get rid of. Laying outside in bright sunlight will photo degrade surface residues with UV light, but sunlight cannot penetrate very deep to neutralize residues locked within the wood. Fluvalinate is soluble in organic solvents to some degree. Coumaphos hydrolyses under alkaline conditions the warmer the faster, or said in another way, it undergoes decomposition by water that is alkaline solution based. Beekeepers around the turn of the century and up to the 50,s – 60,s used to clean old equipment for reuse with lye baths (potassium hydroxide / caustic potash, sodium hydroxide / caustic soda). 1 pound lye to 10 gallons water minimum to 25 gallons maximum, depending upon how fast they would want to clean. It is an alkaline solution so it would help neutralize residues, and it penetrates wood, but to use, one must take care to protect eyes and wear rubber gloves and apron, and not allow to soak for too long (about 15 minutes with higher concentration of 1 lb. lye to 10 gal water, as mixture softens wood if allowed to soak too long).

Following cleaning, the equipment is rinsed well with clean hot water. Equipment should then be laid out in sunlight to both dry and use UV sunlight to photo degrade the surface of the bee equipment for any remaining residues, making sure frames, supers, tops, bottoms, etc., are turned, exposing all surfaces and sides to the sunlight for at least 48 hours. Aside from being labor intensive for which there is no way to avoid, the process is dangerous in that lye can burn skin and is corrosive when splashed into eyes and much care must be taken using this process. Today it is considered not very friendly relative to the environment, food safety, nor handling by people. Another way is needed!

Many beekeepers erroneously believe that processing wax cappings / slumgum into wax cakes, by either use of solar-wax melters, water-bath presses, or steam-heated melters, they can reuse wax reclaimed without fear of residue problems. But most substances applied as preventative treatments for parasitic mites and secondary diseases migrate both into beeswax and honey, and are extremely hard to get out; and in reality, although approved for treatment, actually these compounds have no known way (published) of being decontaminated completely out of beeswax in one fast processing cycle, nor removed from honey.

Many residues that contaminate wax are insoluble in water and require an antidote to hydrolyze according to various Farm / Agri Chemicals Handbooks. Here many times, the current effect, in decontaminating the wax residues, is that the wax itself is destroyed. Some residues high heat cannot destroy without destroying the wax itself. To destroy residues by burning also destroys wax. But in one instance this is alright as candles are meant to be burned. If one processes beeswax into candles and sells, as the residues are burned (possible fluvalinate and coumaphos here in the USA), a product is saved as a monetary source.

But caution, here in the USA it is known in industry practice, that some beekeepers have also used chlorine bleach within hives to fight secondary diseases by adding to liquid feed or spraying diluted, lightly on brood combs, or even bleach their wax to obtain a higher color grade. This is dangerous practice, in that chlorine is readily absorbed by beeswax (or honey) and when recycled and made into candles, chlorine gas is given off when candles are burned. As for other residues destroyed, well this is okay as long as the beekeeper knows for certain that no harmful residues are escaping the burning process, that might take the lice out of someone's hair (reputable testing needs to be done to ascertain that no harmful residues are being given off)! As for sub-lethal residues returned to the hive in the form of wax recycled into foundation base, well the problem is still there - contamination!

In the December 1995 issue of the American Bee Journal, Klaus Wallner writes in his article about a laboratory test to show the migration of residues from wax into honey. "A thin layer of contaminated wax was put into petri-dishes and a layer with approx. 2.5mm of honey free from residues was put on the surface. The closed petri-dishes were stored in an incubator with 86° F (30° C) for 30 days. Then aqua dest. (distilled water) was added into the honey layer and after 24 hours the honey solution was poured off and analyzed." *Again – this is a very important fact to know.* It shows the penetration of residues from wax into honey per say, that is, if there are residues in the wax it indicates there *will be* residues in the honey.

It must be understood to decontaminate! *Why should pesticides which are wax-soluble and water-insoluble migrate into the watery medium of honey?* This, Klaus Wallner showed, with the chemicals fluvalinate and coumaphos, which are now both approved for treatment here in the USA for parasitic mites and believed to be safely held in the wax with no fear to the beekeeper of residue contamination of honey if used properly.

Obviously the belief of no residue contamination with contact applied pesticide strips is in error, but to what degree? On the other hand, we are looking here for possible decontamination possibilities for beeswax , and how it can be applied to wax combs. It does show that honey pulls chemicals held in wax out, by migrating it into the honey itself, thus lowering the residue level in the combs, but how?

Let's look closer at this enigma as to why pesticides which are wax-soluble and water-insoluble, in actuality, migrate into the watery medium of honey. One would think that this should not be happening, but as proven by Klaus Wallner and published in ABJ in December 1995 it is. We know that fluvalinate is acid based when we look at various farm chemical handbooks that say among other things it is corrosive to eyes. We know that coumaphos is also acid based in that it hydrolyses under alkaline (opposite of acid) conditions. We as beekeepers know that honey is acid based. This being the common link then one might suppose that a honey-water bath (but not strong enough to make a wax emulsion) might be used to migrate residues out of wax during decontamination processing in a wax press to reduce the residue level lower, once wax combs have topped-out and can no-longer support life. But what could beekeepers use for a second water-bath to lower the residues even further?

Let's look again at the old turn of the century water-bath method of washing beekeeping equipment in lye. We know that lye is in reality a strong alkaline solution of potassium hydroxide (caustic potash) or sodium hydroxide (caustic soda), and an alkaline solution over time will decompose even coumaphos. But beekeepers must also remember that anything that will wash wood, penetrate it and decompose it if left soaking too long, and shine rusty nails, will probably destroy the wax within the combs too (that's why they are removed first)! Further, honey bees probably would not want to live in these combs even if they could be treated with a very mild solution. So what do we have available that would be food-friendly, and alkaline (to neutralize what is left after a honey-water bath initial processing of combs into wax), and not burn fingers, or be corrosive to eyes if splashed into, when used?

Further, it must also be cheap because we are beekeepers, and be environmentally friendly with the growing emphasis on food safety. The only thing that comes to mind when looking at known alkalis (a hydroxide of any of the alkali metals, soluble in water and capable of neutralizing acids i.e. lime, lye, magnesia, lithium, sodium, potassium), or alkaline compounds (earths such as the oxides of calcium, strontium, barium and magnesium), is maybe that Baking Soda (sodium bicarbonate) would work. We use it in cooking, for medicinal purposes, as a laundry aid, to absorb odors, to clean, and use it also to neutralize battery terminals. Perhaps in putting the two baths together, one following the other, we can come up with a decontamination process for our hives that includes both woodenware and wax combs, though it might take a little work.

Suggested decontamination process

We have talked about saving woodenware using a lye bath. Now suggest to beekeepers to use a sodium bicarbonate bath to clean their equipment with. Both are alkaline solutions but the sodium bicarbonate is user friendly and should not destroy the woodenware if left to soak too long, further, it is environmentally friendly. The results with AFB and EFB fears should be the same. – *eliminated* – for there will not be enough spores left for bees to find and digest to re-infect thus starting a problem. Further, bees will not be repelled by the sodium bicarbonate solution, like they are with lye, if the frames have to be reused right away within a few days. Following cleaning, the equipment should still be rinsed well with clean hot water and laid out in sunlight to both dry and use UV sunlight to photo degrade the surface of the bee equipment for any remaining residues, making sure frames, supers, tops, bottoms, excluders, etc., are turned to make sure all surfaces and sides are exposed to sunlight for UV photo degrading of at least 48 hours each side.

We still cannot talk about honey decontamination because there is in actuality none, other than heating to try to remove objectionable smell by volatilization, and/or filtration of solid particles to remove contaminated solid matter (residues held in wax particles, pollen, and propolis, bee parts, wood, etc.). Residues which are locked within the honey itself so far as we know cannot be removed without destroying the honey.

Decontamination processing for old combs, cappings wax, slumgum wax salvage, to recycle into wax for market (candles or bulk wax buyers) or for recycling into new foundation for either personal use or sale is recommended as follows:

1. Since both fluvalinate and coumaphos now approved in the USA are acid based to some extent, and honey is acid based; and we know that traditionally honey taken before meals helps older persons, stomach acids digest foods better showing that acids mix together, then recommend wax be processed initially in a honey-water bath. (Caution: Reference Coggshall and Morse in "Beeswax" that state, "Beekeepers occasionally report the formation of wax emulsions during rendering; these are sometimes difficult to break. The most common emulsion is a wax-in-water type, described in the beekeeping literature as granular wax. Granular wax is found most

often when cappings or combs in contact with *considerable honey* are melted without the addition of sufficient water to dilute the honey. Remelting the wax with a large quantity of water usually breaks the emulsion and the wax solidifies in firm cakes." We have also found that letting the emulsion set and dry, then rewater-bathing also works.)

Further suggest double-sacking within burlap bags tightly closed, layered between pressing plates under water in a Kelly wax press. Recommend lowest grades of honey be used (also create a market) and heating of honey-water mixture to 190° F (88° C) and held there for a minimum of 24 hours to slow cook rather than rapid boil. At this temperature residues which are volatile should release, residues which are acid friendly should migrate somewhat into the acid based honey-water with the help of heat, dirt should release from the wax to help lighten, and when pressed, - cocoons, pollen, and small solid impurities should all remain within the double burlap sacks, which in itself will reduce contamination still contained within solid matter by separating from wax. Drain Kelly wax press from bottom after wax solidifies on top after most wax has been poured off into cakes.

2. Since coumaphos is what many beekeepers in the USA consider hard-core pesticide treatment where the line is crossed (equate fluvalinate usage to marijuana and coumaphos usage to heroin), suggest second water-bath to further neutralize coumaphos residues (fluvalinate has a CF3 besides Cl; Terramycin has nitrogen and carbon to neutralize). Suggest baking soda (sodium bicarbonate) be used in a second water-bath (1/2 cup) in the Kelly wax press, to create an alkaline condition within which to hydrolyze residues at 190° F (88 ° C) and hold there for a minimum of 24 hours.

(Caution: Reference Coggshall and Morse in "Beeswax" that states, Beekeepers occasionally report the formation of wax emulsions during rendering; these are sometimes difficult to break. Less common is the water-in-wax type of emulsion, where the solution of water, with whatever impurities it contains, is held in droplet form by a layer of wax that surrounds it. On cooling, the wax holds a considerable quantity of liquid. If this emulsion is more complete, the wax does not become a solid cake but instead a mushy mass, from which a large volume of solution can be squeezed by hand pressure. Remelting the wax with a large quantity of water usually also breaks this type of emulsion. We have found that too much sodium bicarbonate causes the wax to lighten considerably into a fine paste emulsion. This was then allowed to set out to evaporate water, after which when dried was reprocessed into a wax cake.)

Suggest Kelly wax press be filled with wax processed and drained from step #1 in loosely layered cakes or cut-out chunks. Then press is refilled with hot water with baking soda added and wax allowed to melt. Stir floating wax occasionally throughout the time period to keep separation point between wax and water co-mingling. You will notice wax lightening in color during the period as more dirt and impurities drop out into the water and sink to the bottom. Cool and drain.

3. Repeat step #2 with plain water bath to rinse (liquefy in hot water stirring occasionally for a few hours to rinse) and drain into thin trays to cool (could be large shallow cookie type sheets or trays) and set-out into bright sunlight to photo degrade any remain surface residues with UV light 1/4 to 1/2 thick. Flip over after 2-3 days and continue for another 2-3 days. If unable to set outside in bright sunlight consider UV black lights inside. Set up lights in a series shining onto a 4 x 8 sheet of

plywood on legs (or large table) for holding thin trays of wax.

4. Market to bulk wax buyer, or recycle into candles or foundation base.

If unable or unwilling to process contaminated combs to decontaminate, then beekeepers must eventually junk them or send them out to professionals for processing when their bees can no longer live upon them, or buy new clean processed wax or foundation. However, bear in mind as combs and equipment become more contaminated, their value due to the contamination will rapidly depreciate in value, so to recoup loss, many beekeepers will be forced to learn to either process or trash every comb they own or possibly go out of business.

It should be obvious that decontamination will probably be a long hard process, but when combs will no longer support life and resistance cannot be avoided, it's either replace with new or recycle the old. Many beekeeper will probably not want to go through the process more than once. To junk combs completely and replace with new clean combs is costly. To recycle is time consuming and costly also. To stay on an artificial oversized bigger is better system aided by the use of chemicals, drugs, and essential oils to control parasites and secondary diseases, until ones spirit is broken or one becomes acclimatized to a perpetual time loop of changing wrong size combs is frightening! Unless beekeepers are willing to make changes to go back to beekeeping's traditional methods, used before today's current problems started in the first part of this century, they will not be able to solve their problems. Indeed it will be a hard long way back to biological beekeeping without the use of drugs, chemicals, and essential oils, but it must be done if our worlds industry is to survive!

That beeswax is different in composition than honey is obvious to beekeepers. That most pesticides for Varroa control used in Western Europe and the USA are fat soluble substances and will have residues accumulate in the wax combs, to give off sub-lethal dosages of these same pesticides, is not so obviously understood by beekeepers. Even less understood are the consequences of mixtures of sub-lethal dosages coming together, that could be devastating to a hives well-being 5-7 years into treatment with a pesticide with no prior use of chemicals, and 2-3 years into treatment with the second pesticide used, once mite resistance has totally developed to the first pesticide used. If mite resistance to the chemical develops 2-3 years into the second pesticide used, the beekeeper very well may be left hanging to handle decontamination of a bad chemical reaction internally, locked within the treated brood combs themselves (residues migrated into the wax and held there from each pesticide used in succession), now spread throughout the whole hive structure by contact transfer migration that will continue to give off chemical reactions effecting his colonies well-being, until removed by decontamination procedures from the hive. Beekeepers should have plans on how to handle this situation, for it is coming, if not already here.

Fluvalinate has been approved for use in the USA for several years now to fight Varroa mites. The most current usage being in the form of Apistan control strips. Now

Coumaphos is being rapidly approved in the USA for use under a Section 18 authorization only, sold as Checkmite control strips, to combat both Varroa mites and small hive beetles! *Question*: What do beekeepers know about these two pesticides, in-hive effects of sub-lethal dosages used either singly or mixed in synergism, locked into wax combs, that they will be forced to deal with later on relevant to effecting their colonies, health, plus decontamination?

Symptoms of chemical effects that show the pesticides are working

Fluvalinate

USA beekeepers should be aware and I am sure many of them remember reading in The Varroa Handbook, by Bernard Mobus and Larry Connor, that "however inviting the treatment may seem, we should heed a warning given to German beekeepers who want to play as researchers, that *fluvalinate* could have neurotoxic effect in humans (ADIZ, April 1987). Handle paper, wood or plastic strips, or any other formulation of *fluvalinate* with great care and gloves. In Agricultural Chemicals Book 1 Insecticides, 1992 Revision, we learn that fluvalinate is a synthetic-pyrethroid compound used as a selective contact and stomach poison insecticide. It may cause eye and skin irritation. It suppresses spider mite populations. Further, it maintains its activity under high-temperature conditions.

Fluvalinate is a Class 2 synthetic-pyrethroid compound that is unique, in that besides maintaining its activity under high-temperature conditions, it is a pyrethroid that works to the negative, in that it gets stronger as the temperature gets colder (Chaney, 1988 PHD Thesis). What happens in this inverse relationship is that the fluvalinate stored in the hive may not singularly cause colony mortality, but will act in conjunction with other factors i.e. temperature or interaction with other chemicals bees are exposed to, to increase stress within a beehive and cause a decline in population of adult bees in an overwintering situation. If there are insufficient bees left coming out of winter to begin hive build-up in the Spring, then colonies quickly crash with inclement fluxional changes in the weather due to the chemical inverse relationship to colder temperature (Note: Look for bees dying around the outer layers of the cluster, like peeling the outer skin off fruit or leaves dropping off of a tree, with each successive cold snap, dropping in waves to the cold).

Reports of hives crashing from mite populations should therefore also be tested for adverse chemical inversion to see which is the real culprit! Chaney's PHD thesis showed that the relative toxicity to adult honey bees of fluvalinate was shown to increase at 18° C (64° F) and 12° C (54° F). At 25° C (77° F) the LD50 was 800ppm and observation of bee behavior was acceptable. At 18° C (64° F) the LD50 dropped to 615ppm and observation in bee behavior changed. At 12° C (53° F) the values for fluvalinate were not shown due to the great difference in magnitude as temperature dropped and continued to drop, and the observation of bee behavior was bees not clustering; no normal behavior observed and feeding was reduced significantly.

In the Varroa Handbook, by Mobus and Connor it was also pointed out that bees exposed to fluvalinate had a memory loss in forager bees, rendering them useless for nectar / pollen foraging (Beekeepers not knowing what this means could equate this to short term memory loses, like Alzheimer's disease, that gets progressively worse as residue levels rise). Wolfgang Ritter pointed out in Chemical Control: options and problems in Living with Varroa, Edited by Matheson, that "the effectiveness, especially of pyrethroids, persists for several months the mites emerging from the brood are also killed, thus enabling a successful treatment for colonies with brood...residues cannot be avoided, especially in wax, if colonies are treated for a long period.

Question: What happens when resistance sets in and mites stop being killed, but the residues in the combs are still there reacting against the bees, and now are contaminating the honey; or worse yet, to fight the resistance – another chemical is added? In Pollinator Protection, by Johansen and Mayer we learn that mixtures of more than one pesticide are a special hazard to bees as relates to certain insecticide-specific miticide combinations. In such cases, the miticide appears to have a synergistic effect which causes the mixture to be more hazardous.

Coumaphos

The first thing beekeepers need to learn now about coumaphos is that the Agricultural Chemicals Book 1 Insecticides, 1992 Revision, by Thomson says "Do not use before or after application of natural or synthetic pyrethrins or compounds used to synergize them." This means that they react together so that 1+1 does not equal 2. It will probably equal much more. Coumaphos is a systemic, organic phosphate (nerve gas). What this means according to Pollinator Protection by Johansen and Mayer is that "The term organophosphate is a generic term used to cover all the toxic organic compounds containing phosphorus." They kill animals, including insects, by inhibiting cholinesterase, a vital enzyme of the nervous system. Constant disruption of nervous activity occurs at the nerve endings. Insects literally jump their nerves to death.

Organophosphorus compounds *are likely to cause the bees to become agitated and aggressive* (Just what Calif. is looking for with AHB syndrome on hand for public perception). They also cause paralysis, abnormal jerky, wobbly, or rapid movements (Equate to motor dysfunction in higher animals). Bees slightly affected by some organophosphorus compounds will crawl up the walls of the hive and fall to the floor over and over. Severe poisoning (or high residue build-up) leads to lack of young workers. Poor housecleaning is another sign typical to look for. Nectar is often deposited in empty brood cells and queens may stop laying simply because there is a lack of clean cells to receive the eggs (propolis coating the brood cells has a high affinity for chemicals).

So what does this mean? It means according to the chemical books, we have two approved incompatible chemicals coming together with effects outlined above to create new effects of the unknown. Add then to this the fact that Florida in the USA, one of our milder climates, is where coumaphos was first approved. This is a relatively warm climate, even in the winter. Now it is approved for much colder climates without prior pre-testing here in the USA in those colder climates. It is known that chemicals applied during cool weather retain a longer residual hazard and regional differences in the hazard of a given pesticide can often be explained in terms of differences in climate. If one equates this with temperature variation and fluvalinate is known to get stronger with its LD50 as the temperature goes down, then in about 2-3 years if not already this year, strange things could start happening in our Northern USA states in beekeepers hives, as coumaphos is added now to the scenario and these two chemicals mix with residual action within the wax combs, giving off sub-lethal dosages for the bees to contend with. I hope our industry is ready!

On another note briefly, even when using essential oils attention must be paid to residues, because some of them downgrade the scent and flavor of honey or reduce honey production or are even injurious to human health. In the December 1999 issue of the American Bee Journal, is an article beekeepers should read by Heather R. Mattila and Gard W. Otis, titled "Trials of Apiguard, a Thymol-based Miticide. Part 1. Efficacy for Control of Parasitic Mites and Residues in Honey." In the article they write, "Although not statistically significant, honey production was reduced by 30% during the Apiguard treatment period, a result which warrants further study."

Question: Would not this be highly significant to beekeepers pocket-books? Further stated in the article was, "The influence of temperature on the evaporation of volatile oils is difficult to regulate. High temperatures cause rapid vaporization of thymol, exposing bees to lethal concentrations, while low temperatures reduce vaporization and result in ineffective control of mites." They further referenced that Cox et al (1989) found that colonies treated with menthol during the summer registered significantly lower colony weight gain and honey production. Bees were repelled by strong vapors in hot weather; few bees were seen inside the honey supers while large numbers of bees covered the fronts of the hives. So much for approved alternate control in the USA.

Question: What happens when equipment is mixed as hives die and outfits go out of business or sell off equipment downsizing? Who really knows what is going on within our beekeepers hives? Now add to this the added compounded problems of illegal spray or spraying per say during crop pollination and one can now really say do the bees have a chance and why is there not more comb rotation here in the USA.

Making Foundation by Hand

Recycling one's own beeswax into foundation is not hard. It is the other side of off-season maintenance of woodenware supers, frames, tops and bottoms. After all, once you make the woodenware, don't beekeepers need something to put into it, and there is nothing wrong with homemade. Homemade lets beekeepers know what their own beeswax purity is, besides letting them know they did right by their bees biologically without the use of harsh chemicals, antibiotics, and essential oils, or artificial feed, to produce a quality food safe product for their own personal use or sale.

Equipment for making foundation

Since equipment for making foundation is currently not sold on the market, other than foundation mills, beekeepers will have to make their own which really is not hard. Three vats will be required: 1) one for dipping wax sheets, 2) one for holding dipping boards, and 3) one for melting a steady supply of wax. 4) Also, required will be pine dipping boards, and 5) a hot-box for warming wax sheets.

Making equipment

Vats for dipping and melting wax are double vats made by using 15 gallon barrels 27″ high x 14.5 – 15″ wide tops, with 7.5 – 8 gallon smaller barrels mounted inside

20 -22″ high x 10″ wide tops. Use an electric immersion heater mounted in the side of the 15 gallon barrels near the bottom. (Walter T. Kelly sells electric immersion heaters, supplied with 1″ pipe threads, thermostat controlled with graduated off-on switch adjustable from 60° – 250° F (16° - 121° C) with 4 foot capillary tube which can easily be connected with 12-ga. wire) Over top of the immersion heater a shallow stand approximately 5″ high needs to be used to support the smaller 7.5 – 8 gallon inner barrel. The inner barrels will need tight fitting metal lids (ideally – the original that were made for the barrels to close), to use during heating/setting overnight, to alleviate condensation of moisture into the wax there contained. Make the 5″ stand from a metal 5 gallon bucket by cutting off the bottom 5″ with a metal cutting blade. Cut a rectangular cutout to allow placing the stand over the heating element approximately 4″ x 4″. Then punch a series of holes at roughly 1″ intervals around both sides of the welded bottom circular seam with a series of smaller circles punched on the circular bottom itself to allow air to escape from heating the water in the jacket in the vats. Simple lids for the double vats can be made from plywood with handles, with an under coating of aluminum coated stiff insulating foam, to help retain heat when the double vats are turned off at night. Around the outside of the double vats a simple hot-water heater jacket can be used to retain heat and also thus lower heating costs. Recommend 220 vs. 110 voltage also as an economy measure.

Alternate use for double vats

If inner smaller barrels are stainless steel (can be found if one looks around) the double vats can also be used for making long taper candles, besides foundation base which can also be used to make candles, when your own hive foundation needs are met. Also beekeepers could use for bulk pouring of wax ornaments for sale, as rubber molds are relatively easy and inexpensive to make, verses buying them.

Beekeepers might want to make extra outer vats from 15 gallon barrels with immersion heater elements mounted within, for use in recycling combs from culled frames. The wax would be accumulated as a combination wax/slum block for further processing. The wooded frames could be easily scraped clean in the hot bath for further soaking later in large lots for further decontamination if necessary. Remove wax with cocoons, pollen, etc with a metal sieve made from a wide-top shallow can with holes punched in a bottom corner to allow water to drain-out, then wax/slum is dumped in pans to accumulate until fill. When sufficient pans or blocks of salvaged wax/slum are gathered, then process in a Kelly wax press with decontamination processing if necessary. Upon completion, either further process into foundation, candles or sell.

Making equipment

A hot-box for warming wax sheets in stacks can easily be made from an old refrigerator, that is gutted inside with the exception of the shelves, which will be used for holding sheets of foundation (could also be used for liquefying honey in various sized containers). Inside on the floor of the refrigerator place a fan forced electric heavy-duty space heater that has been linked into an automatic thermostat with settings between 60° - 150° F (16° -66° C), with dial settings mounted on the outside. Wax will be warmed using temperatures between 100° - 110° F (38° - 43 ° C) approximately (use a setting of approximately 120° F (49° C) to liquefy honey slowly in containers without over heating to preserve natural vitamins and minerals). Remember to key the automatic thermostat to key readings of temperature to the middle shelve of the inside box (top shelf will then register automatically slightly higher and bottom shelf will then register automatically slightly lower, with thermometers being placed on each shelf.)

Dipping boards for making wax sheets should be made from single-ply pine boards and measure 9″ x 16 1/2″ long, with as few knots as possible. If knots are unavoidable, make sure they are real tight knots that will not come loose later on with successive soaking and dipping, sanding, etc. Recommend having a minimum of 3-4 pine dipping boards on hand at all times.

Miscellaneous equipment

Other equipment to have on hand for use in making foundation is 1) an instant hot-water maker pot, 2) a paper-cutter for making sheets of paper to put in between sheets of stacked wax, 3) round pizza-cutter, 4) thin masonite boards cut to size foundation wanted, i.e. 8″ x 16 1/2″, 4″ x 16 1/2″, 9″ x 16 1/2″, etc., 5) flat cutting station or cutting board for foundation, 6) plastic-bristle scrub brush for cleaning foundation mill surface, 7) roll of 30 x 36 plastic garbage baggies, and phosphate free biodegradable ingredient liquid lemon-scent dish soap.

Dipping wax into usable sheets

If beekeepers are making foundation for their own personal use, it is not necessary to have yellow wax to make foundation with. In fact, we have found that the darker grades of wax foundation are more readily accepted by the bees than lighter wax grades and mill easier into foundation. We believe that the difference is the propolis content which gives the wax foundation better strength and ductility. (Note: Proper warming and cooling of wax sheets also helps to give foundation better ductility)

Prior to dipping wax sheets for foundation, the two double-vats of wax are heated and liquefied overnight to be ready the following morning. The double-vat to be used for melting wax, to have on hand a continuous supply of dipping wax, is heated to 190° F (88° C) and held at that temperature. Beekeepers will find that by doing this, following completion of the first days dipping and turning off the wax vat at night, by having the vat insulated so that it will not lose too much heat, the vat to some degree will act as a clarifying tank at night and give impurities within the wax (mostly dirt suspended) a chance to settle to the bottom of the container (many times you will find the mornings wax lighter in color than the afternoon/evening before when you stopped for the day). Periodically drain or dump the wax out of this inner vat into wax pans (every 2-3 months). Let solidify and scrape underside of wax cakes to remove dirty residue accumulation (also to remove any inadvertent water accumulation, from a loose fitting cover on the inner wax vat, during heating, etc.). Further clean out the bottom of the wax vat as a fine slum of dirty wax residue will accumulate there. Reprocess the dirty slum residue in burlap sacks, to reclaim wax contained or throw out. Each morning prior to dipping foundation, figure on turning on the wax vats approximately 2 – 3 hours ahead of the actual process of dipping wax for foundation, so everything is liquid again prior to starting the day's work.

The double-vat to be used for dipping wax should be heated to between 165° – 180° F (74° - 82° C), depending upon the actual climatic temperature outside (time of the year). The operation of making wax sheets is simply the dipping of a 9″ x 18″ single-ply pine board, pre-sanded and smooth, into a deep vessel of melted wax (When buying wood for dipping boards try to find center-cut or heart wood boards with the gain of the wood centered as much as possible. These type boards will hardly warp and the grain is not as liable to fray, peel or split catching the wax sheets as they are being peeled off). Prior to dipping, the board is sanded smooth and soaked for about 1 hour

(15 gallon barrel filled with water), in water-vat barrel for dipping boards, and after each dip into the dipping wax-vat. A film of wax will adhere to the board, which after the last dip into the water-vat is peeled off, after the sides are cut-off all around the board with a blunt straight knife, exposing the two sheets dipped (one on each side of the board.)

Here, with experience, beekeepers dipping wax must learn to feel and recognize their wax as to how it is acting. The wax in the dipping vats if too cold will leave little ripples (waves) on the sheets dipped on each side of the board and the sheets will be wavy with irregular thickness bottom to top. If the wax gets cooled too fast in the water vat, it can crack apart in a fast line across the dipping board. If the dipping board is too bluntly (hits the water flat rather than correctly slightly angled about 5%), too fast, fed into the dipping-vat of wax, the dipped sheets of wax will not adhere properly to the dipping board and then when first dipped into the water-vat, can explode the wax sheets off of the dipping board or rip them apart. The dipping boards must be dipped straight (horizontal while holding) and held straight (horizontal while holding) after each successive dip into the dipping tank, before immersing again, to allow all the wax ripples to flow smoothly off the dipped board back into the dipping-vat without creating waves or lines across the wax, that when milled later on will create a non-uniform thickness in the foundation sheets made. Or in other words, the boards must be held straight when draining fresh dipped wax, so that the wax does not flow to either the right or left side of the dipped sheet. (Note: As the smooth ripples form and the wax flows, draining, a grain within the wax is created, like a grain within wood. If too much either right or left, then the sheet of wax created, when fed into the mill may not mill straight in its path, which could create unwanted milling problems later on.)

Beekeepers must recognize that the temperature of the wax, and the quickness of the plunge of the dipping board with its angle of trajectory, just like the temperature outside, and the adjustment of the mill later on, all have their influence upon the finished product, namely the foundation being made. Further, many beekeepers will find, it may be an advantage to reverse the dipping board end with each dip and do a two-ply dip, to create an evenly made sheet of wax at a higher temperature of wax (180° F or 82° C), rather than dip the board once and have the sheet thicker on one end than the other (165° F or 74° C), or dip the board once, with a second fast dip on the opposite end to even just the opposite 3-4 inches of the wax sheet (165° F or 74° C). This repeat dipping and rotation of the dipping board becomes even more important when even more layers of wax are added to the dipped sheets, for those beekeepers who want thick sheets of wax run through adjustable-thickness mills for easier wire embedding later on into frames.

(Note: Though this might be easier foundation for beekeepers to embed, it really is a waste of wax to the bees, and is a disadvantage in decontamination procedures, because whenever the base of the cells is made thicker than the bees make it, they will rarely take the trouble to thin it down; but, no matter how thick the cell-wall, they will thin it down to nearly or the same thickness as the natural feral. What this means is, that if you have just lowered your contamination level of residues down within the wax by processing, and intend to keep bees with no further doping of chemicals, beekeepers really need to run the sheets thin, so that more fresh uncontaminated wax is drawn-out by the bees, thus lowering the overall contamination level, i.e. 90% new wax should be the goal here. Further, by milling thinner and not attempting to load the mill to the hilt, in the processing, thus getting less cell walls to match the old diamond-match pattern of the early part of this (1900) century, the bees will have a better chance of adapting 4.9mm foundation to either slightly smaller or slightly bigger, to acclimatize the bees better to their own local geographic area without the restraints of the cell walls, that act to artificially keep big bees bigger without breaking wax pattern, when drawing the wax base out. Further, by bees having to thin the cells walls thinner, you are in actuality creating even bigger inside cell diameters, which will only enhance the Varroa and tracheal mite problem with still slightly larger bees yet.)

Following dipping, stack sheets of warm wax, peeled off dipping board (after cutting around sides of board with knife), neatly on another board or plate, by layering between sheets of pre-cut paper so that the wax sheets do not stick together. The neater the stacks, the better warmth will be retained within the stack of dipped sheets of wax. This is very important! The longer the wax sheets stay warm, the easier they are to mill, if milling into foundation right away. Further, the longer it takes for the wax sheets to cool if left alone to set on a shelf, the greater the ductility of the sheets will be on handling. Ductility of the wax sheets must be maintained throughout the whole process from beginning to end!

Milling wax into usable foundation

Milling foundation by hand is not hard and is accomplished by placing warmed sheets of wax into a sleeve of plastic so that *no stretching* of the foundation takes place in any direction, *distorting the cell size*, while lowering the concentration of release agent used. (It also helps to break in a new machine not seasoned to constant milling.) Every time a sheet of wax is fed through the foundation mill, a water-bath with a 2% bio-degradable mild soap-suds must be used to keep the wax from sticking while also acting to keep the mill rollers seasoned

(Note: wax contains minute amounts of acid which act on the metal rollers to make them stick). If sheets of wax fed through the foundation mills are not sleeved, *great care must be taken not to stretch the foundation thus giving a distorted cell size,* as the wax is ejecting from the milling process. It is important to let the sheet of wax eject at normal turning speed without pulling on the sheet of wax to help it along. If wax should severely stick to the surface of the mills *great care must be taken not to damage the milling surface of the rollers*, in removing it! The easiest way to remove stuck wax on rollers is to pour boiling water over them and gently brush with a plastic or animal bristle brush until gone. *Never use metal type brushes no matter how soft you think they are to clean with.* After cleaning, then seasoning with clean release agent must start again on the stubborn spot.

Prior to milling, the sheets of wax need to be pre-heated to between 90° – 110° F (32° - 43° C), uniformly throughout. To do this they can be either placed in a thermostat controlled water bath at constant temperature and soaked until warm, or merely placed into a thermostat controlled hot box and left to set until thoroughly warmed. (Note: If working with processed decontaminated wax, choose the warmed water-bath method, because one cannot be too sure that full decontamination has been accomplished 100%. Around 1891, it was common practice to soak dipping boards in brine water for a few hours, the proportion of salt in the water being about a teacupful to about three pails of water. The salt served a double purpose: It acted somewhat as a lubricant in facilitating the removal of the sheets from the dipping boards, and as a preventive against the grain rising in the board and consequently, roughening. In the warmed water-bath method, back then also briny water of about 110° F (43° C) was used to soak and warm the sheets for milling.

Considering decontamination today, this might be an old practice worth following because an alkaline solution is used to hydrolyze coumaphos over time and briny water is alkaline! Thus after decontaminating in the Kelly wax press and photo degrading in UV rays of the sun, foundation making could continue the process by briny water bath by both dipping to sheet wax and soaking to warm prior to milling to make foundation.) While the sheets or stacks of wax are warming, the room you are milling in should be heated to a minimum of 80° F (27° C), 90° F (32° C) being much better, to allow for rolling and embossing the foundation, without fighting the wax trying to cool too quickly, once removed from the either the warm water-bath or the hot-box it is being stored in, until milling actually takes place. Just prior to beginning milling, take boiling water and pour it over the mill rollers to pre-heat them up, so that you will not have warm sheets of wax meeting cold rollers to create a big milling mess. You can also pre-heat the rollers by placing a heating-pad around them for a few hours before hand. At any rate the rollers should be pre-heated to about 95° – 100° F (34° - 38° C) prior to beginning to roll wax sheets through the mill rollers.

During milling foundation, make sure the immediate milling area is clean of debris that could end up accidentally adhering to a warmed wax sheet and end-up fed into the foundation mill, only to damage the rolling/embossing surface permanently. Also make sure the gears of the mill (teeth on either end of the embossing rollers) are greased often with a heavy, thick, gear grease, to keep the gears running tight and lessening wear, that over the long haul can effect timing of the gears and the actual impression that is embossed. If you are working with a pre-set spaced foundation mill i.e. Tom's Mighty Mini Foundation Embosser, plan on obtaining an average of 7 sheets of foundation to the pound for an average 8″ x 16 1/2″ embossed sheet of wax. If you are working with an adjusting foundation mill where you can change the thickness of the wax sheets you are embossing, be very careful not to screw down the mill rollers too much, thus damaging the embossing heads on the foundation rollers

(Note: If the foundation is feeding to one side and not rolling straight through the rollers, before you try to adjust the foundation mill to an adjustment that could make things worse, look at how you are dipping the wax sheets. The ripples or waves of the grain should run 90° to the sides. If the ripples or waves run on the diagonal any way, even 5 - 10%, you are holding the dipping board wrong when you are dipping the wax sheets and creating the problem by careless dipping procedures). Stop here then and go back and redo your wax sheets until you learn to do it right! Properly made sheets of wax should feed true and list neither right, nor left, when being fed through the embossing rollers of the foundation mill.

Remember when milling foundation, with every sheet of wax fed, the rollers must be lightly soaped with biodegradable soap suds (that bubbly soap part!). To make an exception is to have the rollers stick wax and then need cleaning. To make plastic sleeves to put the wax sheet in prior to milling to avoid stretching the cells embossed, use plastic garbage baggies 30 x 36 in size, bought from your favorite Grocery Store. Each plastic garbage baggie will make three sleeves. Each sleeve will produce anywhere from 20-40 embossed foundation sheets before needing to be changed, giving an average of 60-100+ sheets embossed before having to use another new garbage baggie at an average cost of 5 cents each divided by 3. Following milling, stack embossed sheets of warm foundation neatly on another board or plate, by layering between sheets of pre-cut paper so that the wax sheets do not stick together. The neater the stacks, the better warmth will be retained within the stack of embossed sheets of wax. The longer the embossed sheets take to cool the greater the ductility of the foundation will be for handling later on when wiring into frames and taking to the field.

Cutting foundation to size

Cutting foundation to desired size is not hard. First beekeepers must take a Masonite board (1/4) and cut out the desired size they wish to have foundation made into, i.e. 8″ x 16 1/2″, 4″ x 16 1/2″, etc., to use as a template to cut foundation with. Beekeepers will also need a flat-wooden top cutting station or cutting board to place the foundation and template upon while cutting wax embossed sheets to size. Then take a sponge and soak it into some of the mild soap solution on hand, and wet the cutting board before placing milled foundation sheet of wax to be sized upon it. Then place template over the milled foundation sheet and align the sides of the template along the rows of cells milled on the foundation and cut straight rows.

Cut wax using the pizza cutter, also pre-dipped into soapy solution, prior to starting. (Make sure room you are cutting in is pre-warmed to about 80° F (27° C) and wax is also of similar temperature to allow for easy and quick cutting.) Then re-dip as needed to avoid sticking. (Note: Try not to cut foundation with rows of cells aligned on a diagonal, as bees could have a tendency to draw the cells out wrong, i.e. drone or hodgepodge pattern.) Save scraps, wash, then re-melt in wax melting pot to make more wax sheets for foundation. When finished cutting foundation, box and seal, storing for future use until needed.

Year #3 in the Field

We've talked about the process followed the first year for catching feral bees or retrogressing domesticated colonies from oversized brood foundation. We have also talked about the process to be followed the second year and how to look for visual signs that colonies of honey bees are cleansing themselves of parasitic mite and disease problems, to stabilize themselves. Beekeepers should continue catching feral swarms as a renewable resource for both honey bees and clean uncontaminated wax for recycling into foundation. Beekeepers need to plan how many hives to get ready for production the third year to end the fall season with colonies a minimum of 3 deep supers full of bees, pollen and honey, with all drawn 4.9mm foundation. They also need to plan how many hives to set aside for the production of seed-frames to continue shake-downs each Spring, until all hives maintained are in the process of retrogressing back onto a natural biological system. It is suggested that newly caught feral swarms be allowed to do the balance of the drawing-out of foundation work each year, along with forced-splits (swarm cells in colonies, necessitating divides rather than having bees go to the bushes), until all hives maintained have been converted.

The third year is a big year for stabilization, for by now what hives are going to die, have. Further, what hives are going to survive, are. It is the year for beekeepers to start making limited honey production again. It is also the year to get 4.9mm comb foundation drawn-out, because beekeepers reaching this year with their bees should have whole boxes of brood to work with, including accompanying stores of honey and pollen, to start Spring build-up. By not having the first box of comb to draw out, beekeepers will find their hives will now start sooner in the Spring to brood and whiten comb; allowing for populations to build faster pot-progressive by at least one brood cycle.

It is critical this third year of hive retrogression to continue culling all combs not drawn-out properly. *Cull any combs with more than 10% drone cells drawn on any one frame side. make this a mandatory field management technique. also cull any combs drawn-out in a broken pattern helter-skelter.* There is valid reason for doing this, i.e. 1) to prevent reproduction of mites in drone brood; 2) to prevent laying of worker brood in too large a cell, therefore preventing/limiting reproduction of mites in these oversized cells too; 3) to help limit the spread of secondary diseases associated with mites, by their wound infection bites upon the exoskeleton of the bees (Varroa mites) or their internal organs (tracheal Mites).

It is important for beekeepers going back to biological beekeeping with accompanying decontamination processing of their wax combs (melt combs down and restart their outfits), that they *refrain from further use of chemicals, essential oils, and antibiotics*, so that it will not become necessary to do it again, due to the excessive time and cost involved, both to themselves and more importantly, their bees. Beekeepers must remember that pesticides and other foreign substances have been shown to be both movable into wax and out of wax into honey. Once the decision is made and action of decontamination processing is accomplished, beekeepers must remember they need their wax to continually become less and less saturated by diluting, with their bees help, what little residues are left in the foundation bases they milled from recycled wax to restart their colonies. To be biologically sound in management, beekeepers need to refrain from reintroducing foreign substances, especially contact poison treatments that allow during the course of their bees movement within their hives, the addition of new layers of contamination. If this would happen, it would have the effect of re-elevating residues they worked so hard to get rid of during decontamination processing, besides recontaminating woodenware (frames, supers, tops, bottoms, etc.). This would have the effect of making it necessary to restart decontamination processing all over again if biological beekeeping is still their end-goal!

Beekeepers need to plan on marking their frames if they can acquire uncontaminated beeswax from either feral cut-outs or purchase on the open market, to recycle into new foundation base, that shows the wax to be fully clean, rather than having gone through decontamination processing that still might contain some residues. Later (figure 4-5 years average), after their hives have stabilized and they are extracting clean wax (no new added residues to the foundation base, making for 90% new wax added by bees, with the caps being essentially non detectable), beekeepers should cull marked decontaminated frames again (those combs that still have some residues locked within the foundation base from the 1st decontamination processing) and reprocess the wax, selling to bulk buyers or process into candles and sell. These frames should be replaced by clean wax from extracting or feral cut-out wax (or both) that has been recycled into foundation base.(Note: Honey bees will rarely take the trouble to thin down the cell base bottom on milled foundation, but they will readily thin down and reuse the wax contained in thick cell walls on foundation base. Therefore, to lock in as much decontaminated wax as possible from the 1st decontamination processing into the foundation base after recycling wax, the foundation must be made thin when embossing, to mimic the old "Diamond Match pattern without the cell walls, so that the bees will add new uncontaminated wax cell walls, while not reusing the wax in the foundation base bottoms. This way, by the time the cells are drawn out with fresh clean wax from the bees own wax glands, the cappings should be safe to cut off for recycling.).

The above marking and sorting of frames within colonies is time consuming, as is the second recycling of comb wax from those frames still containing small amounts of residues left after the 1st recycling, but beekeepers will find marking will be mandatory to go through if they are ever to become fully biological in their beekeeping, with the end purpose of selling *organic honey, pollen, propolis, and wax.* Beekeepers cannot technically manage and sell natural products of honey bees and use substances foreign to a hive that contaminate any of these hive products (Note: This would also include larvae, eggs, brood, etc.). The best way to avoid having to go through this process is to not have used chemicals, essential oils, or antibiotics to begin with; and/or to have access to clean uncontaminated feral cutout wax or purchase uncontaminated wax from someone else. Either way, it's not going to be easy to take the long way back to biological beekeeping!

Beekeepers actively manipulating and working colonies up during the third year, need to learn to manage their bees using traditional-style unlimited brood nests, to end the fall season with colonies a minimum of 3 deep supers of bees, pollen, and honey. This will average out to about a box of pollen, a box of honey, and a box of bees at the start of winter, but not necessarily in 3 separate supers. Above these supers are stacked, when needed, the honey supers without the use of queen excluders to separate them from the brood nest. With this traditional-style brood nest, beekeepers will find that the bees will place a majority of the pollen in the bottom half of the hive, while a majority of the honey will be in the top half. Brood will be throughout with a good flow on, but for wintering, many will center with the cluster, dropping down towards Spring and then quickly expanding upward as the season turns on!

It is recommended to let the bees expand to their full potential in the course of the year, supering new boxes of foundation as necessary while at the same time working it in. Beekeepers are urged to make splits only as necessary when swarm cells appear, to keep the bees from going to the bushes, by setting down a box of bees, with accompanying brood, pollen, and honey, and most importantly – *the old queen*. Leave the old stand (hive) with the majority of the bees and field workers to raise the new one. Keep the hives worked and opened-up at all times. This means leaving empty frames for the queen to lay in or new foundation for the hives to draw-out. *Do not let the hives hanker-down* (force the queen down with frames full of honey and pollen during the flows). Feral bees need plenty of room for expansion *and when cramped they swarm.* If you must, resituate the combs within the hives. Bottom box: 1 honey frame on outsides, next, 2 pollen on outsides coming towards center, equals 6 frames, with the remaining 4 for brood. Second Box: 2 honey on outsides, next 1 pollen on outsides coming towards center, equals 6 frames, with the remaining 4 for brood. Third Box: 3 honey on outsides, next 1 pollen on outsides coming towards center, equals 6 frames, with the remaining 4 for brood. In the fall as the season down swings, let the bees fill out the third box completely for over wintering in colder areas. (Note: above this set-up are the honey supers!) What is traditionally kept in the 3 main boxes of the brood nest belongs to the bees – *do not touch their stores, they will need it in the spring to come on strong, early, and fast.* (Note: Once a hive is strong, if it hankers-down with stores and you are actively working the bees up, give the extra stores to a slower paced hive or a hive started late in the season for carry-over stores, so you will not have to feed them. Once you have enough for the bees, then start extracting for yourself on a limited basis this third year. But remember that the objective is to get the bees back to existing on their own biologically with ample stores. Once you arrive there and secure your bees, survivability, then many things become possible, including extracting honey for profit.

American Mite History Background

FROM USDA BEE LABS QUARTERLY REPORTS

Bee Culture Research Investigations Madison, Wisconsin Period: Jan 1 - Mar 31, 1960 ENT c10-1(C) Biology of diseases and pests of honey bees and development of control methods. Mite Survey - Late March. A sample of 25 live bees from each of 50 colonies was shaken in about 15 cc. of 1:10,000 Triton X-100 solution in a shell vial. About 10 cc. were pipetted from the bottom and examined in a petri plate at 20X. Mites were observed in 40 of the 50 samples. When the 10 colonies from which no mites had been found on the first sampling were resampled, 9 had mites. The colony that exhibited no mites on the first two samples had mites on the third sampling. As many as 10 adult mites were found in one sample, although 47 of the samples had 4 or less. It is evident that external mites, Acarapis sp. infest all or almost all the colonies at the Madison laboratory. A sample was sent to Beltsville for identification - reported spade shape coxal plates. Live bees were sent for studies on focal point of infestation and further study for species identity.

--

Bee Culture Research Investigations Madison, Wisconsin Period: Apr 1 - Jun 30, 1960 with Quarterly Progress Report now labeled Administratively Confidential. ENT c10-1(C) Biology of diseases and pests of honey bees and development of control methods. Mite Infestation. (Acarapis spp.) Fifty colonies were sampled during late March and five during early May. A sample of 25 live bees from each colony was shaken in about 15 cc's of 1:10000 Triton X-100 solution in a shell vial. About 10 cc's were pipetted from the bottom and examined in a petri plate at 20X. As shown in table 1, mites were observed in 40 of the 50 samples taken in late March. When the 10 colonies from which no mites were found on, the first sampling were resampled, 9 had mites. The colony that exhibited no mites on the first two samplings had mites on the third sampling. As many as 10 adult mites were found in one sample, although 47 of the samples had 4 or less. The 5 colonies sampled in early May all had mites, although 3 was the maximum number of adults found in a sample. It is evident that external mites Acarapis spp. infest all or almost all of the colonies at the Madison laboratory. A sample was sent to Beltsville for focal point of attack and identification. External mites were found in all samples having the truncated coxal plate. They were found on the wing bases, a position usually associated with Acarapis vagans. The most common external mite thus far found has been A. dorsalis, showing a deep cleft in the coxal plate. This mite is usually located in the scutellar groove. Colonies will be sampled again during August to determine the incidence of infestation at that time.

ABSTRACT: A 1959 report that bees from California infested with Acarapis woodi had been intercepted in California, which was later proved to be in error, has stimulated an intensive search for the acarine disease causing mite and other mites. An external mite tentatively identified as A. vagans has been found infesting bees of all colonies (50) examined at the Madison laboratory. This mite has been reported from a few other locations but appears to be less common than A. dorsalis. Separation of species is based on point of infestation and shape of the mite's posterior coxal plate.

Progress Report Under Cooperative Agreement # 12-14-100-2362(33) between University of Wisconsin Agricultural Experiment Station and Entomology Research Division, ARS, USDA Period: Oct 1, 1959 - Jun 30, 1960 These studies are concerned with: #9. Evaluation of the incidence of external mite (Acarapis spp.) infestation of bees in the University colonies.

ABSTRACT: A 1959 report that bees from California infested with Acarapis woodi had been intercepted in California, which was later proved to be in error, has stimulated an intensive search for the acarine disease causing mite and other mites. An external mite tentatively identified as A. vagans has been found infesting bees of all colonies (50) examined at the Madison laboratory. This mite has been reported from a few other locations but appears to be less common than A. dorsalis. Separation of species is based on point of infestation and shape of the mite's posterior coxal plate.

--

Bee Culture Research Investigations Laramie, Wyo. Period: Jan 1 - Mar 31, 1961 with Quarterly Progress Report labeled Administratively Confidential. Work Project ENT c10, Line Project Ent c10-1 Biology of diseases and pests of honey bees and development of control methods, #3. External mites on honey bees, Table 7 External Mites on honey bees at Laramie Page 10, Table 8 Sexes or developmental stages of Acarapis dorsalis mites found at Laramie Page 11.

ABSTRACT: Page #1, Acarapis dorsalis mites were found on about 3% of all sample bees examined, from our Laramie colonies in February and March. At least 58% of our colonies and all of our apiaries have these external mites. Eggs, larvae, and both male and female adults were found in the dorsal scutellar groove of the thorax of the honey bee. Their identity was also confirmed from cleared microscopic mounts of adult females, which disclosed the deeply cleft hind coxal plate characteristic of this species. All the same bees were examined for neck mites: A. externus, also, but none was found. This appears to be the first record of external mites on honey bees in Wyoming, but Mr. Revell's notes indicate some were seen at this laboratory in 1944, and by Dr. Burnside at Beltsville, Maryland in 1941. Also, he recalls seeing some at Ames, Iowa in 1917 from honey bees.

3. External mites on honey bees. Page #9, Acarine mites have been found in the dorsal scutellar groove on the thorax of adult worker honey bees from many of our colonies here in Laramie. A few of the adult female mites have been mounted in a polyvinyl alcohol + lactic acid + phenol mounting medium, and a few have been mounted in Hoyer's solution. These media clear at least some of the specimens satisfactorily so as to disclose the coxal plate. In our specimens this is deeply cleft, characteristic of Acarapis dorsalis. Morgenthaler (1934), as recently determined by E. W. Baker of the U.S. National Museum (See Beltsville 4th Qtly. Rpt. 1959: pp. 7-7b). This appears to be their first record in Wyoming (But see below). This breeding place of external mites was first discovered in England by Morison (1931) Bee World 12(4): 40-42 (April) and 12(10): 110-111 (Oct)

Table 7 summarizes a survey made of all our colonies in February and again in March, 1961, for external mites, both in the scutellar groove and in the ventral neck region. No mites were found in the latter region which is the typical breeding place of A externus Morgenthaler (1927), originally observed there by Homann in Germany (1933): Zeitschr. f. Parasitenk 6(3): 350-415. It may be noted that about 3% of all bees examined were found infested with the dorsal back mite, A. dorsalis. A surprising 58% of our colonies were infested with this mite, and since most samples obtained in February contained not more than 20 bees per colony, and in March not more than 30 bees per colony, it is probable that even a larger percentage of our colonies are infested. All six of our apiaries are infested with this external mite.

Page #12, Table 8 summarizes the sexes and developmental stages of Acarapis dorsalis found during the above survey. It may be noted that eggs, larvae, and both male and female adults were found. Eggs were most abundant, and bees infested with eggs only were found more frequently than bees infested with any other stage or combination of stages. Two mite eggs on one bee were common. Three of the four adult male mites found were with an adult female on the same bee. All mites were found by examining individual bees under a dissecting microscope. They were then transferred with a teasing needle to a drop of glycerine on a microscope slide, and their identity confirmed under a binocular microscope at a magnification of 150X. It is surprising that the eggs are nearly as large as the adults.

A few measurements showed the eggs averaged about 130 x 65 microns. Morgenthaler (1932) Zeitschr. f. angew. Ent. 19(3): 449-489, found adult female A woodi (Rennie) (1921), mites varied 106-180 x 65-85 microns, and external mites are indistinguishable by size from it. The eggs are a shiny white, and show up more distinctly than the slightly yellowish adults. The eggs adhere firmly to the tiny branched hairs in the scutellar groove of the bee, and in removing the eggs one nearly always breaks off the tips of these bee hairs also. The larvae cling to these hairs tenaciously, so that one usually finds tiny branched hairs next to the larvae also. The reliability of any washing technique for discovering external mites on honey bees needs further investigation. A survey of 3900 bees from 48 Italian and 12 Caucasian colonies in Laramie was made in February and March 1950, using Morgenthaler,s method of washing 50 or 100 bees in Oudeman's fluid, filtering the supernatant through filter paper, and examining the sediment left on the paper for external mites under a microscope. No external mites were found then (Laramie's 1st Qtly. Rpt. 1950:17-18).

Similarly, no mites were found in 1000 sample bees from 10 Laramie colonies examined by a mass washing technique at Beltsville in May 1959 (Beltsville 2nd Qtly. Rpt. 1959: p. 6). It seems doubtful that any washing technique would wash off any mite eggs which adhere so firmly to the branched hairs of the bees, unless it were allowed to act a long time and was capable of dissolving the adhesive material. Dr. J. E. Eckert of Davis, California, made the verbal statement at the January 1961 Omaha meetings of the American Beekeeping Federation, that the "neck" mite, Acarapis externus, is found in a very "sticky secretion" in the ventral neck region of the bee. Therefore, even the adults of this species probably would be difficult to wash off their host. Dr. Eckert stated that the few mites he had found on the wings or on the fore part of the abdomen of honey bees had this same sticky secretion, so that he suspects that the so-called A. vagans Schneider (1941) reported from these breeding sites may simply be A. externus colonies resulting from a gravid female having to start egg-laying in these other locations before she was able to migrate to the neck region. Any washing technique also has the disadvantage of requiring identification by microscopic examination of cleared specimens of adult females in order to identify the species, or determine its probable breeding place. The wings and fore-abdomen were first found to be breeding places of external mites by Orosi-Pal in Hungary: Zeitschr.

f. Parasitenk. 7(2): 233-267 (1934), and: Deutsch. Inkerfuh. 9(12): 398-400 (1935). Males and immature stages of these four Acarapis species cannot be differentiated morphologically. Morgenthaler (1928, 1932, and 1934) had found biometrical averages of the adult female only could be used to distinguish A. externus by its long hind tarsal segments, and A woodi by the shorter distance between the inner margins of its spiracles.

Page #13, We can find no mention, even in Laramie's Quarterly Reports, of external mites having been found on honey bees in Laramie or Wyoming before. However, Mr. Irven Revell has kept some typewritten notes he made which show that some external mites were found at this laboratory in December 1944. Samples of 50 bees per colony collected from our Collegian Dairy apiary on Dec. 8, 1944, were examined by brushing the top of the head and thorax and the underside of the wings of individual bees with a fine brush while being held over a smooth bond paper, and inspecting the debris falling on the paper, under high-dry magnification (645x) of a binocular microscope.

Eggs and adults were also seen on individual bees (under a dissecting microscope?) by Dr. Burnside and Dr. Sturtevant, but their location on the bee was not recorded. Four of the twelve colonies at this apiary were found to have external mites. About 3% of the bees examined were infested at that time. Burnside and Revell also found external mites on bees of three of the nine colonies at the Fish Hatchery apiary, examined December 16, 1944. As one mite was found under a mandible and others were found on the underside of the wings close to the base, it seems probable that these were not A. dorslalis, but probably were one of the other external species. It is not recorded how many bees from this apiary were examined for external mites, but 34 to 50 live bees per colony were examined for Nosema at the same time, so it is presumed that these same bees were examined for mites also.

Since package bees were purchased annually at Laramie, it was supposed by them that these external mites were imported with package bees from the South. Mr. Revell,s notes also state that Dr. Burnside had found similar external mites on honey bees at Beltsville "in the fall of 1941." Apparently this was not recorded in the Beltsville Quarterly Reports. This seems surprising, since external mites on honey bees in the United States had never previously been recorded by American investigators. Morgenthaler in Switzerland first reported finding external mites on honey bees from North America (Canada) in 1926: Schweiz. Bienenztg. 49(5): 220-224. Brugger in Switzerland first reported finding external mites on honey bees from the United States in 1930 (in a sample from Geneva, New York: Arch. f. Bienenkunde 17(4/5): 113-142 (1936).

In August 1959 Australian authorities notified the California State Apiary Inspector that California bees shipped to Australia had been found infested with Acarapis woodi mites. As this is the internal species that infects the thoracic tracheae of honey bees, causing Acarine disease, which is common in Europe, has never been known in North America, this caused a serious scare. Fortunately, the mites proved to be the external species (or subspecies): Foote, H. L. Amer. Bee Jour. 99(10):415 (Oct 1959); Anonymous, Amer. Bee Jour. 99(12): 490 (Dec. 1959). A dorsalis has since been found on other California bees: Harper, R. W. Coop. Econ. Insect Rpt. 9(44): 968 (Oct. 30, 1959). A. externus also has been found here: Eckert, J.E. and Shaw, F.R. (1960) "Beekeeping": p. 375., Macmillan, N.Y.

Surveys of bees from many states by A.S. Michael at Maryland have shown external mites present in most of the southern states and in several other scattered states also (Beltsville 3rd Quarterly Report for 1959 and subsequent Reports) e. g. from: Louisiana, California, Utah, Maryland, Florida, Georgia, West Virginia, Mississippi, and Tennessee.

Page #14, Incidentally, Dr Echert found A. dorsalis mites on Nebraska bees, during his demonstration of examining individual bees for mites, at the Omaha (1961) beekeepers, meeting. Therefore states adjacent to Wyoming, both on the east and west, also have external mites on honey bees. One of our Campus colonies previously found infested with Acarapis dorsalis mites was moved on February 23, 1961 into a greenhouse to pollinate some special alfalfa plants of the Wyoming Agricultural Experiment Station agronomists. In order to determine whether the warm greenhouse temperature influenced the external mites on the bees of this colony, samples of 100 bees each were examined on three subsequent dates: Feb. 28, March 13, and March 24, or 5, 18, and 29 days after moving the colony, respectively. Mite eggs only were found on 3% of the bees examined February 28th; no mites of any stage were found at the two later inspections.

Mr. Revell states that in the spring of 1917 when he was a student at Ames, Iowa, his zoology professor demonstrated the method of brushing bees, bodies to look for external mites, and that some microscopic mites were found on honey bees then. This precedes the European investigation of external mites, such as those of Morgenthaler and his associates in Switzerland, and even the British investigations of Rennie and his associates on the internal Acarapis woodi of honey bees, both in the early 1920's. We wonder if such early records of finding external mites on honey bees in the United States were ever published. Manuscript published: Hitchcock, J.D. A bee mite (Acarapis dorsalis) — Wyoming. Coop. Econ. Insect Report 11(11): 157 (3/17/61).

--

Bee Culture Investigations, Laramie, Wyo. Period: Apr 1 – Jun 30, 1961 with Quarterly Progress Report labeled Administratively Confidential. Work Project ENT c10, Line Project ENT c10-1: Biology of diseases and pests of honey bees and development of control methods, 4. External mites, Table 4. External Mites Found on Individual Bees at Laramie, Wyo. : 2nd Quarter 1961, Page 9; Table 5. Numbers of Bees Infested with Different Stages of External Mites, Page 10.

ABSTRACT: Page #1, Acarapis dorsalis mites were found in the dorsal groove on the thorax of about 3% of the bees and in about 60% of the colonies over wintered at Laramie. Acarapis dorsalis mites were found on about 8% of the bees in 100% of packages received from Louisiana. The same bees were infested with Acarapis vagans on their wings. About 2% of the bees and 40% of the packages were infested with these wing mites. Thoracic mites were also present on about 7% of the bees in 68% of packages received from Georgia. The same bees were infested in their neck region with Acarapis externus. about 1% of the bees and 20% of the packages were infested with neck mites.

Page #6, 4. External Mites., The first observation of external mites on honey bees in Wyoming was reported in the Laramie 1st Quarterly Report for 1961: pages 9-14, including Tables 7 and 8. These were the "Back" mites: Acarapis dorsalis, living in the dorsal scutellar groove on the thorax of the bee. About 60% of the colonies, and about 3% of the bees sampled were found infested. Our bees were also examined for "Neck" mites: Acarapis externus, known to live in the neck region of the bee, but none were found.

Page #8, Another sampling of bees from all our apiaries was made in April, 1961, for the dorsal "Back" mites only. Results are summarized in Table 4. All apiaries were infested, showing 20 to 60% of the colonies in different apiaries infested: average 44% of all colonies. From 1.1 to 4.7% of the bees sampled from different apiaries were infested: average 2.9%. This is about the same percentage of bees infested as during the previous two months. Among package bees received from Mitchell's apiaries of Bunkie, Louisiana, sampled in May, we observed both "Back" and "Wing" mites: Acarapis dorsalis and Acarapis vagans: see Table 4.

The "Wing" mites occurred both on the upper side of the hind wings and the lower side of the front wings. Three individual bees were found infested with both "Back" and "Wing" mites. "Back" mites were found on 8.4% of the bees and in 100% of the packages. "Wing" mites were found on 2.1% of the bees and in 40% of the packages. these bees were examined for "Neck" mites also, but none were found. Among package bees received from Rossman's apiaries of Moultrie, Georgia, sampled in June, we observed both "Back" and "Neck" mites: Acarapis dorsalis and Acarapis externus: see Table 4. "Back" mites were found on 7.2% of the bees and in 68% of the packages. "Neck" mites were found on 1.4% of the bees and in 20% of the packages. Once an individual bee was found infested with both "Back" and "Neck" mites. These bees were examined for "Wing" mites also, but none were found.

The "Back" mites observed on Louisiana package bees included 64 eggs on 37 bees, 42 immature stages on 34 bees, 26 adult females on 25 bees, and 6 adult males on 6 bees. The "Back" mites observed on Georgia package bees included 45 eggs on 23 bees, 22 immature stages on 16 bees, 24 adult females on 20 bees, and 3 adult males on 3 bees. As many as 5 eggs were found in the dorsal groove of a single bee. As many as 4 immature stages were found in the dorsal groove of a single bee. As many a 3 female mites were found on a single bee. Of the 9 males observed in the dorsal groove of both groups of package bees, 6 occurred on bees found to have a female mite in the dorsal groove also.

The "Wing" mites observed on Louisiana package bees included 30 eggs on 16 bees, 9 immature stages on 5 bees, and 3 adult females on 3 bees. As many as 3 eggs or 3 immature stages were found on the wings of a single bee. Of the 19 infested bees, 8 had mites on both the front and hind wing (on the same side), 7 bees had only a hind wing infested, and 4 had only a front wing infested. "Wing" mites were found on the right wings of 12 bees and on the left wings of 7 bees. The "Neck" mites observed on Georgia package bees included 12 eggs on 5 bees, 12 immature stages on 6 bees, 4 adult females on 4 bees, and 1 adult male on 1 bee. As many as 3 eggs, or 3 immature stages were found on a single bee. Table 5 summarizes the numbers of bees infested with different stages of these external mites: eggs, immature, and adults. Most individual bees, 68%, were infested with only a single stage of external mites. These included eggs only: 28%, immature stages only: 25%, and adult mites only: 15%, of the infested bees. however, 30% of the infested bees were infested with two stages of mites, and 2% of the infested bees were infested with all three stages of the mites.

--

Bee Disease Investigations, Laramie, Wyo. Period: Oct 1 – Dec 31, 1961 with Quarterly Progress Report labeled Administratively Confidential. Work Project ENT c10, Line

Project ENT c10-1: Biology of diseases and pests of honey bees and development of control methods, 4. External Acarapis mites, Table 2., Survey of external mites on adult worker honey bees at Laramie: December 1961: Summary, pages 4-5. Page #4, 4. External Acarapis mites. A survey was made of sample bees from all our colonies for external mites. Usually 30 bees were examined from each colony. The back (dorsal groove and adjacent areas of the thorax), underside of the neck, and the wing bases–both right and left sides, of each bee were examined with a dissecting microscope for the presence of external mites, in various stages of development.

Results are summarized, by apiaries, in Table 2. All apiaries were infested; on the average, 46% of the colonies, and 1.6% of the bees sampled, were infested with external mites. All stages of development: eggs, immature, and adults, were found. External mites of all three species found in package bees at Laramie last spring (2nd Quarterly Report 1961: page 8, and Tables 4 = 5), were still present on the bees in December. These include the back mite: Acarapis dorsalis, the neck mite: Acarapis externus, and the wing mite: Acarapis vagans. These are believed to be the first reports of their occurrence in Wyoming.

--

Bee Disease Investigations, Laramie, Wyo. Period: Apr 1 - Jun 30, 1962 with Quarterly Progress Report labeled Administratively Confidential. Work Project ENT c10, Line Project ENT c10-1: Biology of diseases and pests of honey bees and development of control methods, 7. Paralysis, page 10, 8. External Mites, page 11.

ABSTRACT: About 2% of package bees received from Louisiana were infested with external mites. Eggs, immature, and adult mites were found of both the back mite, Acarapis dorsalis, and the wing mite, A. vagans.

Page #11, 8. External Mites. Samples of 30 bees were examined from each of 12 queenless packages received from Louisiana, for external Acarapis mites. The back mite, A. dorsalis, was found on 4 individual bees, and the wing mite, A. vagans was found on 4 other individual bees. The neck mite, A. externus was not found. Thus, about 2% of the bees were infested with external mites. All stages: eggs, immature, and adult; and both sexes: male and female, of the mites were seen. They are considered of no economic importance.

--

Bee Disease Investigations, Laramie, Wyo. Period: Jan 1 - Mar 31, 1965

with Quarterly Progress Report labeled Administratively Confidential. Work Project ENT c10, Line Project No. and Title: ENT c10-1(c), Biology of bee diseases and development of control methods for diseases and pests., 6. External Mites: On honey bees from California and Wyoming, Page 6.

ABSTRACT: External mites: Acarapis dorsalis and Acarapis vagans were found on the back or wings, respectively, of a small percentage of honey bees from Fresno, California, and Laramie, Wyoming. No neck mites were found. Page #6. 6. External mites: On honey bees from California and Wyoming. Sample bees from all 6 packages from dwindling colonies received from Fresno, California, in mid-February were examined for external mites. "Back mites": Acarapis dorsalis, were found in the dorsal groove of 3 out of 30 bees examined from package #5, and also "Wing Mites": A. vagans, (eggs and larvae only), from 2 other bees of the same sample, and also from 2 bees out of 30 examined from Package #4. No "Neck mites": A. externus, were found. An additional 189 bees from Package #5 were then examined individually for external mites and 2 bees were found with back mites, plus 7 with wing mites. Thus a total of 14 bees out of 219 examined, or only about 6%, were infested with external mites. Samples of 30 bees per colony were also examined March 16, 1965, for external mites from all 13 colonies in our Canal apiary at Laramie. Again, wing or back mites were found on 1 to 3 bees per sample in 5 of the colonies, and 1 adult mite with 1 egg, presumably A. vagans, was also found on the base of the abdomen of a single bee. No neck mites were found.

--

Bee Disease Investigations, Laramie, Wyo. Period: Oct 1 - Dec 31, 1969 with Quarterly Progress Report labeled For Official Use Only, with following: This progress report includes tentative results of research not sufficiently complete to justify general release. Such findings, when adequately confirmed, will be released promptly through established channels. Therefore, this report is not intended for publication and should not be referred to in literature citations. Work Unit/Project No. 00 ENT C1005 5301 00. bee Disease Control, External mites on honey bees, page 6.

ABSTRACT: Acarapis dorsalis mites were found breeding in the dorsal scutellar groove on the thorax of adult worker honey bees sampled in December from an apiary established last spring with package bees from California. The same apiary also had bees infested with immature stages of a similar mite, presumably Acarapis vagans, breeding on the wings or on the front part of the abdomen. All bees were individually examined under a dissecting microscope, also for Acarapis externus, known to breed in the ventral neck region, but none was found. Twelve of 22 colonies in this apiary were found infested with external mites. These included 3 colonies infested with the "back" mite, 2 infested with the "wing and abdomen" mite , and 3 infested with both species.

Another apiary of 23 colonies, previously over wintered in Laramie, also sampled in December, was found to be free of infestation by external mites.

Page #6. External mites on honey bees (Hitchcock). External mites of presumably 2 species were found among sample honey bees collected in mid-December from colonies established last spring in Laramie, Wyoming, with package bees imported from central California. thirty bees were examined individually under a dissecting microscope from each of 22 colonies. Each bee was examined for adult mites, or their eggs or immature stages, on 8 possible breeding sites: 1. Dorsal scutellar groove of the thorax, 2. Under side of left fore wing, 3. Upper side of left hind wing, 4. Under side of right fore wing, 5. Upper side of right hind wing, 6. Anterior-dorsal broad surface of the apparent 1st (true 2nd) abdominal segment, 7. Ventral neck or cervix, 8. Feste of the proboscis (or groove into which tongue is folded back). Ten of the colonies were negative for mites, but the other 12 colonies had from 1 to 3 bees infested with external mites, out of the 30 bees examined from each colony.

A total of 29 bees had eggs, immature, or adult mites in the scutellar groove on the thorax: undoubtedly the "back" mite: Acarapis dorsalis. eggs or immature mites were found on the wings of 3 bees and on the abdomen of 3 bees. These are presumed to be the "wing and abdomen" mites: Acarapis vagans, but unfortunately no adult mites were found on those breeding sites at this time of the year. All the bees wee also examined in the neck regions for the "neck" mites: Acarapis externus, but none were found. 5 colonies had bees infested with the "back" mites, 2 colonies had bees infested with the "wing and abdomen" mites, and 5 colonies had mixed infestations with both these mites. Even one individual bee had a mixed infestation of mite eggs: both in the scutellar groove on the thorax and on the underside of its left forewing. External mites were found about 11% of the bees examined from the infested colonies, and on about 6% of all bees examined from this package apiary. No external mites were found on honey bees sampled from 23 colonies, previously over wintered in Laramie, from 1 other apiary, sampled in early December.

--

Bee Disease Investigations, Laramie, Wyoming. Period: Jan 1 - Mar 31, 1970 with Quarterly Progress Report labeled For Official Use Only, with following: This progress report includes tentative results of research not sufficiently complete to justify general release. Such findings, when adequately confirmed, will be released promptly through established channels. Therefore, this report is not intended for publication and should not be referred to in literature citation. Work Unit/Project No. 00 ENT C1005 5301 00 Bee Disease Control, External Mites on Honey Bees (Hitchcock), page 8, Table 6. External Mites on Honey Bees in Over wintering Colonies at Laramie, Wyoming, page 9; Table 7. Distribution of External Mites in Different Breeding Sites in individual Colonies, page 9; Table 8. Distribution of External Mites in Different Breeding Sites on individual Bees, page 10; Table 9. External mites found in the scutellar groove on the thorax of honey bees (Sagebrush apiary): Distribution and frequency of developmental stages, page 10; Table 10. External mites found on the wings of honey bees (Sagebrush apiary): Distribution and frequency of developmental stages, page 11; Table 11. External mites found on the abdomen of honey bees (Sagebrush apiary): Distribution and Frequency of developmental stages, page 11.

ABSTRACT: External Acarapis mites were found on honey bees in 62% of our colonies, and in 7 out of 8 of our apiaries. In the infested colonies, about 10% of the bees were infested. The mites are of at least two species: Acarapis dorsalis breeding in the scutellar groove near the posterior dorsal part of the thorax, and A. vagans breeding on the wings or on the front portion of the abdomen. All bees were also examined under the neck for A. externus, known to breed there, but none was found. Many colonies (37%) had multiple mite infestations on the thorax, wings, and abdomen. Individual bees were infested most frequently on the thorax (40%), and least frequently on the abdomen (13%).

The average numbers of mites of all developmental stages occurring on individual bees were about 1 1/2 on the wings, 2 on the thorax, and 3 1/3 on the abdomen. The maximum number observed was 2 adults + 4 eggs + 11 immature equals 17 mites on the abdomen of a single bee. It is interesting that the severely Nosema diseased colony also had an exceptionally high percentage (60%) of its bees infested with external mites Acarapis spp. on its thorax, wings, and abdomen. This severely Nosema diseased colony died within 3 weeks after sampling, and the equipment was removed from the apiary.

Page #8, External Mites on Honey Bees. (Hitchcock). A survey was made for external mites on honey bees in over wintering colonies in all our apiaries at Laramie, Wyoming. A sample of 30 bees from each colony was obtained, and the bees were examined individually under a dissecting microscope, on 8 possible breeding sites (Laramie 4th Qtly. Rpt. 1969, p.6). The data are summarized in Table 6. External mites were found in 81 of 131 colonies (62%), in 7 of 8 apiaries. Mites were found on about 10% of the bees in the infested colonies of all apiaries. However, the average infestation was as high as 15% in one (Harnden) apiary, where one very weak colony had a high mite infestation (60%), and also very severe (93%) Nosema disease. Perhaps such internally parasitized bees do not have the strength to attempt to brush off external parasites. The mites were apparently of two species: Acarapis dorsalis, which breeds in the scutellar groove on the thorax, and A. vagans, which breeds on the wings and abdomen. All bees were examined under the neck also, for A. externus which is known to breed there, but none was found.

The distribution of mites in different breeding sites on the bees in different colonies is summarized in Table 7. This shows that colonies most frequently had bees infested on the thorax, and least frequently infested on the abdomen. Multiple infestations of more than one breeding site were very common, occurring in 37% of the infested colonies. The distribution of mites in different breeding sites on individual bees is summarized in Table 8. About 40% had mites on the thorax only, 35% had mites on the wings only, 13% had mites on the abdomen only, and about 10% had multiple infestations of more than one breeding site. Infestations of the wings and abdomen, on the same individual bees were more frequent (about 75), then infestations of either the thorax and wings or the thorax and abdomen (1.6 each). This may be another indication that the mites breeding on the wings and abdomen are the same species.

Page #12, It was observed that the mites on the wings always occurred on the underside of the front wings or the upper side of the hind wings. When the wings are folded over the bee's abdomen, as normally occurs when inside the cluster of bees within the hive, the mites would be between the front and hind wings, and thus would be protected from getting rubbed off by other bees. Mites on the abdomen occurred on the front vertical portion, normally held close to the back of the thorax. Thus they were between these two main body segments, and again were protected from rubbing against other bees of the cluster. Mites on the thorax breed in the comparatively deep scutellar groove near the posterior portion of the top of the thorax, in which groove they cannot be rubbed off by contact with other bees.

Tables 9, 10, and 11 summarize the numbers of each development stage of mite, found on individual bees, on each of the 3 breeding sites: thorax, wings, and abdomen, for the bees from our largest (Sagebrush) apiary. The average number of mites of all stages (eggs, immature, and adults) on a single honey bee was about 1 1/2 on the wings, 2 on the thorax, and 3 1/3 on the abdomen. The maximum number of mites on all stages on a single bee was 4 on the thorax, 4 on the wings, and 8 on the abdomen. However, in another apiary, 2 adults + 4 eggs + 11 immature, or a total of 17 mites were found on the abdomen of a single honey bee! The frequency of infestation of individual bees by mites of different developmental stages is also shown in Tables 9-11. Bees infested on the thorax were most frequently infested with eggs only. Bees infested on the wings were infested about equally with eggs only or by immature stages only.

Comparatively few bees were infested on the abdomen, but a high proportion of these (11 out of 19) had infestations by two or three stages of mites simultaneously. It was observed that the eggs and immature stages of mites occurring in the scutellar groove on the thorax were tightly "glued" to the adjacent hairs. When they were removed with a fine needle, many hairs broke off, but the eggs and immature stages were easily floated off the needle into a drop of water or of 20% glycerin. By contrast, the eggs and immature stages of mites occurring on the wings or abdomen were "glued" very tightly to the membrane wings or the comparatively smooth surface of the front of the abdomen. It was very difficult to pry them off with a needle, and when this was accomplished, they adhered very firmly to the needle, so that it was difficult to transfer them to a drop of fluid on a microscope slide.

Page #13, A number of adult mites from each of the above three breeding sites on honey bees were preserved in Hoyer's,s medium on microscope slides. Those from the scutellar groove have the characters distinctive for Acarapis dorsalis: namely, an apodeme the full length of the propodosoma, and the rear margin of the coxal plate (between the hind legs) having a deep indentation between two broadly rounded lobes (Michael, 1962). The mites from the wings or abdomen have a short apodeme (about 2/3 the length of the propodosoma).

The shape of the rear margin of their coxal plate is very difficult to determine, and has not been clearly seen in enough individuals to determine its exact character. It appears to be almost truncate, like that of A. externus, the neck mite. However, in some individuals it appears to be a slightly rounded lobe without any indentation. In most specimens it is obscured by internal structures or elements–perhaps including ova or feces. If it is identical to A. externus, it seems strange that not a single mite of any developmental stage was found breeding under the neck of any of the nearly 4000 bees examined individually in this survey, where the neck mite is known to breed. Polaroid photographs have been taken which show the almost transparent eggs or immature mites attached to each of the three breeding sites: (1) scutellar groove of thorax, (2) beside a vein near the base of the wings, or (3) on the front of the abdomen.

--

Bee Culture Investigations, Laramie, Wyo. Period: Apr 1 – Sep 30, 1970 with Quarterly Progress Report labeled For Official Use Only, with following: This progress report includes tentative results of research not sufficiently complete to justify general release. Such findings, when adequately confirmed, will be released promptly through established channels. Therefore, this report is not intended for publication and should not be referred to in literature citations. Work Unit/Project No. 00 ENT C1005 5301 00 Bee Disease Control.

ABSTRACT: Page #2, Only a slight amount of Nosema disease was found in 5 out of 25 package colonies sampled in early May. The same package bees, received from California, had external mites on bees from 92% of the package colonies, and on about 15% of the bees in the infested colonies. These mites were found on 3 locations on the bees, bodies: the scutellar groove on the thorax, the wings, and the front portion of the abdomen. eggs, immature stages, and adult mites were found at each location. There was an average of 2 mites per bee on the thorax or wings, but about 3 1/2 mites per bee on the abdomen. The mites in the scutellar groove undoubtedly are Acarapis dorsalis. those on the wings and abdomen are believed to be A. vagans. The underside of the neck of each bee also was examined for A. externus, known to breed on that location, but none was found. Single colonies occasionally included different bees infested with mites on all 3 body locations. Individual bees occasionally had mites on both the thorax and wings, or both the wings and abdomen, but not on both the thorax (scutellar groove) and abdomen.

Page #20, External Mites on Honey Bees (Hitchcock). Samples of 30 worker honey bees from each colony were obtained on May 5, 1970, from 25 package colonies, obtained from Davis, California, established in our Gunnerson apiary. each individual bee was examined under a dissecting microscope to determine the number and distribution of mites on various body regions of the bees. The data are summarized in Table 17. External mites were found in 92% of the packages. Mites occurred on 14.5% of all the bees examined from the infested colonies.

Page #21, Table 18 summarizes additional data on the numbers of honey bees having various numbers of each stage of the mites on various body regions. While comparatively few bees were found infested with mites on their abdomens, these mites were more abundant per bee, than mites on the thorax or wings. The average number of mites of all stages: eggs, immatures, and adults, occurring on the front part of the bees, abdomen was nearly 3 1/2 mites, compared to averages of about 2 mites per bee for those on the thorax or wings. Photographs were taken of mites on each body region. The mite species occurring in the dorsal scutellar groove on the thorax is undoubtedly Acarapis dorsalis. The mites occurring on either two wings or abdomen are believed to be A. vagans, but more detailed observations of their microscopic anatomy, on more specimens, are needed to confirm their identify, and to ascertain that mites on both the wings and abdomen are the same species. The underside of the necks of all bees sampled from these package colonies was also examined for the neck mites: A. externus, but none was found.

--

Bee Culture Laboratory, Beltsville, Md. Period: Apr 1 - Jun 30, 1959. Page 6, A survey of Acarapis mites on honey bees in the United States was initiated during this quarter. Twenty-three routine samples from Connecticut, Florida, Georgia, Idaho, Kentucky, Maryland, Mississippi, New Mexico, Pennsylvania, Texas, Virginia, and North Carolina were examined for the presence of external and internal mites. In addition, thirty special samples from Wisconsin and ten samples from Wyoming were also subjected to the same examination. No external or internal mites were found in the total of 63 samples. Additional samples from other geographic areas will be solicited during the next quarter.

--

Bee Culture Laboratory, Beltsville, Md. Period: Jul 1 – Sep 30, 1959. Forty-eight samples of bees were examined for the presence of Acarapis mites. No Tracheal mites were found but a mite was recovered from one sample by washing. This mite has been sent to the National Museum for identification. Page #5, The survey of Acarapis mites on honey bees in the United States was continued during this quarter. Forty-eight samples of adult bees were examined for the presence of external and internal mites. In excess of 1,000 individual dissections for tracheal mites were made with negative results. A mite was recovered by washing from a sample of bees obtained from Louisiana. This mite has been sent to the National Museum for final identification. additional samples have been requested from the apiary from which the mite was obtained.

--

Bee Culture Laboratory, Beltsville, Md. Period: Oct 1 – Dec 31, 1959. One hundred and seventy-three samples of bees were examined for the presence of Acarapis mites. One hundred and eighteen of these samples were found to be infested with external Acarapis. The positive samples were from the states of California, Utah, and Maryland. No tracheal mites were found in over 5,000 dissections of bees from these samples. A sample of Acarapis woodi infested dead bees in alcohol was received from Italy. These bees wee sent by Dr. Giordani for demonstration purposes.

Page #6, II. REPORT OF PROGRESS (Continued). The survey of Acarapis mites on honey bees in the United states was continued during this quarter. One hundred seventy-three samples of bees were examined for the presence of external and internal mites. More than 5,000 individual dissections for tracheal mites were made with negative results. However, examinations revealed 118 of these samples to be infested with external acarapis. One hundred fifteen of these samples were from Shasta County, California, and were examined for confirmation of the findings of the California workers. One sample of bees from Utah and 2 samples of bees from the laboratory apiary at Beltsville, Md., were found to be infested with external Acarapis. A change in technique for external mites has resulted in easier detection of the presence of these mites on bees. Approximately a dozen bees are placed in a petri dish and wetted with a 1 to 10,000 solution of triton X-100. This wetting agent is forcibly ejected from a plastic squeeze bottle directly onto each individual bee. The bees are then liberally washed with distilled water forcibly ejected from a plastic squeeze bottle directly onto each individual bee. This can be done rapidly until approximately 1/4 of an inch of fluid has collected in the petri dish. The presence of the Triton X-100 eliminates currents in the fluid and also causes the mites to settle to the floor of the petri dish. Examinations therefore can be made in a single plane under the dissecting microscope at a magnification of 20 times.

Page #7, The mites when located are easily removed from the fluid with a capillary pipette and transferred to a microscope slide for examination at higher power. Dr. Edward Baker, of our insect identification group at the National Museum, working with materials supplied by us and material obtained from the California workers has succeeded in locating a definite morphological difference between Acarapis woodi and the external mite under study which is apparently Acarapis dorsalis. Drawings demonstrating this difference are attached. A sample of dead bees infested with Acarapis woodi has been received from Italy. These bees were sent by Dr Giordani for demonstration purposes. Tracheal dissections are being made to obtain material for the preparation of microscope slides for our own files and for distribution.

--

Bee Culture Laboratory, Beltsville, Md. Period: Jan 1 – Mar 31, 1960. I. SUMMARY OF PROGRESS. A total of 32 accessions were received for diagnosis. External mites were found to be present on bees obtained from Florida. An intensive survey of the diseases of adult bees in Maryland has been initiated.

Page #6, II. REPORT OF PROGRESS (Cont'd). The survey of Acarapis mites on honey bees in the United states was continued during this quarter. External mites, Acarapis dorsalis, were found to be present on bees from Florida. Mites other than Acarapis that have been recovered from samples of bees are as follows: Carpoglyphus, a widely distributed genus, found on dried fruits, milk products, glucose, decaying potatoes, flour, and many other food products. One species of Carpoglyphus has been recorded as breeding in large numbers inside bottles of wine in Paris, maintaining itself on floating pieces of cork and drawing nourishment from the wine. Glycyphagous domesticus, a species found in dried fruits, and organic matter such as skin and feathers, and is often found in enormous numbers in homes and stores. This mite also causes the "grocer's itch" when highly infested material is handled. It has also been reported as the intermediate host of Catenotaemia pusilla, a cestode parasite of rodents. tydeus, mites that are worldwide in distribution and appear to be predaceous on small insects and mites and their eggs and Oribatid. Tyrophagus; Tarsonemus; Typhiodromus.

Page #8, II REPORT OF PROGRESS (Cont'd). Fifty-five microscope slides of Acarapis dorsalis and twenty-eight slides of Acarapis woodi have been completed. Upon receipt of additional A. woodi material from Italy, completion of these slides for distribution can be accomplished. The late winter and early spring in Maryland were unusually severe with resulting losses of approximately 10% of the colonies. Time has been devoted to apiary cleanup and preparation of equipment for spring management.

--

Bee Culture Laboratory, Beltsville, Md. Period: Apr 1 – Jun 30, 1960. I. SUMMARY OF PROGRESS. A total of 244 accessions were received for diagnosis. Laboratory examinations totaled 948, and 3,810 individual dissections were made for tracheal mites. External mites were found on bees from the States of Georgia and West Virginia.

Page #6, II. REPORT OF PROGRESS (Cont'd). The survey of Acarapis mites on honey bees in the United States was continued during this quarter. Acarapis dorsalis was found to be present on bees from Georgia and West Virginia, two states from which they had not been previously reported. Dissections for the internal mite, Acarapis woodi, totaling 3,810, were made on 127 samples and all have been negative. A survey for adult bee diseases in the State of Maryland covering 103 apiaries revealed that 16.5% of these apiaries were harboring Nosema disease in detectable amounts; 4.9% of these apiaries were found to contain detectable infestations of external mites; and one apiary was found to contain Septicemia. It is planned to survey at least 400 apiaries in this study.

--

Bee Culture Laboratory, Beltsville, Md. Period: Jul 1 – Sep 30, 1960. I. SUMMARY OF PROGRESS. A total of 197 accessions were received for diagnosis. Laboratory examinations totaled 790, and 3,150 individual dissections were made for tracheal mites. External mites were found on bees from the States of Mississippi and Tennessee. Sixty-five apiaries in Maryland were surveyed this quarter for adult diseases. Nosema disease was found in 25% of these apiaries, external mites in 7.5% of these apiaries, and Septicemia was found in one apiary. A spore-forming bacillus and an unidentified coccus were isolated from Wisconsin bees showing symptoms of paralysis. Both organisms demonstrated some pathogenicity for adult bees.

Page #6, II. REPORT OF PROGRESS (Cont'd). The survey of Acarapis mites on honey bees in the United States was continued during this quarter. Acarapis dorsalis was found to be present on bees from Mississippi and Tennessee, two states from which they had not been previously reported. Dissections for the internal mite, Acarapis woodi, totaling 3,150, were made on 105 samples and all were negative.

--

Bee Culture Laboratory, Beltsville, Md. Period: Oct 1 – Dec 31, 1960. Page #2, I. SUMMARY OF PROGRESS (Cont'd). Sets of microscope slides of A. woodi and A. dorsalis have been prepared and distributed. A sample of Apis indica received from Oregon contained mites identified as A. woodi.

Page #9, A series of 35mm. transparencies have been completed on the subject of Acarine mites of the honey bee. A set of these slides has been furnished to each of the Bee Culture laboratories. Two sets have also been sent to Mr. Seymour Bailey, President, Apiary Inspectors of America. A set of microscope slides of A. woodi and A. dorsalis have also been sent to each of the Bee Culture laboratories.

Page #10, A sample of Apis indica was received from the Oregon State Department of Agriculture for examination for Acarine disease. This sample was from experimental stock obtained from Japan in a double screen package and was to be maintained in a special bee room. Our examination did not reveal the presence of any mites in the trachea of these bees. However, when muscle sections were made of the anterior portion of the thorax including the anterior tracheae and the muscle mass was dissolved by lactic acid, mites were found with coxal plate configurations identified with Acarapis woodi. These mites were sent to Dr. E. W. Baker who identified them as A. woodi.

--

Bee Culture Laboratory, Beltsville, Md. Period: Apr 1 – Jun 30, 1961. A 5-week's visit was made to European laboratories in Italy, France, Switzerland, Germany, and England to study Acarine disease. Samples of dead preserved bees infested with A. woodi and A. vagans were brought back from Europe for additional study of differential characters of various species of honey bee mites.

Page #9, In Europe diagnosis of acarine disease in the apiary is considered of a presumptive nature. A colony with reduced supplies, a normal complement of young bees but low in numbers of foragers is suspect. A severe infestation results in a colony greatly reduced in strength, especially in the spring. Individual bees with wings unhooked are unable to fly, climb blades of grass, etc., or lay on the ground motionless or with wings vibrating ineffectively. These symptoms are mostly observed during the spring. In summer not many diseased or dead bees are found but more frequently weak colonies with reduced foraging activity are observed.

USA Paralysis – Vicious Bee Breeding Historical Background

FROM USDA BEE LAB QUARTERLY REPORTS

USDA Bureau of Entomology and Plant Quarantine, Division of Bee Culture Madison, Wisconsin Period: Jul 1 – Sep 30, 1942 An unusually vicious temper was observed in only a few of the hybrid groups which reached its highest level in a few colonies of Group GxHxS.

--USDA Bureau of Entomology and Plant Quarantine, Division of Bee Culture Madison, Wisconsin Period: Apr 1–Jun 30, 1945.

ABSTRACT: Page #2. Two groups of resistant queens proved to be so vicious that it is almost impossible to handle them in full-strength colonies. Queens of this stock crossed with Caucasian drones produced workers of satisfactory disposition.

Page #17. Work Project t-1-5. Management of Bees for the Production of Bee Products. Groups Ma, Ra, and K deserve special mention. The Ma queens produced bees so vicious that it is almost impossible to handle them in full-strength colonies. The level of brood production from these queens is so low that it is unlikely this stock will prove practical for production. The Ra group contains representative colonies equally vicious as the Ma but on the whole they are somewhat easier to handle. Colonies headed by K queens, which are sisters of the Ra, show desirable characteristics, both in behavior and expected production. We are convinced that unless resistance can be retained upon top crossing with a gentler strain of bees, the present resistant lines will be unacceptable by and dangerous to the beekeeping industry. Large apiaries of bees as vicious as the two lines in question are almost certain to cause unsuitable relations between beekeepers and people of the community in which they operate. Future plans for the selection and breeding for resistance must take these factors into consideration. From an experimental standpoint it would be worthwhile to raise queens from one of the better yet more vicious Ma queens to be top crossed with drones from gentler stock, both for the purpose of testing their resistant character and temper.

--USDA Bureau of Entomology and Plant Quarantine, Division of Bee Culture Madison, Wisconsin Period: Jul 1 – Sep 30, 1945

Page #8. Lines S-36, Hr, and Ra appear superior in production. Lines F, Fa, and K show high production and are considered to have characteristics desired in good stock. Lines G-Ga and H-Ha showed a marked tendency for swarming, and this lowered their average production. Some attempted swarming and some actual loss of swarms occurred in all stock groups. Consequently, some of the better colonies were handicapped at some period during the flow whereas some of the retarded colonies had an advantage because of more storage space in proportion to their population. this situation will prevent a strict analysis of differences in the productive capacity of the several stock lines. This year's tests indicate that artificially inseminated queens are as dependable as those naturally mated. We should, therefore, plan to use as many artificially mated queens as possible in future tests. Other noticeable stock differences were evident. The Hr group was exceptionally slow in sealing honey and used excessive burr comb. Several colonies in the H and Ha groups cut some of their brood combs down to the midrib without building them. The vicious temper of the Ma and Ra groups was almost intolerable. However, these lines built practically no burr comb and sealed their honey sooner than any of the others.

--USDA Bureau of Entomology and Plant Quarantine, Division of Bee Culture Madison, Wisconsin Period: Jul 1 - Sep 30, 1948

Page #18. The 123 queens were less prolific than the other lines tested. The bees were cross and production was low. It had many of the characteristics of the RA 3-way resistant stock of the same component lines tested in 1945 that showed good average production but extreme viciousness. This hybrid line has few desirable characteristics to offer. (Note: Ref (W39xW64)x(A18)

--USDA Bureau of Entomology and Plant Quarantine, Division of Bee Culture Madison, Wisconsin Period: Apr 1 - Jun 30, 1949

ABSTRACT: W.C. Roberts has been at Kelleys Island since April 5 making preparations for the queen-rearing project this year. Stephen Taber and William N. Edwards are carrying on the work this summer.

Page #2. Progress and Status of Work. W.C. Roberts has been stationed at Kelleys Island since April 5 making preparations for the queen-rearing project this year. A 20′ x 40′ x 8′ portable insulated steel building has recently been erected in one of the queen-rearing yards. The portable steel grafting house erected last summer was moved to this location. Mr. Roberts returned to Madison June 25 to carry on his work in bee breeding. Stephen Taber and his assistant William N. Edwards have taken over the queen-rearing work on the Island this summer.

--USDA Bureau of Entomology and Plant Quarantine, Division of Bee Culture Madison, Wisconsin Period: Jul 1 – Sep 30, 1949

ABSTRACT: At the start of the season the bees on Kelleys Island were in excellent condition. However, there was little nectar available throughout the summer, handicapping the production of queens severely. Approximately 1,500 queens were produced and distributed for testing. Large quantities of sugar syrup will need to be fed to insure satisfactory wintering of the colonies. More cases of paralysis were observed this summer than during any previous year. With one exception, all colonies suffering from paralysis were headed by (S-10 x W39)x(A18 x 16-3) queens, which suggests a stock weakness.

Page #2 Progress and Status of Work. "Kelleys Island Queen Production Project" A 20′ x 40′ portable insulated steel building was erected late in June to facilitate the project. The bees were in excellent condition at the start of the season. Queen rearing go off to a good start. However, except for a brief honey flow from sumac, there was little nectar available throughout the summer, which handicapped the production of queens severely. Very little rain fell during June, July, and August. Robbing bees were serious throughout the summer but special cages for putting over colonies while they were being worked were of some help. The production schedule fell below planned estimates. Approximately 1,500 queens were produced and distributed for testing. Honey reserves are too low for satisfactory wintering, which will make it necessary to feed a large quantity of sugar syrup before the 1st of November.

Page #4. "Paralysis" (Floyd Moeller) Paralysis became so severe in one colony in the Slotten yard that the colony was of no value. The condition was first noted June 3.

After adding a quantity of brood from a triple nuc on June 20, the colony made some improvement but due to queen trouble, the colony was finally disposed of. Two other colonies in the same yard later developed paralysis, which was never severe, and recovered. In July two more colonies in this yard developed bad cases of paralysis. Both of these are very severe at this time. Two cases of paralysis appeared in early August in the Primrose yard – one is severe and the other mile, but both persist as of this date. A further observation was made to the effect that all five of the colonies severely infected with paralysis were of the stock line (S-10 x W39) x (A18 x 16-3). The two colonies that showed light infection and then recovered were of the line (Cau. x B149 x E-182 x Sh) None of the (S-10 x W39) x (A18 x 16-3) cases has recovered. This would seem to indicate some degree of resistance and susceptibility to the disease among the lines of stock. The means of spread is not well understood, but may be in large part by crawling bees. It has to this time been restricted to the two yards – Slotten and Primrose.

Page #6. Work Project I-g-2. Bee Management Investigations. I-g-2-5. Stock selection and inbreeding. (W.C. Roberts) The production phase of the Kelleys Island project was assigned to Stephen Taber and two assistants in late June. Thus the work of production and maintenance in inbred lines of bees could be renewed at Madison. Established lines were inbred for another generation and new lines were started from some of the outstanding hybrid crosses. Pedigrees of the nine lines (which we will winter this year) are shown in the diagrams. These pedigrees give a history of the lines for the last 4 years and show the matings made each year.

Page #15. Some of the principal distinguishing characteristics of the lines under test are given below.

Detailed study of the season's data has not been made at this time. The Kelleys Island line (S-10 x Cau) x (A-18 x 16-3) has bees of gentle disposition, seal their honey white to the bottom bar, and do not produce bur or brace comb even when heavily crowded. The top producing colonies in three of the four yards were of this line. In the other yard, the top producing colony produced only 5 pounds less than the highest yielding colony. In some colonies the tendency to propolize the entrance was shown, but not to excess. From the standpoint of handling characteristics and honey production, this is the best of the 1949 series. Some of the queens were lost due to an unexplained rather abrupt cessation of egg laying at various times of the season, which necessitated replacement of those queens.

The Kelleys Island line (S-10 x W-39) x (A-18 x 16-3) has some temper and is of a rather nervous temperament, probably due to the W-39 influence. The bees tend to build bur and brace comb, often to excess. These colonies produced crops well in the upper range, indicating superior performance of the queens. Five of the colonies of this line were the only ones to come down with paralysis and seem unable to recover, at least at present. This would indicate unusual susceptibility to this disorder.

The Kelleys Island line (S-10 x W-39 x Cau.) x (A-18 x 16-3) shows nervous "runny" tendencies, probably also due to the W-39 influence. The honey is well sealed and white capped. The tendency to build bur comb is not present, and they do not build much brace comb. Queens are large and prolific. Production is superior, except in the Slotten yard where only two queens are represented.

The Kelleys Island line (D182 x B149 x Cau.) x (A-18 x 16-3) is of mild disposition, does not build much bur or brace comb, and seals honey well. Production from this line is also superior. All four lines of queens from Kelleys

Island were introduced in the fall and overwintered in the colonies to be tested.

The commercial line S-229 x Sh was average or below average in production. The queens were not as large as most Short stock but were of good conformation and quite uniform. The bees showed some temper. Line 50 x Sh had large light colored queens of remarkable uniformity with bees of some slight temper. Honey was not sealed as well as some of the other lines. Production was average or slightly below. The 51 x Sh line was one of the better Short crosses as far as temperament is concerned. Production was good. The 53 x Sh cross produced bees of mild disposition with a tendency to propolize the hive entrance heavily. (Caucasian characteristic) Production was average or above. The queens were darker in color than the other Short hybrids and somewhat smaller than the 50 x Sh queens.

The 1 x Sh hybrids showed bees of some slight temper. The tendency to build bur comb was notably absent. Entrances to some of the colonies were heavily propolized. Production was average or slightly above. The 3 x Sh cross produced bees with some nervousness, but not excessive temper. The bees built bur comb freely even when not crowded, but this characteristic was not clearly defined among all the test queens of the line. Production was well above average. Sister queens mated to Harrell drones (3 x Hr) produced gentle bees. Again the tendency to build bur comb was present. Production, however, in this cross was not outstanding. 5 x Hr and Hr x Hr queens were not outstanding, and produced average or below average crops.

The 111 and 112 lines were tested last year, and the queens representing these lines were in their second year. Production for both lines was well above average. The 112 x Sh hybrids produced bees with some temper. Production was good. Some of the colonies had a tendency to propolize the entrance. Queens of this line were of very good size and conformation and produced excellent quality brood. All of the colonies are being prepared for overwintering and necessary weight adjustments made to insure adequate winter stores. The four remaining cases of paralysis are being kept under observation, and, if their condition does not improve, they will have to be disposed of rather than to attempt wintering them. Test queens reared on Kelleys Island will be introduced this fall to the colonies they are to head next season.

--USDA Bureau of Entomology and Plant Quarantine, Division of Bee Culture Madison, Wisconsin Period: Oct 1 – Dec 31, 1949

ABSTRACT: Five colonies headed by (S-10 x W39) x (A18 x 16-3) queens still showed paralysis late in October, although of much less severity, and the colonies are being overwintered. A sample from one of these colonies is being observed for longevity study.

"Paralysis": Paralysis in colonies headed by (S-10 x W39) x (A18 x 16-3) queens referred to in the Third Quarterly Report (Plus one additional colony developing late in the Zweifel yard) persisted late in October. Colony conditions appeared to have improved and the colonies are being overwintered. Only 62 queens were shipped during the 4 week period between August 15 to September 12. The reasons for this are many and interlocking, but, in general, our failure to control or prevent robbing during the dearth of honey flow in August and September is perhaps the most important.

We again experienced a shortage of drones in August, but this was remedied with only a short period of no matings in contrast with the 4-weeks' period of 1948 in which drones were exceedingly scarce. Plans are being made to distribute cell building and drone colonies in several small yards instead of one yard. The number of stock colonies on the Island will be reduced next year to lower the population of bees on the Island so that competition for floral plants will not be so great. Any suggestions on methods to prevent or control robbing will be appreciated as we consider that our number one problem.

The exceedingly hot weather in July and August (Maximum 102′) on several occasions caused poor acceptance of cells in the swarm boxes, but this was not as serious as the sudden complete robbing out of finishing colonies that occurred several times. These factors reduced the number of cells available to put out but an epidemic of absconding in the mating yard (probably due to nucs being robbed out) during the last half of august almost completely stopped the shipment of queens. This occurred at the time we expected to ship queens to the Laramie, Logan, and Columbus laboratories. The result was that Laramie received only half of its order and the other laboratories none. When things got back to near normal again, we were able to produce two-thirds the number of queens promised the Madison laboratory and then closed down queen rearing to get the colonies in condition to winter.

After taking up the nuc yard and consolidating the equipment, we were fortunate in getting a fall honey flow. Seven-hundred pounds of sugar were fed in September but robbing continued even during the very good honey flow from golden rod and aster during early October. Seventy-four colonies are being wintered on the Island and 23 on the mainland. All colonies were requeened with Island raised queens. Only 20 of these colonies have the proper queens for next season's drones so some requeening will be necessary in March and April of 1950.

Page #8. "Stock Testing Reports" Reports on the performance of queens produced at Kelleys Island in 1948 are slow in coming in and in general very incomplete. Some cooperators gave good reports on individual colonies and supplied a comprehensive summary of general conclusions. Other cooperators just wrote a letter of regret that colony records were not taken. It is evident from the tests that the S-10 x Cauc (Red) queens were considered best by most cooperators. The other lines were in most cases inferior but in a few cases the S-10 x W39 (yellow) or S-10 x W39 x Cau (orange aluminum) queens were considered superior. In the majority of cases where the S-10 x Cau (Red) queens were best the D182 x Bur X Cau (aluminum) queens were in second position with orange aluminum and yellow following in that order.

In most cases the best Kelleys Island lines were considered superior to commercial stock in the same test yards. No general summary can be prepared at this time and a statistical analysis of the data may never be made because the records are incomplete. However, a few general impressions are indicated by a study of the records of the various tests. The S-10 x Cau (Red) line gave the best record of honey production, brood production and gentleness. The other lines were more variable and were often criticized for temper, nervousness, or slow to build up. In most cases the S-10 x W39 (yellow) was reported to be more vicious than the other lines. Some beekeepers considered this line gentler than their own stock, whereas other reported them to be very vicious.

Page #9: The records showed a general tendency for the S-10 x W39 (yellow) and the S-10 x W39 x Cau (orange aluminum) to be slow in building up in the spring. This is in general agreement with our observations of the queens in the cell builders at Kelleys Island. The cooperators who expressed a preference for the yellow or orange aluminum queens were located in areas having a honey flow that started later than in areas of cooperators who found the red or aluminum queens to be best. Although the records are only approximate, it appears that the queens that were one-half or one-fourth W39 had a tendency to reach the peak of brood rearing later than those queens having no W39. This seems to indicate that certain hybrids build up a population quicker than others. Location and time of honey flow may thus be very important in the selection of the most desirable hybrids.

It may be assumed that S-10 x Cau is a 75-day hybrid and S-10 x W39 x Cau is a 90-day hybrid. This conjecture, however, is based upon scattered observations. Cooperators often commented on the high quality or compactness of the brood from the Kelleys Island queens. Individual colonies, however, were often recorded as having spotted brood. These records occur mostly in S-10 x W39 (yellow) queens and indicate that the S-10 or W39 lines may have one sex allele that is the same as one of those in the inbred A-18 or 16-3 lines. S-10 and 16-3 were originally Short lines, while W39 and A-18 are both resistant lines and may be somewhat related or have a common sex allele.

Page #16. Due to its unusual susceptibility to paralysis, stock (S-10 x W39) x (A18 x 16-3) was hampered considerably in production. (Six colonies showed paralysis: 3 in the Slotten yard ((one lost)), 2 in Primrose, and 1 in Zweifel.) On the basis of relative yields, it ranks about eleventh among the sixteen lines tested. This line evidently is the more inferior of the four Kelleys Island crosses. Temper and nervousness is also associated with this line, probably due to the W39 breeding.

--USDA Bureau of Entomology and Plant Quarantine, Division of Bee Culture Madison, Wisconsin Period: Apr 1 – Jun 30, 1950

ABSTRACT: Fifteen cases of paralysis have been noted among test colonies. The disorder is confined to five lines of stock. One colony at the Hill Farm had American foulbrood. European foulbrood has been serious at the Sherwin and Hill Farm apiaries. European foulbrood in the queen yard has seriously affected bee breeding progress. Queens and drones from all lines have been produced so that each line will be further inbred one or two generations this season. A shortage of bees on Kelleys Island has curtailed the production of queens for distribution to cooperators. It is anticipated that the number of queens produced will be somewhat less than expectations.

Page #1 Progress and Status of Work. "Diseases" – Paralysis. It is interesting to note that the five cases of paralysis encountered last year were all of the stock line (S-10 x W39) x (A18 x 16-3) All hybrids received for 1950 tests from Kelleys Island were mated to S-10 x W39 drones. If stock has any relationship to the occurrence of paralysis, this would indicate that we can expect trouble.

Page #2. Already fifteen cases of paralysis have been noted among test colonies this year. The disorder is confined to five lines of stock at present. The (S-12 x Bur) x (S-10 x W39) line developed 6 cases of the disease among 20 colonies, 2 of which were very bad. The (A-18 x Bur) x (S10 x W39) line developed 4 cases among 18 colonies, 2 of which became severe. The (Hr x Gaf) x (S-10 x W39) stock shows paralysis in 3 colonies out of 20, one being severe. The (A-18 x Gaf) x (S-10 x W39) and (Bur x Hr) x (S-10 x W39) lines each have one case of the disease. Strangely, the (S-10 x W39) x (A18 x 16-3) colonies that had severe paralysis and recovered last year are excellent colonies this year and show no signs of the disease.

--USDA Bureau of Entomology and Plant Quarantine, Division of Bee Culture Madison, Wisconsin Period: Jul 1 - Sep 30, 1950

ABSTRACT: Page #2. Adult bee paralysis reached epidemic proportions in two yards, was severe in a third yard, and spread to all other yards before the end of August. With the exception of two colonies, the most devastating cases were confined to stocks containing at least 50% S10 x W39 blood lines. See Table 4.

Page #6. "Paralysis" Adult bee paralysis reached epidemic proportions in two yards, was severe in a third yard, and spread to all other yards before the end of August. With the exception of two colonies, the most devastating cases were confined to Kelleys Island stock. One Short and one Harrell colony represented severe cases. However, in these and the milder cases among commercial stocks the disease developed slowly. With Kelleys Island stock, they contained S10 x W39 either through the queen or drones.

Page #7. The first indication of paralysis usually was indicated as mild and within 5 to 7 days the majority of bees were affected. In many colonies from 1 to 3 gallons of dead bees would accumulate in front of the hive within a few weeks. In most cases queens were superseded, although in one of the worst cases brood rearing was maintained at an excessive level, yet the colony population remained static or decreased. In a few cases where infection and supersedure occurred early, the colonies appear free of paralysis.

Two of five (S10 x W39) x (A18 x 16-3) colonies overwintered showed severe infections during December 1949 that lasted until the close of brood rearing. These survived the winter in excellent condition and built up splendid populations. They did not show paralysis this year until midsummer. The cases changed from mild to severe in 4 or 5 days. In order to handle 245 queens for test next year, they were established in nuclei above main colonies. The nuclei were made irrespective of the presence of paralysis in the colonies. We intended to unite the nuclei to the colonies below early in September but further consideration of the relationship between stock and paralysis suggests the advisability of allowing the nuclei to develop new populations.

By so doing we may get some indication of differences in stock susceptibility by comparing the bees in the nucleus with those in the colony below. By delaying the uniting of the nuclei, we will accomplish two purposes – (1) a check between stock lines and paralysis and (2) possibly provide populations more capable of overwintering where paralysis was severe. Especially noted was one colony divided the first week in August to establish a young laying queen on top. The nucleus showed no symptoms of paralysis the middle of September whereas the main colony below developed a severe case.

Page #10. During August and September, queens produced at Kelleys Island for 1951 tests were introduced. The plan calls for 9 lines with 20 queens in each line. To date, all but two of the lines have been introduced. Five commercial lines were also secured for test – 20 queens from Davis, 20 queens from the American Bee Breeders Association (ABBA), 26 queens from Short, 20 queens from Rossman and Long (G42 xS10 breeding), and 20 queens from Kelley (Hr x S10 breeding), making a total of 180 Kelleys Island and 106 commercial or 286 queens in all for 1951 test. Some losses are expected but the numbers should be adequate to provide 15 to 20 queens from each line. Losses incurred at introduction or shortly after are being compensated for in part by Dr. Robert's replacement queens of the breeding (MxE) x F, (MxE) x Naturally Mated, and (YFxD) x T. Table 2 shows the queen losses to date.

A new method of queen introduction was tried this fall. One brood chamber with an auger-hole entrance, turned to the rear was used to establish a nucleus above the inner cover with the escape hole screened. All older bees than drifted back to the old colony below, leaving younger bees above – a desirable condition for queen introduction. The Ashurst paper push-in cages were employed. They worked satisfactorily and eliminated extra work and disturbance experienced when removing the wire push-in cages. Introduction loss other than for cause was 1.6 percent. Closer observation might have explained the loss of the four queens recorded as not accepted. The majority of nuclei were made 1 to 3 days prior to caging the new queens. Seven losses resulting from virgins or queens in nuclei were due to apparent queenlessness or unobserved queen cells.(Three mother and daughter colony units were observed, two of which involved losses of queens caged.)

The greater loss of queens this year over that experienced in 1948-49 can in part be explained by the epidemic of bee paralysis and a shortage of help in manipulating colonies. In late July when extracting was begun, colony manipulation was neglected, allowing any queen cells present to mature and emerge. Colonies afflicted with paralysis invariably commenced queen cell building. This is largely responsible for the average of 36% of original queens surviving among the Kelleys Island stocks mated to S-10 x W39 drones, as compared to 57% of original queens of the commercial stocks surviving and 53% of queens of the Short top-crosses carried for a second-year test. Of the 25 Kelleys Island queens carried over for a second-year test, 32% of the original queens survive.

--USDA Bureau of Entomology and Plant Quarantine, Division of Bee Culture Madison, Wisconsin Period: Oct 1 – Dec 31, 1950

ABSTRACT: Page #2. Data on queen cell and queen production at Kelleys Island for the season of 1950 is summarized. Over 1,600 test queens were produced but only 1,225 were shipped. A comparison of the 3 years of queen production at Kelleys Island shows that progress is being made in solving the problem of losing virgin queens due to robbing and insufficient drones during the last half of the season. Data is given to indicate that the number of test queens produced each season can be increased most readily if more nuclei are stocked in the months of May and June.

Paralysis was the major factor in the poor showing of the test queens produced in 1949 and tested by commercial producer in the summer of 1950. Where paralysis was not present, some test colonies had remarkable records of production and exceeded commercial stocks by large margins. A study of the survival of original test queens was made and the results tabulated. Survival among all stocks is much lower than normal. This is explained by a higher incidence of swarm preparations and the occurrence of paralysis disease among the Kelleys Island stocks top crossed to (S10 x W39) drones. A total of 241 test queens were successfully introduced to colonies for 1951 test.

Paralysis disease persisted in late October but may have been subsiding. Nine cases of the disease were found in late September, making a total of 77 cases for the season. Nuclear divisions into which 1951 test queens were introduced during August had fewer paralyzed bees than their respective parent colonies below. This strengthens evidence that susceptible stock was the basis cause for the high incident of the disease.

Page #12. Most cooperators reported some paralysis in their test colonies. Some beekeepers, however, were unable to differentiate between paralysis and Nosema. Several cooperators reported differences between yards in the incidence of the disease, although the queens in each yard were of the same breeding.

Page #14. Paralysis was not noticeable or appreciably evident in the stock colonies or in the cell building and finishing colonies on the Island in the summer of 1949. Since these colonies were the source of the S10 x W39 drones that mated with the test queens, it must be concluded that the question of heritability of susceptibility to the disease is not established. We, however, do not plan to repeat these identical crosses for large-scale tests again in the near future.

--USDA Bureau of Entomology and Plant Quarantine, Division of Bee Culture Madison, Wisconsin Period: Apr 1 - Jun 30. 1951

ABSTRACT: Robert Kleba was employed under letter of authority at Kelleys Island starting April 2. W. C. Roberts was at Kelleys Island on May 21 to 24 and June 25 to 28 to direct the queen rearing work by Mr. Kleba. Similar to the epidemic of paralysis encountered in Kelleys Island stock last year and due to the unfortunate selection of S10 x W39 paralysis - susceptible drone stock for securing matings, we are now faced with an epidemic - this time being VICIOUS bees due to another unfortunate selection of A-18 x Bur - drone stock for mating on Kelleys Island last year. There are apparent differences between the lines, both in temper and in other characteristics, so an effort will be made to carry the tests through to completion.

Page #7. I-g-2-6 Production of hybrid queens for testing under commercial conditions.(W.C. Roberts). Production of queens on Kelleys Island has been satisfactory during the quarter, and all order have been filled according to schedule. Approximately 175 queens were mailed out previous to June 26 and another 100 queens will probably be sent out before June 30. Four-hundred-eighty nuclei were established previous to June 25, and adequate numbers of bees and brood are available to establish all the remaining nuclei within the next 2 weeks. The delay in getting the yard completely established before the end of June was due to difficulty in obtaining adequate numbers of queen cells for establishing nuclei.

Mr. Kleba has had some difficulty with acceptance of cells in the swarm boxes, but this situation is rapidly improving. I-g-2-7 (Farrar) The unfortunate mating of all Kelleys Island queens to drone source A18 x Bur has resulted in bees so vicious that this is no satisfaction in working them. It has been necessary to use bee gloves for the first time. In past years where a vicious line was experienced, the number of colonies represented only a small part of the total. There are apparent differences between the lines, both in temper and in other characteristics, so an effort will be made to carry the tests through to completion. Ribbands of Rothamsted reported that nurse bees could be converted into field bees by treatment with CO2 (Changes in Behavior of Honey bees Following Their Recovery From Anesthesia, The Journal of Experimental Biology, Vol. 27, Nos. 3 & 4, pp. 302-310, December 1950.).

To determine the effects of CO2 on the behavior of bees, four 2-pound package colonies were established at the Hill Farm. Two of the packages were anesthetized with CO2 for 5 minutes at the time of introduction and again 4 days later. The other two were handled in the usual manner to serve as checks. The gassed colonies gained approximately 1 pound during the first 4 days compared with 4 pounds for the two checks. Queens in the former remained shrunken and produced few, if any, eggs, whereas those in the checks started laying immediately. Prior to the second treatment with CO2, 500 bees were marked with distinctive color in each of the four packages. Bees marked in the treated colonies were observed in colonies throughout the yard even though the test units were well isolated.

Marked bees identifying the check colonies have not shown up in other hives. Several questions may be raised. Did the queens fail to lay because the bees in the gassed hives failed to feed them? Did the bees drift because treatment with CO2 caused them to lose their sense of orientation or because the queens were not laying? Similarly, did the lower gain of the gassed colonies result from the fact that their queens did not lay or from the change in behavior of the individual bees? Further studies along these lines seem desirable.

Page #12 " Temperament of Bees" Similar to the epidemic of paralysis encountered in Kelleys Island stock last year and due to the unfortunate selection of S10 x W39 paralysis - susceptible drone stock for securing matings, we are now faced with an epidemic, this time being VICIOUS bees due to another unfortunate selection of A18 x Bur drone stock for matings on Kelleys Island last year. The (A18 x Bur) x (S10 x W39) stock tested last year was outstandingly vicious, so the choice of A18 x Bur drones for Kelleys Island last year was expected to be unfortunate and it was. All Kelleys Island stocks being tested this year are vicious, with some variations in the queen lines. The temper of these bees is such that unless a honey flow is in progress, they cannot be manipulated without the use of gloves and adequate clothing. Even with a flow on, they are far from gentle.

--USDA Bureau of Entomology and Plant Quarantine, Division of Bee Culture Madison, Wisconsin Period: Jul 1 – Sep 30, 1951

ABSTRACT: An abstract summary is submitted in lieu of the Third Quarterly Report due to pressure on field work, resulting from a shortage of help for completion of the field work.

Page #2.The honey flow ended approximately July 25. Colony yields appear to average 150 pounds surplus and range from less than winter stores to in excess of 300 pounds surplus. European foulbrood infections and extremely vicious bees that increased difficulties in colony management largely contributed to the wide spread in production. Stock differences will be summarized when data is completed and analyzed. Extracting of honey was made difficult by inadequate space for handling the large volume of super equipment, no satisfactory method for uncapping honey, and considerable amount of granulated honey in the comb affecting straining. The finishing of extracting had to be postponed as it stimulated robbing among the breeding colonies and nuclei located in the immediate vicinity. Considerable loss of queens resulted from the excess robbing.

Page #3. Attendant bees from approximately 300 cages of queens received were examined for Nosema. Approximately 40% to 50% contained infected attendants. An analysis of this problem is being undertaken by caging queens removed from colonies with varying numbers of inoculated attendant bees for 4 to 8 days. Over 200 queens will be examined to determine what infection results from this association. Two colonies of bees removed from the refrigerator showed 100-

percent Nosema infections. Cause or significance of this has not been clarified. I-g-2-6.

Data on production of queens on Kelleys Island for the year has not been summarized, but indications are that it will exceed that of any other previous year. All test colonies are being requeened by uniting nuclei that had been established on top after first removing the old queen. Over 300 queens were introduced into the nuclei during August and September. Losses among the introduced queens were approximately 10% and appear to be explained by (1) extremely vicious bees; (2) presence of the colony's queen in the nuclei just prior to caging the new queen; (3) the condition of queens on arrival as indicated by high Nosema infection among attendants, dead attendants in certain groups from Pelee Island (most accepted Pelee Island queens required several more days to lay than from other sources), and notation as to the poor quality of queens at the shipping point; and (4) certain queen lines seem more difficult to introduce. The problem of vicious bees appears to be of greatest importance.

Page #4. Vicious bees necessitated wearing of two layers of clothing, bee gloves, and removing honey by means of bee escapes. These procedures still were not adequate to prevent considerable stinging. All the field operations were materially slowed up as a result. I-g-2-9. Little work on this project was possible due to European foulbrood and demands of other work in handling the test colonies.

--USDA Bureau of Entomology and Plant Quarantine, Division of Bee Culture Madison, Wisconsin Period: Oct 1 – Dec 31, 1951

ABSTRACT: During the season 1,262 hybrid queens were shipped from Kelleys Island. This compares favorably with previous years. Over 250 of these were shipped to Madison and Laramie. twelve different hybrid queen combinations – mated to the (M x E) drone population at Kelleys Island are available for test at Madison. Five of these hybrids will also be tested at Laramie.

Page #2. Excessive temper characterized the stock from Kelleys Island produced in 1950 and tested this summer. This temper was so bad that all colony manipulations were seriously impeded, resulting in incomplete colony notes. Many of the best colonies went out of condition due to swarm preparations that seriously affected yields. European foulbrood reached epidemic proportions, greatly influencing yields and distorting the general conclusions. Forty cases of European foulbrood were recorded, with possibly again that many that were not recorded. Production records, in view of the above conditions, were used only to generalize on the potentialities of the stock lines. All Test colonies were weighed in October and final production records tabulated. All test colonies showed an average yield of 157.5 pounds – maximum 363, minimum -72. Only eight cases of paralysis disease, four of which were bad, were observed this season. All of these cases were either Short stock or contained the S-10 line.

Page #14. Detailed observations on queens and brood were restricted to fewer occasions than in the past due to the excessive temper characterizing most of the stock being tested this season. Colony manipulations were modified somewhat, also due to this cause. Bee gloves and coveralls, frowned upon by progressive beekeepers as cumbersome and unnecessary, had to be used even during the height of the honey flow to avoid unbearable amount of punishment. Paralysis disease has been much reduced this season, only eight colonies showing any evidence of the disease. Five of these were commercial

Short stock, two of (W42 x S10) x (M x E) stock, and one was a (Hr x S10) x (A18 x Bur) colony. Four of the cases were very mild and four were bad. Three of the bad cases were of commercial Short stock and one was of the (W42 x S10) x (M x E) stock. All of the cases were either Short stock or contained the S10 line. It will be recalled that the S10 x W39 matings in past years resulted in unusual susceptibility to paralysis. Experience this year points strongly to the Short stock portion of this cross as carrying much of the susceptibility.

Page #15. Test queens for 1952 were introduced during August and September. The program of testing has been modified for next year to get a better evaluation of stock before it is released as well as to broaden the scope of stocks being evaluated.

--USDA Bureau of Entomology and Plant Quarantine, Division of Bee Culture Madison, Wisconsin Period: Jan 1 – Mar 31, 1952

Page #6. I-g-2-6. Production of hybrid queens for testing under commercial conditions. (W.C. Roberts) This project is in a state of uncertainty. There appears to be little likelihood that any work will be carried on at Kelleys Island this year due to the decision of the Cooperative to conduct queen rearing in South Georgia. The move was necessary to keep the project on a self-sustaining basis. Twelve queens for the production of drones were removed from Hill Farm stock colonies and mailed to South Georgia January 31. According to reports, the Cooperative will have hybrid queens available for distribution about April 15.

--USDA Entomology Research Branch, Section of Beekeeping and Insect Pathology, Madison, Wisconsin Period: Oct 1 – Dec 31, 1954

Report now labeled Administratively Confidential, with following: This report is not for publication in whole or in part without prior approval by the Chief of the Branch. General stock characteristics are summarized in table 11, page 18.

--USDA Bureau of Entomology and Plant Quarantine, Division of Bee Culture, Intermountain States Bee Culture Field Laboratory Period: Apr 1 - Jun 30, 1945

R.L.P. 12. rearing of daughters from queens showing resistance to American foulbrood (H.B. Parks, Collaborator, Texas Agricultural Experimental Station).

Page #6. Artificially inseminated queens, 1945 testing. Table 2 gives a comparison of the 1944 results with the 1945 preliminary results. The W64 line again showed 100% negatives plus recoveries but only 46% negatives. The W39 line results were complicated by the fact that 8 of the 12 colonies under tests were killed on Sept 10, because of the extremely vicious temper then exhibited. This viciousness was a source of increasing difficulty all season. It is very probable however that this group would have been 100% negative plus recoveries if they had gone to the end of brood rearing since none of the non-recovery colonies showed more than 1 or 2 diseased cells at the time the colonies wee killed on September 10.

--USDA Bureau of Entomology and Plant Quarantine, Division of Bee Culture, Intermountain States Bee Culture Field Laboratory. Period: Apr 1 - Jun, 1946

The three queen lines that have been under observation and included in this analysis are as follows: The W64 (Stugar, Yugoslavia, Carniolan) 1939-1945; the W39 (Piana, Italy, Italian) 1937-1945; the A18 (Iowa, Mraz hybrid) 1937-1944.

--Bee Culture Research Investigations, Southern States Bee Culture Research Lab Baton Rouge, La. Period: April 1 - Jun 30, 1960

Quarterly Progress Report labeled Administratively Confidential.

Page #13.BR 14-60. To obtain information on semen and eggs of the honey bee (S.Taber,3rd) Experiments on Semen -shipment: The work this season on this problem has continued some of the work started in the fall, and summer. Two shipments of semen were received from Dr. W. E. Kerr, in Brazil. The first arrived in very poor condition and no inseminations were possible. The second made according to my instructions was quite satisfactory and semen arrived in good condition. Four queens were inseminated, all are laying fertilized eggs. According to Kerr this particular stock (Apis mellifera adonsonii) is quite vicious but a tremendous honey producer. At this time virgins are being reared from these queens and Kerr has been asked to send additional semen from this stock. Three tubes of semen mailed to Kerr in one shipment were received but queens inseminated with the contents of the tubes died. Two additional shipments of semen, 3 tubes each have been made to Kerr, one sent airmail special delivery and the other sent regular mail special delivery.

The results of these shipments are not known yet. Four shipments of semen have been received from Dr. F. Kohler, Wursburg, Germany. He has been working on this problem independently and his shipping tubes are somewhat different but show promise of improvement over the method used here. Semen is placed in very fine capillary plastic (polyetgelene) tubes, the tubes are sealed at both ends. These are then placed in a larger aluminum tube to provide protection. The whole package is quite light and virtually damage proof. However, the semen which he has shipped has all arrived in very poor condition. I think the reason is because of the various diluents he has been using and the inclusion of mucus. Three inseminations have been made with very poor quality semen, results of the inseminations are not available at this time. One shipment of semen has been made to Dr. Kohler, but no results are available. One additional shipment of semen has been made to Illinois, and the results of this shipment are unknown.

Page #14. There has been no chance to study the implications of light eggs versus heavy eggs, in so far as a practical aspect of the problem is concerned. The logic that a small queen might lay small eggs and a large queen larger eggs or that a queen of equal size but laying smaller eggs than another could for this reason lay more eggs, is not necessarily so. These questions and others will have to be examined. Two shipments of semen were received from Kerr during the quarter. The first was especially unsatisfactory, causing death of all queens. The second has just been used and the queens have just started laying so that it will be some time before success or failure is known. If these are successful, it will mean that we now have bees which are 87.5 per cent adonsonii.

--Bee Culture Research Investigations, Southern States Bee Culture Research Lab Baton Rouge, La. Period: Jul 1 – Sep 30, 1960

Quarterly Progress Report labeled Administratively Confidential.

ABSTRACT: Experiments with septicemia disease in artificial insemination showed several antibiotics effective in retarding growth in agar plate cultures; however, various tests with Chloromycetgin and other experiments with this disease, suggest limited usefulness of antibiotics and that the disease is best kept under control by sterilization of instruments and hands between insemination operations. Egg development times were determined for a number of queens and for several samples of the same queen.

These sample averages varied from about 70 to 80 hours. Sample variation from one queen varied in their extremes 10 hours or more in development times. Egg development times were measured in hive and laboratory incubation. Some differences may exist between the two. Semen was successfully sent to Germany, Iowa, and California. Semen was successfully sent to and returned from Austria. Reports of other shipments are not complete. Egg transfers and queen rearing experiments are on extension of the 2nd Quarterly Report. Over the season 2,310 eggs have been transferred and of these 1,582 were accepted. Queens raised from eggs are generally good if the eggs hatch first in a swarm box. Eggs which hatch in queen cups placed in usual queen rearing colonies develop into poor queens, intermediate forms or even workers.

Page #11. 7-Temper and Manageability. Temper, quietness on combs, and similar factors which influence handling of bees were noted at most observations. The rating, except in instances where a colony was noticeable different, were broad comparisons. No accurate measure of temper was developed.

Page #16. "X" Characterization of Line. General: This line, in spite of its nervousness and mean disposition, was the outstanding line under test. It was also the least inbred. It had the highest honey production, was second only to "B" in pollen storage, had the best B/H and E/B ratios, the highest brood average and the best brood pattern of any group. Of the six colonies set up for test, one superseded prior to the data taking period and one failed, queenless, on 5/19/60. Queen cells in various stages were observed in all colonies at various times but there was no serious effort to swarm of to supersede in splits of severe crowding, except in the supersedure colony 36 which tried to swarm about 5/19. The colonies were above average in population throughout the test period. Burr comb was perhaps more plentiful in this line then in any other but it is doubtful that this can be considered a line characteristic since these colonies were stronger and possibly more crowded than some others. There was no disease in any colony during the test, except possibly Nosema which was doubtless present in all colonies.

Page #17. "B" Characterization of Line. General: Not an outstanding line in any particular. Colonies were average or below average in development and production, inclined toward nervousness and meanness and susceptible to European Foulbrood. Honey production was closely comparable in all colonies but exceeded by "I" and "D". The ratio of brood to honey was above average, exceeded only by "X" and would have been better except for colony 37 which produced more brood than any colony of the group but was third in honey production. The line rated 7th in ratio of pollen to brood but the maximum amount of pollen stored was in B-17. There were no supersedures in the line and only one colony failed during the test. In spite of general resistance to wax moth infestation Colony B-1 was finally almost destroyed by moths.

Page #18. "B" line cont'd. Wax Moths. Moths were found in combs in Colonies 10, 17, and 25. Bottom boards, with the exception of 25 were generally clean and no wax moth larvae were found there. Reaction to Smoke was interesting in this line. Bees had a tendency to collect on top bars or to boil over the edges of bodies during manipulation and especially in Colony 37 could not be driven down by smoke. Similar to the "S" line in this respect.

Page #21. "Y" Characterization of Line. General: In spite of the fact that these colonies were good in population throughout the test the line excelled only in the P/B ratio and possibly in resistance to wax moth. Of the 5 colonies set up for test, two colonies superseded prior to the data taking period and one failed on 4/6/60. One test colony #35, and one supersedure colony, #8, had E.F.B. The colonies were gentle, quiet and possibly as easy to handle as any group. However, colony 12, a supersedure colony, was nervous and nasty; qualities which may be forgiven in part since it was one of the strongest colonies and the highest producer in the yard.

Page #22. "L" Characterization of Line. General: This is a difficult line to evaluate. The colonies divide into two groups: 15 and 16 which were alike and uniformly poor throughout the test period and colonies 21, 24, and 33 which were average or above. There were no supersedures prior to the data gathering period and no colony failures during the test which is similar to the "B" and "T" lines. The line is susceptible to "paralysis" with all colonies showing SHINY bees to a certain extent during most of the test and severe symptoms in Nos. 15, 16, 21, and 24 until 4/6. European foulbrood was present in Colony 33 during most of the test period and in Colony 21 on 4/20. In spite of disease the line ranked 4th in honey production and only one colony had any wax moths. The line is inclined to be cross, equal to "B" and exceeded only by "X", but is fairly quiet on the combs. There is a definite indication that stronger colonies would be harder to handle and Colony 21 was definitely MEAN.

Page #26 "S" Characterization of Line. General: Of the three colonies started, two were superseded early and only one colony was available for test. Although the colony was below average in most respects it had some distinctive characteristics. The bees were nervous and irritable but not exactly cross. They did not respond to smoke, flew from frames easily and could not be driven down from top bars or edges of hives. Sometimes a trail of bees would boil over from the second brood chamber, run down the side of the hive and go in the entrance. Presumably they ran up the inside wall and repeated the performance. Both supersedure colonies were downright nasty to handle. Another characteristic different from most lines was the presence of "Lace" comb, a thin line of white wax along the edges of the top bars. The behavior of this line shows a resemblance to the old "black" or German bees. No disease was found in Colony 27 but both supersedure colonies showed E.F.B. at one or more observation periods.

Page #33. Experiments on shipment of bee semen (continued from 2nd Quarter) Dr. F. Kohler, of Wursburg, Germany, reports that he used semen sent to him to successfully inseminate queens. Semen sent to Illinois and California was successfully used to inseminate queens. Four tubes were sent air mail to Dr F. Ruttner, in Austria, with instructions for him to use two tubes for inseminations of his own and to return two to Baton Rouge. Upon receipt of this returned semen from Ruttner, one tube was immediately used to inseminate 4 queens. Three of these produced fertile workers. The other tube was kept for 3 weeks so that it would have been possible to have inseminated daughters of the first insemination with the second insemination. Inseminations with the second tube were unfortunately all unsuccessful. However, they were made at a time when we were having a great deal of trouble with septicemia infections killing inseminated queens, so that the disappointment in the second tube cannot be blamed entirely on the semen. Additional shipments of semen were received from Kohler as described in the last Quarterly Report, but no successful inseminations were made from them. Other shipments have been made from Baton Rouge to various people, but reports are still not complete.

--Bee Culture Research Investigations, Southern States Bee Culture Research Lab Baton Rouge, La. Period: Apr 1 - Jun 30, 1961

Quarterly Progress Report labeled Administratively Confidential, with following: This report is not for publication in whole or in part without prior approval by the Chief of Investigations.

ABSTRACT: page 1. Work Project ENT c10 - Bee Culture Investigations, including other pollinating insects. BR 4-61. To develop methods of measurement of "temper" in honey bees as a basis for genetic study. BR 7-61. To improve individual artificial insemination. BR 11-61. To obtain information on semen and eggs of the honey bee.

Page #9. A shipment of one tube of semen of A. melliferan adonsonii was received from Dr W. Kerr, four inseminations were attempted, 3 queens died immediately, the fourth is still in doubt.

--Bee Culture Research Investigations, Southern States Bee Culture Research Lab Baton Rouge, La. Period: Jul 1 – Sep 30, 1961

Quarterly Progress Report labeled Administratively Confidential.

BR 4-61. To develop methods of measurement of "Temper" in honey bees as a basis for genetic study.

Page #2. In cooperation with the Ontario Agricultural College stocks of bees are being established from immature stages brought to Baton Rouge from England by Dr. Smith, of that Institution. Queens and drone pupae in gelatin capsules, young larvae on royal jelly, and fertilized eggs were included. Three stocks were represented: Adam 1, Adam 2, and Russian. Good sexually mature individuals were obtained from all categories in all stocks except the Russian drones produced very little semen. Virgin queens of all lines were shipped to Ontario for natural mating there and drone production. Artificial matings were made between the virgins and drones of each Adam line and some of those shipped to Ontario and some to Madison for virgin production.

The Russian imported drones were unusable but a few queens were inseminated with semen shipped from England. Virgin daughters of all of these were mated artificially with sons of queens of imported larvae, and this stock can be considered secure in this country. Similar queens have been shipped to Ontario and Madison for establishment of stocks at these places. Although all methods of importation were successful, the most practical method of establishing a stock appears to be by importing larvae or eggs and then semen when the virgins are ready to be mated.

Page #3. BR 4-61. To develop methods of measurement of "Temper" in honey bees as a basis for genetic study. (Roberts) During the season we have produced over 25 two-way hybrids and 20 four-way hybrids for temper studies. The single hybrids are established in nuclei and the multiple hybrids are now in colonies. Variances in temper of these bees are observable but difficult to measure. So far we have not obtained a "yardstick."

--Entomology Research Branch, Apiculture research Branch, Southern States Bee Culture research Lab Baton Rouge, La. Period: Jan 1 – Mar 31, 1962

Quarterly Progress Report Labeled Administratively Confidential.

BR 4-62 Temper of Bees, page 3. BR 9-62 Studies on semen and eggs of the honey bee. (Taber) Several new ideas have been developed to further extend the usefulness of sperm shipment and storage. As yet these trials are in a very preliminary stage and no report will be made on them at this time. The Apis mellifera adonsonii. Stock that was lost during the late spring because of neglect has been re-imported with two successful shipments of semen from Dr Kerr, in Brazil. This stock will be available for tests by the various interested people by the end of the summer. Stock of over 90% adonsonii is now available and with a little inbreeding this will be taken to over 95%.

--

Entomology Research Division, Bee Management Investigations Madison, Wisconsin. Period: Apr 1 - Jun 30, 1966

Quarterly Progress Report labeled Not For Publication, with following: Not for publication without prior approval of the Entomology Research Branch of the Agricultural Research Service or for use in sales promotion or advertising which expresses or implies endorsement of the product by the Branch, Service, or the USDA.

Page #7 As a first step in evaluating the colonies in this test group, a brood count was made June 29. This is shown in table 3. The colonies will be moved into a 50 acre alfalfa field at Arlington Farm in August. Undesirable handling qualities of runniness and some temper were noted in most of the APC stock, both high and low lines.

--Apiculture Research Branch, Bee Management Investigations Madison, Wisconsin. Period: Oct 1 - Dec 31, 1969

Quarterly Progress Report labeled For Official Use Only. This progress report includes tentative results for research not sufficiently complete to justify general release. Such findings, when adequately confirmed, will be released promptly through established channels. Therefore, this report is not intended for publication and should not be referred to in literature citations.

Page # 4. Temper in the CrPC stocks (for which we did not keep production records) was excessive. Because of this unwanted temper, we will discontinue further cranberry selections in these lines.

--Apiculture Research Branch, Bee Management Investigations Madison, Wisconsin. Period: Apr 1 - Jun 30, 1970

Quarterly Progress Report labeled For Official Use Only. This progress report includes tentative results of research not sufficiently complete to justify general release. Such findings, when adequately confirmed, will be released promptly through established channels. Therefore, this report is not intended for publication and should not be referred to in literature citations.

Page #7. Paralysis. One colony of miscellaneous stock had severe paralysis June 4, lost most of its population, went queenless, and had to be replaced. This is the first case of severe paralysis seen for many years at Madison. (F. E. Moeller).

Bee Breeding in the Field: Part 3

Back To Basics

With increasing problems of parasitic mites, secondary diseases, and scavengers within our domesticated honey bee hives causing considerable damage, while today's controls of antibiotics, chemicals, and essential oils are more and more not effectively working, beekeepers should be asking –How did we manage to get ourselves into this situation on so broad a scale industry wide, not only within our own country, but around our whole global world? This feat, should be no easy accomplishment and yet it seems, as an industry, we have managed to pull it off! *Why?*

What ever happened to basic biological beekeeping? How has deviation from thousands of years of beekeeping traditional practice caused it to virtually cease, within the short time span of approximately 100 years? It is a known fact that both honey bees and mites have been on this earth and have co-existed for many millions of years. Parasites cannot survive if they kill their host. Something has evidently gone wrong! Colonies of honey bees do not naturally succumb from Acarapis mite infestations with their accompanying secondary stress diseases, without cause and effect transpiring. Could today's modern thinking of bigger is better, and negotiated modern bee breeding ideas and practices be the real culprits?

We have gone over honey bee comb size and ramifications, with instructions on how to retrogress colonies back onto a biological system approximating the feral in size. This is a necessary and mandatory step on the long way back to biological beekeeping. This industry cannot survive without being able to breed back and forth within the feral population, to recapture for our bees, lost survival and characteristic traits thrown aside, as our honey bees were artificially mutated bigger in search of "Bigger Golden Honey Crops," which in the end were not to be found without the extraction from our industry of a huge price, – namely, possible extinction as a beekeeping industry.

Just as too big artificially beyond that designed by nature is wrong, so is too loose with artificial insemination. This is not to say that artificial insemination does not have a place within beekeeping, but that place has gone beyond permissible parameters, when beekeepers believe that they can actually select better for all attributes necessary to the survival of our industry, but end-up with the culmination we see all around us of industry disintegration instead.

If bee breeding as taught today, and bigger is better philosophy would work, beekeepers would not have today's industry's problems so devastating on a worldwide basis. Therefore, my husband and I stand on the principle that only retrogression back onto a fully biological system of beekeeping without the in-hive use of chemicals, essential oils, and antibiotics will overcome today's ever increasing problems of parasitic acarapis mites (no matter what the species), their associated secondary diseases, and internal hive scavengers. That retrogression must be both physical pertaining to (1) the beekeeping equipment used; (2) the way honey bees are bred. We believe that our industry cannot have one without the other.

Therefore, just as we have gone over retrogression of combs with accompanying shakedown of honey bees to resituate them and acclimatize them back onto a naturally sized biological system of beekeeping approximating the feral, we will now go over bee breeding. (Note: Bee breeding will be gone over as it relates to commercial levels of colonies. Hobby and Sideline guidelines should still be applicable, though modified in actuality because of fewer colonies.)

The first year of retrogression is extremely hard, for to succeed, the combs mandatorily must be drawn-out properly without blown-out cell patterns. Beekeepers will find that honey bees that cannot draw out comb properly will die or abscond. Most of this will be from acquired disease problems that will not clear up due to the misdrawn comb.

As colony numbers increase making the transition surviving their first critical overwintering and self-requeening by supersedure, beekeepers will find that their mental outlook will change from one of defense to one of offense in working field management strategy. Beekeepers should look at their first 100 hives successfully retrogressed and surviving their first winter as a milestone. Colonies found dead going into or coming out of their first winter by absconding or not overwintering, need to have their combs absorbed as a regular field management practice into surviving colonies still trying to retrogress, to make their retrogression quicker and easier, by having less foundation to draw out. This will give the colonies involved, still trying, an edge on turning critical brood cycles easier. It will also give colonies going into winter a chance to store needed honey and pollen in combs that otherwise would be in short supply.

Bee breeding is not recommended until year four in the field or until the number of colonies successfully retrogressed is approaching a minimum number of 500 colonies. There is reason for this stipulation. Beekeepers must remember that certain principles must be in place (objectives obtained in the field) to enable a successful breeding program to take place.

First, survivability must be obtained and demonstrated over the course of the preceding years of retrogression without the aid (crutch) of purchasing queens from an outside source; or the use of artificial insemination; or hands on grafting for new queens either virgin or mated locally (whether done by self or purchased).

Second, once survivability is gained and held; and the numbers of colonies are increased, variability comes into play as a part of field management. *It is only through the attainment of sufficient numbers and variability that bee breeding becomes an attainable reality. at this time. beekeepers wishing to retrogress their bees back onto a natural biological system must remember and keep in mind, what price the color of their bees will play, as to whether they will succeed or fail, based upon local and regional requirements for maintenance of desired characteristics while maintaining the ability to survive.*

Field Breeding Basics

Honey bees can be controlled by working in harmony with their natural instincts. How honey bees behave, both individually and as a whole colony working-unit, depends upon the field temperatures and the weather conditions. Colony thermodynamics, which means working with nature's natural temperature rhythms and climate as it relates to honey bees, controls the behavior of the colonies relative to brood-rearing, swarming, honey gathering, wax production, queen rearing, etc., throughout the year. Beekeepers can create an environment for their colonies to build up strong populations for breeding and honey gathering, etc., by working with colony thermodynamics and learning to remove adverse hive conditions through sound field management practices using retrogression back onto a natural biological system of beekeeping.

The queen is the heart of each colony. However, the life of each colony depends upon temperature. In cold weather, the honey bee activity slows-up and finally completely stops each winter. If the winter cold is too severe, the colony may die from cold or starvation. In warm weather, the honey bee activity increases up to a certain point and then colonies may die from heat. It does not take a very high temperature to kill an entire colony.

To manage honey bees successfully means, therefore, controlling their behavior with sound field management on a year-round natural biological program. Honey bees always react in the same way to the same conditions relative to temperature and climate. If beekeepers learn to understand how these conditions work relative to honey bees, then they can anticipate and control their behaviors within the framework of a sound year-round natural biological management program.

Queen breeding should rank as the most important activity in a sound program of biological honey bee management without the use of chemicals, antibiotics, and essential oils upon honey bees in the field. Queen breeding is simply an increase in the number of queens a beekeeper manages, thus increasing colony numbers. Yet, it is not merely a question of reproduction for numbers only. Breeding implies an improvement of the honey bee's performance capabilities by the augmentation of the best attributes and the elimination of negative attributes, the final result being the production of colonies which are uniform in all aspects and have above average production performances.

Some beekeepers mistakenly believe that acarapis mite resistance (whether internal or external mites) must be bred for, having been told that the solution lies within the artificial control of internal genes of the honey bee. *Unfortunately, this is not true in real world control of the problem and its accompanying secondary diseases.* So to explain this fallacy that acarapis mite resistance must be bred for, we will divide this section into two distinct parts, namely bee breeding and biological manipulative treatment for control of honey bee mites (basically retrogression explanation as to why and how it works). Since we are now going over bee breeding in its pure sense, we will continue, finishing up with the later towards the end. (Note: Let us now state, that commercial beekeepers wishing to see with their own eyes the field may do so. Several have already done so and we expect more will follow. It is important to see it to understand, so adaptation can be made for others to follow working their bees their own way, but following basic field management principles.)

Continuing bee breeding therefore, the major limiting factor of the start of queen breeding is the rearing of sufficient drones and nurse bees. Insufficient numbers of either will doom most operations attempting requeening to unsatisfactory results (the exception being breeding to raise the incidence of thelytoky). Beekeepers using colony thermodynamics relative to local area breeding cycles within the framework of year-round biological field management, geared to nature's natural temperature rhythms and climate, can greatly improve overall colony performances in a period of 3-5 years. Beekeepers need to learn that queen breeding is progressive and retrogressive in results and can even hold status-quo, as in the case of cloning.

Beekeepers should know both the main flow-breeding and stress-breeding times of the year in their local areas. Main flow-breeding mainly hybridizes and/or breeds honey bees forward progressively, while stress-breeding when used at either the beginning or the end of selected breeding cycles can retrogress bee stocks, like separating oil from water (yellow bees from black bees), so that they may be re-hybridized again and again to re-infuse hybrid-vigor for increased colony production standards.

Basic Colony Thermodynamics

1. A cold-blooded animal is one that has a body temperature below 80° F (27° C), and that takes on the temperature of the air, water, or other element in which it lives. One bee, or a few bees, do take on the temperature of the air around them and cannot protect themselves against the loss of heat or cold

2. A warm-blooded animal is one having a relatively high and constant body temperature relatively independent of the surrounding environment. The bee cluster can keep itself warm against a temperature of sub zero temperatures of -100° F (-73° C) or cool against a temperature of over 135° F (57° C) by metabolic activity mimicking warm-bloodedness by working together as a whole harmonious unit to provide an optimum and constant body temperature relatively independent of the harsh surrounding conditions of temperature and humidity.

3. With an internal ambient temperature of approximately 106° F (41° C) both bees and brood die without some measures of heat regulation.

4. When the ambient temperature inside the hive drops to 45° F (7° C), bees normally cease work, cluster loosely,

and maintain the cluster temperature at 57° – 58° F (14 ° C).

5. The cluster is most nearly dormant at 57° – 58° F which still allows the bees to be able to break cluster and move to a new store of honey when all within the cluster has been consumed.

6. Honey bee clusters generate 12° – 13° F (7° C) heat by their normal and natural bodily metabolism or activity incidental to living.

7. The brood rearing temperature is approximately 93° F (34° C) to stimulate both the queen to lay eggs and the worker bees to feed and care for larvae.

8. Once the brood rearing has begun, bees must generate whatever heat it takes to maintain the brood nest temperature at approximately 93° F (34° C) until the brood emerges.

9. If the temperature of the outside air rises to 90° F (32° C) or higher, bees normally carry water into the hive, evaporating it by forced air circulation and thus removing the excess heat from the hive (Evaporation of water cools the hive because the specific heat of water is more than 4 times that of air).

10. Pure hybridization occurs where hot-weather bees (yellow) and cold-weather bees (black/brown) come together naturally by either latitude or altitude with a mean monthly temperature of 75° F (24° C). (Note: Now refer back to Saga #8 picturing in your mind the world as a basic map of honey bee thermal/cell size zones based on composites of hot and cold land area masses. Read and recap here, then proceed on.)

11. As the inside ambient temperature approaches and/or exceeds both 45° F (7° C) and 106° F (41° C) small black bees approach the breeding condition of thelytoky (Have

not been able to accomplish with either yellow-mix or large dark castes.)

12. Humidity in the brood chamber should be about 60% relative humidity, while in the supers where the honey is being ripened it should be 10% relative humidity.

Other basic guidelines for bee breeding

(Note: Now refer back to Saga #8 again picturing in your mind the world as a basic map of honey bee thermal/cell size zones based on composites of hot and cold land area masses. Read and recap here, then proceed on.):

1. Dark (brown/black) cold-weather bees exist naturally below 30° latitude where higher altitudes permit.

2. (Yellow) hot-weather bees exist naturally above 30° latitude where warm thermal areas permit.

3. Small caste races/strains of hot-weather (yellow) bees exist at the Equator and large caste races/strains of cold-weather (brown/black) bees exist as they approach the poles.

4. As all races/strains of bees advance towards temperature transition-zones at near 30° latitude, hot-weather (yellow) bees hybridize more (larger), while cold-weather (brown/black) bees hybridize less (stay more towards pure races/strains and smaller).

5. Nature breeds constantly and constant when all optimum basic evolutionary needs are met i.e. water, food, shelter, and temperature.

6. Mongrel hybridization (man made) is not an evolutionary progression for it separates when artificial stimuli are removed i.e. inappropriate artificial bigger comb size, surrogate geographic areas, and forced climatic breeding.

7. Nature breeds evolutionary change that is progressive, retrogressive, or cloned, when race/strain survivability is at stake.

8. Each race/strain of honey bees has its own separate breeding cycle in Nature providing, an evolution separate from all others, enabling it to exist.

9. Large caste bees on a natural system equate with: 1) fewer bees per brood comb, 2) slower developmental time, and 3) slower mating flight speed.

10. Small caste bees on a natural system equate with: 1) more bees per brood comb, 2) faster developmental time, and 3) faster mating flight speed.

11. Drones take mating flights only on days when bees are able to break cluster and fly outside.

In queen rearing, not the outside air temperature itself is the focal point which beekeepers must consider, but the temperature of the skin surface of the artificially boxed hive where exposed to the sun or the chill-factor of cold winds, which may reach 135° F (57° C) or sub zero -100° F (-73° C), or even more depending upon latitude and altitude, and time of the year. This heat or cold passes through the wall/entrance of the hive to its interior, thus increasing or decreasing it to far above or below the outside temperature. Beekeepers seriously breeding bees can help colonies thermoregulate by maintaining tight and painted equipment, and leaving full frames of honey surrounding the brood nests to act as insulation against extremes of cold and heat.

By natural metabolic cluster reactions, honey bees thermodynamically overcome these effects of unfavorable weather conditions within the hive during cold winters and hot summers. However, to bees, the temperature of the skin surface of the artificial box is a trigger mechanism to which they must react, to average the maximum and minimum temperatures of each day. Day in and day out, bees must manipulate natural weather conditions to approach and provide optimum mean temperature conditions for brood rearing and colony survival.

An ambient temperature lower than about 80° F (27° C) inside the colony results in one of two things. Either the brood rearing within the colony decreases and cuts back or, if seasonal conditions cause the bees to react favorably (fresh pollen and/or nectar coming in), the bees will increase their metabolic activity and produce the necessary heat to offset any short-term decrease in temperature, adding a minimum of 12° - 13° F (7° C) of their own body heat to raise brood, if there is a supply of pollen and reserve honey stored. As soon as the brood rearing temperature of 93° F (34° C) is reached, the queen begins to lay eggs and the brood is reared and cared for by the colony.

In spring, when most beekeepers think of rearing queens, they think of progressive breeding techniques, waiting until colonies produce sufficient drones and nurse bees before beginning their queen rearing. M*any wrongly believe that hybridization is progressive breeding. it is not!* In today's world, *hybridization is for the most part mongrel breeding that produces only a short burst of hybrid vigor and then quickly falls apart with each succeeding generation.*

For most beekeepers, there should be no breeding from hybrids since it is beyond most beekeepers to control it (We will talk about this control and the field mechanics involved before we end the bee breeding portion of this saga.). The final result is nearly always total mongrelization of local area bee stocks and an uncontrolled mixture of overly aggressive honey bees which makes beekeeping more and more impossible in today's urbanizing world. (Note: Could this be an underlying causative effect giving rise to a myth of "Africanized Killer Bees" as being uncontrollable, when the real cause is probably mongrelization with beekeepers mixing uncontrolled variances of races/strains of honey bees together, as each one uses the type of honey bees he/she prefers?)

In a long-term stock improvement program, artificial insemination and various closed-population breeding methods should be avoided, as they lead to severe inbreeding, resulting in poor brood patterns, poor product averages, weak winter cluster carry-over, and colony collapse over a period of 20 – 30 years. This is not to say however, that artificial insemination methods for honey bee queens does not have its place in bee breeding, but the technique is definitely over used in today's world. In skilled hands, the technique of artificial insemination can save many years work in development of properly field-managed stock lines of several hundred colonies, when used in conjunction with a modified open-mating system.

Nature breeds evolutionary changes that are progressive, retrogressive, or cloned, when race/strain survivability is at stake. To accomplish either of the three, beekeepers must remember that all breeding begins with the selection of notable breeding stock of above average overall colony performance. Beekeepers should look for and select honey bee breeder colonies based on a whole-bee theory of field characteristics. To do anything else will, in the long-term, doom the breeding program to problems and necessitate retrogression before being able to proceed further. (Note: We will talk about a whole-bee theory of field characteristic selection and the field mechanics involved before we end the bee breeding portion of this saga.)

Retrogression in a bee hive is not a simple process. We have talked about cell size retrogression and what it involves in physically sizing honey bees back down to natural feral sizing for control of all acarapis mites and their accompanying secondary diseases. This necessary process sets the stage for bee breeding as survivability and variability are achieved. But, just what is progressive breeding, retrogressive breeding (not to be confused with retrogression relative to size), and cloning (thelytoky) as pertains to breeding honey bees?

Progressive breeding

Is the production of uniform progeny within the framework of a fully naturalized breeding program which will true breed and the results of which can only be obtained from uniformly bred colonies. Permanent results can only be achieved by the use of naturally occurring races/strains of honey bees. Since a bee by any other name is still a bee, then beekeepers must use individual or combinations of large or small caste races/strains of hot (yellow) or cold-weather (brown/black) bees to accomplish this.

Artificial hybrids may then be created by mimicking natural hybridization, when two of these races/strains are assimilated. Nature does not produce complex mongrels. Nature transitions in and out from one race/strain to another, with a brief transition-zone between, that is a mixture of each, while always maintaining compatibility to localized geography and climatic thermodynamics.

Retrogressive breeding

Is the reversal of either natural or artificial hybridized combinations of large or small caste races/strains of hot or cold-weather bees, resulting in the production of uniform progeny within the framework of a fully naturalized breeding program, which will then result in each separation achieved, breeding true to their own hot or cold-weather characteristics and large or small caste delineations.

Results can only be achieved by the use of stress-breeding at either the beginning or the end of the selected race/strain breeding cycles where no overlap occurs, one projected breeder-cycle to the other(s). Artificial races/strains can then be created by mimicking natural races/strains where complex mongrelization has taken place, to gain uniformity of characteristics then necessary for the advancement of desirable traits i.e. gentleness and production.

Cloning (thelytoky)

Is the holding constant of race/strain genetics from one generation to the next naturally or by artificially increasing the propensity of worker bees to lay viable brood, to raise queens as an alternate survival system to supplement normal queen mating in case the virgin queen is lost during the mating flight.

Results can only be achieved by using severe stress-breeding, by using the temperature outside, the beginning or the end of selected race/strain breeding cycles where no overlap occurs, one projected breeder cycle to the other(s).

It is a short-duration phenomenon initiated by extreme stress to allow perpetuation of species, until the first available normal mating can be accomplished, to allow, the colony to permanently requeen itself in the normal manner of mating. (Note: Simply put, thelytoky is a natural state in Nature found in low incidence in Italian bees and in higher incidence in natural sized small black/brown honey bees that allows for worker bees to be self-fertile for short periods of time so when hives go queenless during hot dearth summers or extremely cold winters, they can requeen themselves with a temporary queen to carry on until season is such that the colony can requeen themselves naturally again.)

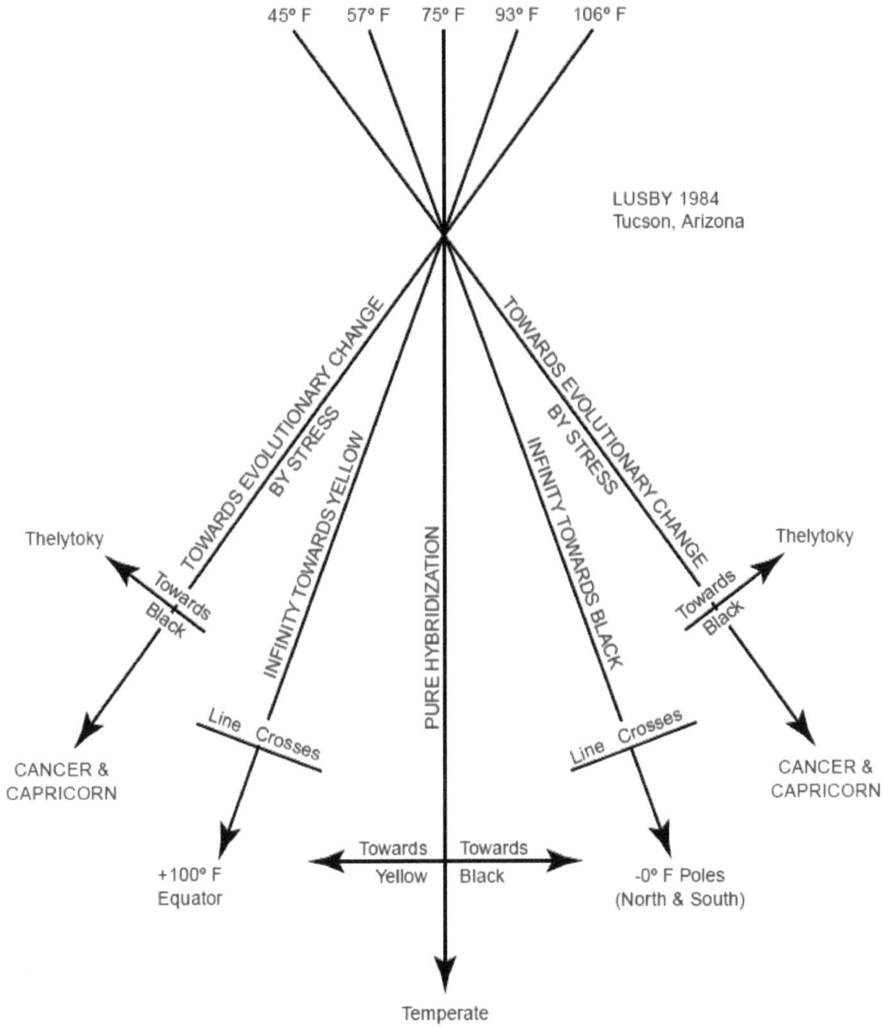
45° F
57° F
75° F
93° F
106° F
LUSBY 1984
Tucson, Arizona
TOWARDS EVOLUTIONARY CHANGE
BY STRESS
INFINITY TOWARDS YELLOW
PURE HYBRIDIZATION
INFINITY TOWARDS BLACK
TOWARDS EVOLUTIONARY CHANGE
BY STRESS
Thelytoky
Towards
Black
Thelytoky
Towards
Black
Line Crosses
Line Crosses
CANCER &
CAPRICORN
CANCER &
CAPRICORN
Towards
Yellow
Towards
Black
+100° F
Equator
-0° F Poles
(North & South)
Temperate

Figure 1

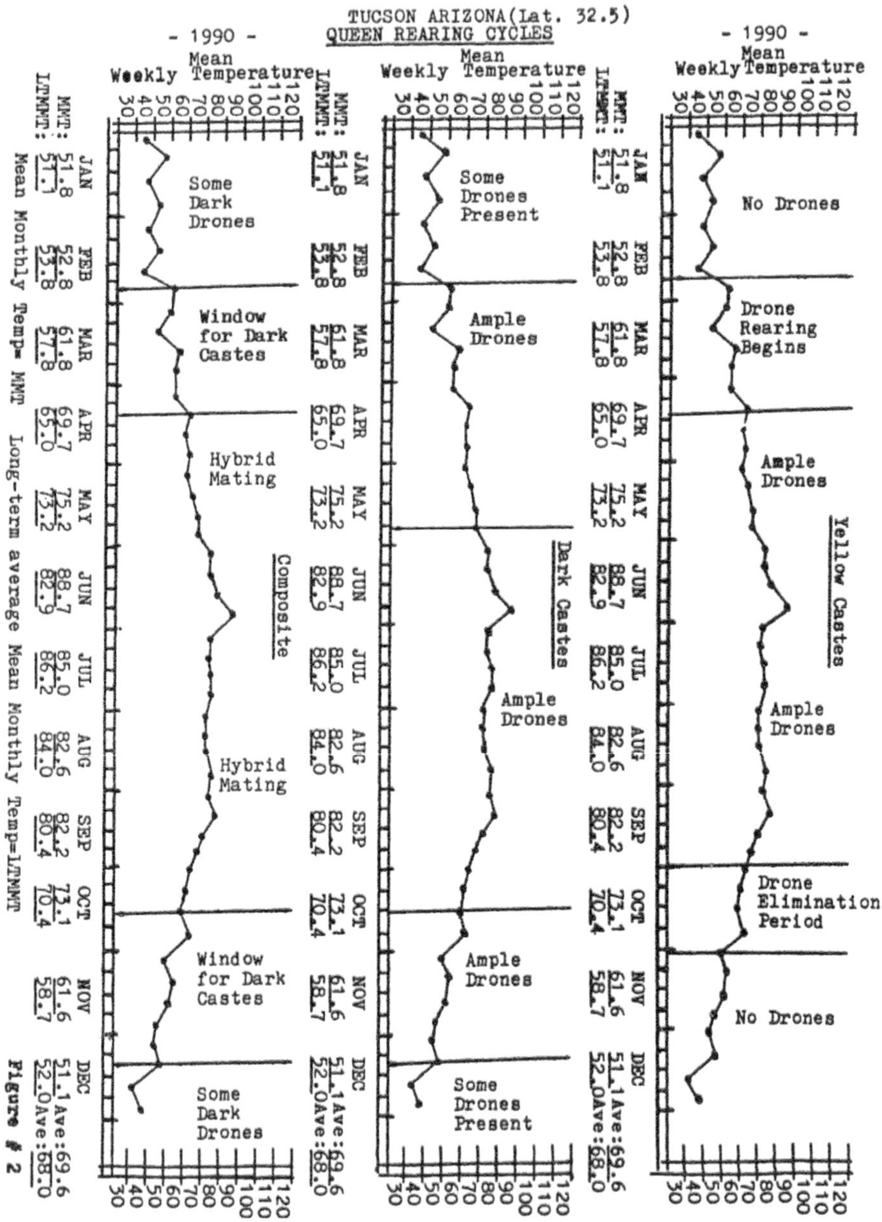
TUCSON ARIZONA(Lat. 32.5)
QUEEN REARING CYCLES
- 1990 -
- 1990 -
Weekly Mean Temperature
Some Dark Drones
Window for Dark Castes
Hybrid Mating
Composite
Hybrid Mating
Window for Dark Castes
Some Dark Drones
Some Drones Present
Ample Drones
Dark Castes
Ample Drones
Ample Drones
Some Drones Present
No Drones
Drone Rearing Begins
Ample Drones
Yellow Castes
Ample Drones
Drone Elimination Period
No Drones
MMT:
LTMMT:
JAN FEB MAR APR MAY JUN JUL AUG SEP OCT NOV DEC
Mean Monthly Temp= MMT Long-term average Mean Monthly Temp=LTMMT
Figure # 2

Figure 2

Projecting Breeding Cycles

Beekeepers should remember when projecting breeding cycles that the color of the exoskeleton is only of significance as a distinguishing character for the purpose of racial analysis where there is the possibility of darker (brown/black) races/strains of honey bees crossing with yellow races/strains of honey bees. In these instances, because the yellow rings of the tergites are so conspicuous, they can be quickly eliminated one from the other. It is because of this that beekeepers from time immemorial have given significance to the coloration of the tergites of the abdomens of honey bees.

Only when more than one race/strain of bees are in a given area do beekeepers need to project breeding cycles to find the best times the drones of their bees have the breeding advantage to maintain racially segregated stock. To project the number of breeding cycle graphs required, beekeepers should first survey colonies in their area that are both domestic and feral (Note – Colonies on oversized artificial brood foundations do not fully correlate with naturally occurring breeding cycles (sometimes by more than 30 days), necessitating that differences be taken into account or excluded from survey.)

Survey information should include:

The number and type of race/strain bees perceived present in the area.

Being specific, the approximate dates the worker bees first begin to either raise or eliminate drones from their colonies.

Being specific, the weeks/months drones are totally absent from all colonies. (Note – If a few drones are present so note, and under what circumstances i.e. laying worker, extra strong hive, etc.)

To plot breeding cycles, beekeepers need to chart month by month, (see the chapter titled "Open-Mating Breeding Charts" Fig. 2 for an example) both the actual "mean monthly temperature" and the "long-term average mean monthly temperature". Beekeepers also need to additionally chart month by month the "mean weekly temperature", noting the approximate dates the worker bees first begin to either raise or eliminate drones from their colonies. Lastly, beekeepers need to chart the week(s)/month(s) the drones are totally absent from all colonies. (Note – "Mean" temperatures are used because Nature does not breed by utilization of daily temperature extremes. Honey combs are Nature's regulator for constant breeding transition, as stored honey/pollen combs each equal stored insulation capacity against extreme heat and cold, that when combined (layered) side by side, help honey bees thermoregulate internal brood nest temperatures thus mimicking warm blooded animals.)

The dominate breeding cycle for the area will be determined by the majority of mean monthly temperature days favoring either right (darker black/brown bees and lower temperatures) or left (yellow/orange and higher temperatures) of 75° F (24° C)on the your now made-up "Open Mating Breeding Chart" from your charted survey on "Queen Rearing Cycles" made-up from your survey of drone flight patterns based on projected mean monthly temperatures.

Beekeepers should then look for either open windows-of-opportunity showing drone breeding advantage, and/or majority-of-temperature dates, to either the cold-weather (dark black/brown) or hot-weather (yellow/orange) breeding side of the Open-mating Breeding Chart (see Chapter titled "Open-Mating Breeding Charts" Fig. 1). Consequently, for raising dark (Black/brown) bees, the closer the breeder can come to the maximum mean monthly temperature, for the warmest month never exceeding 75° F (24° C), the better the results will advantage dark (black/brown) drones. Further, the closer the breeder approaches 57° F (14° C), the darker the results will be.

Beekeepers desiring to raise yellow caste bees should follow the same process, only, the closer the breeder approaches 93° F (34° C) the higher the odds will be for that type of mating.

In areas of complex mongrelization where several races/strains of bees are determined to be present, retrogressive breeding should be a multi-step process. It should start with the separation of yellow races/strains from dark (black/brown) races/strains.

Next, beekeepers should separate color by caste size, to be lastly followed by separation of remaining bees by physical characteristics other than size. (Note - We are talking about honey bees upon a natural biological system only. Nothing mentioned relates to oversized artificially mutated bigger is better honey bees.)

By being able to select how to breed bees, either progressive or retrogressive, beekeepers can initiate methods to return beekeeping back to a sound foundation of natural biological beekeeping without the use of chemicals, antibiotics, and essential oils and a future in the now coming 21st century.

To go forward, beekeepers must learn they sometimes have to go backwards to rectify today's modern bee breeding theories and field management suppositions, that do not stand the test of time eternal as being sound in principle and field application. Beekeeping in the future can only survive and thrive with uniform, well adapted natural biological honey bees on a beekeeping field management system paralleling the feral, thus allowing for the maintenance of a clean sustainable beekeeping agricultural system. It's not an easy road to follow, but it must be accomplished if our industry is to survive. One by one we must all take the long way back to biological beekeeping.

Selecting Breeder Colonies Based on a Whole-bee Theory of Field Characteristics:

What characteristics should we look for as conscientious beekeepers to keep our colonies both profitable and manageable in the field without the use of chemicals, antibiotics, and essential oils, for the control of acarapis mites (all species) and their accompanying secondary diseases?

1. COLOR. Color is of paramount importance in the breeding of honey bees. What price has industry paid for down-playing the importance of color in bee breeding to the detriment of our colonies, for color delineates hot-weather (yellow) bees from cold-weather (black/brown) bees or put in other words, – Tropical Zone bees from Temperate Zone bees. This is a major natural biological division within nature and therefore must be observed in bee breeding. (Note: Now refer back to Saga #8 picturing in your mind the world as a basic map of honey bee

thermal/cell size zones based on composites of hot and cold land area masses. Read and recap here, then proceed on.)

2. LARGENESS OF BROOD PATTERN. This is important because it denotes out-breeding associated with open mating in the field. What beekeepers should be looking for is wall-to-wall solid brood patterns during good flows of nectar and pollen.

3. BODY SIZE/LACK OF DISEASE. These two characteristics must always be linked together because you cannot have one without the other in the natural feral in nature. When body size is correct for the area the honey bees are located in, nature is in "Harmony" in a perfect balanced mix of inter-relationships one element to another. It is only when the basic elements of life change, that created stress begins and problems soon follow.

4. BODY UNIFORMITY. Each race/strain has its own body characteristics that set it apart to be so recognized. Beekeepers must learn to recognize this besides looking at the usual characteristics i.e. formation of legs, wings, thorax, tergites, head, etc.

5. LACK OF CROSS BRACE. Honey bees in harmony with their natural biological surroundings have extremely little cross-brace, if any, and then only for structural support in adhering to places where they build their feral nests/combs.

6. LACK OF SWARMING. While all honey bees swarm that are healthy, it is normally within a natural rhythmic pattern as colonies build throughout the season for perpetuation of species. This rhythmic pattern is to be accepted, but swarming outside of the natural rhythmic pattern in off-season (no honey/no pollen gathering) is not to be tolerated without undue provocation.

7. HONEY GATHERING CAPABILITIES. Beekeepers need to remember here that on a natural system of beekeeping, which normally entails the field management of unlimited broodnests, the honey gathered from the third super/box down belongs to the bees themselves, unless the beekeeper in field management is having to open-up center congestion for continued brood laying during the course of a main flow of honey. What beekeepers should be looking for here is individual colonies capable of drawing-out and filling-up supers up through the 5th deep super in a uniform manner, relative to other desired characteristics. (Note: this characteristic is closely linked with size.)

8. POLLEN GATHERING CAPABILITIES. Beekeepers need to remember here that on a natural system of beekeeping, which normally entails the field management of unlimited broodnests, pollen can be placed by the bees alongside honey anyplace within the hive. Beekeepers should also note that this characteristic is closely linked with size. Beekeepers with their honey bees acclimatized to their own geographical local area should find that their bees will gather enormous amounts of pollen naturally. To not see gathered pollen is to have bees out-of-tune with natural surroundings and this should therefore be a trigger to look for lack of other characteristics.

9. HIVE DEFENSE. This is not a bad trait to have, just not to extremes. So let's put hive defense more into a proper perspective. Beekeepers want natural hives to defend against honey bee robbers from other species; and including their own. This is important during off-season when honey is scarce and needed for winter and/or dry summer carryover. Beekeepers also need natural hives to defend against small scavengers i.e. mice, lizards, ants, moths, etc. Equally important is natural defense against larger scavengers i.e. skunks, coatimundi, etc. It is

important that hives defend when hit or eaten upon. This is natural. It is unnatural for a hive to defend when approached without being touched. It is unnatural for a hive to defend if being opened with smoke (not to say 1 or 2 stings might not ensue due to beekeeper clumsiness) to the extend the beekeeper is sent running for the bushes.

10. HOURS OF FLIGHT. Beekeepers should check their lines of bee stock for colonies that fly earlier in the morning and later at night. With longer hours of flight activity these colonies will many times be better foragers bringing in greater stores of pollen and honey.

11. MANAGEABILITY. Let's equate this with being able to observe, when opened and frames taken out with use of smoke, natural bee movement upon combs. This means, no runny bees, no bees boiling over the sides of the supers either right or left, or bees boiling over both sides at the same time. It also means no absconding simply because the hive was opened.

12. ROBBING CAPACITY. This is technically a good trait to have because poor robbers are poor honey gathers. It is also ideal to have linked with other whole-bee field characteristics like good hive defense and hours of flight where colonies are bred for earlier and later flight times. An example would be where one bee colony is still in cluster, while an early breaking cluster colony with earlier flight times, could send field workers in to strip honey from around the sides of the cluster within, but still out of reach of it to alert defense. Beekeepers would be amazed at how many feral colonies actually do this to artificially oversized domesticated colonies, thus shortening their stores of pollen and honey in times of dearth, whether dry summer or winter.

13. EARLY/PRE-FLOW BUILD-UP OF BROOD. Honey bees that turn their 1st brood cycle earlier anticipating

honey flows and seeking them out are good to have. Mostly this trait is found with the darker races/strains (black/brown) of bees. The objective being to have as much brood ready when the main flow starts, then the queen shuts off and all attention is giving to the gathering of stores. Many times this trait is erroneously looked upon, causing good queens to be killed by beekeepers not knowing that this is a natural characteristic of such black races like caucasian. Then when ample stores of pollen and honey are assured for winter carry-over, the queen restarts laying again. It is a trait of temperate zone honey bees. It is not a trait of tropical zone honey bees, that perpetually build brood with no queen shut off until the flow is actually over, and then these natural bees naturally swarm to another site and start again. Linked with other whole-bee field characteristics such as honey and pollen gathering, high quality production stock can be obtained.

14. CLUSTER AND FANNING ABILITY. Beekeepers need to look for honey bees that can move the cluster in winter consuming pollen and honey stores as needed, for those that cannot, do not carry-over through severe winters. The same is true concerning fanning ability. Beekeepers need to look for honey bees that can cool the colony during hot summers. This is a trait that is linked closely with body size and lack of disease, early/pre-flow build-up of brood, and largeness of brood pattern. This is because small natural sized bees within a given area in a cluster cover more brood cells per square inch, thus giving greater return on each brood turn made. This can be critical in short flow years. Further, naturally sized honey bees have denser flight muscle, allowing for better generation of heat by shivering over artificially enlarged domesticated honey bees, allowing for better clustering and fanning ability for greater colony survivability. These traits when linked with body uniformity help bees to have

less parasitic mite problems, because denser flight muscle from greater use, makes for a tighter thorax with less air pockets and reduced cavity for tracheal mites to attempt to live in.

15. PROPOLIZING ABILITY. This is a trait that should be sought by beekeepers because honey bees that cannot gather and use propolis cannot maintain the sterility of the broodnest against disease. Further they cannot coat the inside walls of the comb cavity against intrusion by weather, scavengers, fungi, molds, bacteria, etc.

Beekeepers will find that when they try to identify colonies to graft and breed from that contain as many characteristics above as possible they will perpetuate better honey bees more naturally adapted to their surroundings.

Field Breeding Basics for Honey bees Using Colony Thermodynamics within the Transition Zones

Apiacta XXX, 20-29 (1995), Dee A. LUSBY, E.W. LUSBY

Honeybees can be controlled by working in harmony with their natural instincts. How honeybees behave, both individually and as a whole colony working-unit, depends upon the field temperatures and the weather conditions. Colony thermodynamics, which means working with Nature's natural temperature rhythms and climate as relates to honeybees, controls the behavior of the colonies relative to brood rearing, swarming, honey gathering, wax production, queen rearing, etc., all around the year. Beekeepers can create an environment for their colonies to build-up strong populations for breeding, honey gathering, etc., by working with colony thermodynamics and learning to remove adverse hive conditions through sound field management practices.

The queen is the heart of each colony. However, the life of each colony depends upon temperature. In cold weather, the honeybee activity slows-up, and finally, completely stops each winter. If the winter cold is too severe, the colony may die from cold or starvation. In warm weather, the honeybee activity increases up to a certain point and then colonies may die from heat. It does not take a very high temperature to kill an entire colony.

To manage honeybees successfully means, therefore, controlling their behavior with sound field management on a year-round program. Honeybees always react in the same way to the same conditions relative to temperature and climate. If beekeepers learn to understand how these conditions work relative to honeybees, then they can anticipate and control their behaviors within the framework of a sound year-round management program.

Queen breeding should rank as the most important activity in a sound program of honeybee management. Queen breeding is simply an increase in the number of queens a beekeeper manages, thus increasing colony numbers. Yet, it is not merely a question of reproduction. Breeding implies an improvement of the honeybee's performance capabilities by the augmentation of the best attributes and the elimination of negative attributes, the final result being the production of colonies which are uniform in all aspects and have above average production performances.

The major limiting factor of the start of queen breeding is the rearing of sufficient drones and nurse bees. Insufficient numbers of either will doom most operations attempting requeening to unsatisfactory results (the exception being breeding to raise the incidence of thelytoky). Beekeepers using colony thermodynamics relative to the local area breeding cycles within the framework of year-round field management, geared to Nature's natural temperature rhythms and climate, can greatly improve overall colony performances in a period of 3-5 years. Beekeepers need to learn that queen breeding is progressive and retrogressive in results and can even hold status quo, as in the case of cloning.

Beekeepers should know both the mainflow-breeding and stress-breeding times of the year in their local areas. Mainflow-breeding mainly hybridizes and/or breeds honeybees forward progressively, while stress-breeding when used at either the beginning or the end of selected breeding cycles can retrogress bee stocks, like separating oil from water, so that they may be rehybridized again and again to re-infuse hybrid-vigor for increased colony production standards.

Basic colony thermodynamics for bee breeding

1. A cold-blooded animal is one that has a body temperature below 80° F (27° C), and that takes on the temperature of the air, water, or other element in which it lives. One bee or a few bees do take on the temperature of the air around them and cannot protect themselves against the loss of heat or cold.

2. A warm-blooded animal is one having a relatively high and constant body temperature relatively independent of the surrounding environment. The bee cluster can keep itself warm against a sub zero temperature of -100° F (-73° C) or cool against a temperature of over 135° F (57° C) by metabolic activity mimicking warm-bloodedness by working together as a whole harmonious unit to provide an optimum and constant body temperature relatively independent of the harsh surrounding conditions of temperature and humidity.

3. With an internal ambient temperature of approximately 106° F (41° C) both bees and brood die without some measures of heat regulation.

4. When the ambient temperature inside the hive drops to 45° F (7° C), bees normally cease work, cluster loosely, and maintain the cluster temperature at 57° – 58° F (14° C).

5. The cluster is mostly nearly dormant at 57° – 58° F (14° C) which still allows the bees to be able to break

cluster and move to a new store of honey when all within the cluster has been consumed.

6. Honeybee clusters generate 12° – 13° F (7° C) heat by their normal and natural bodily metabolism or activity incidental to living.

7. The brood rearing temperature is approximately 93° F (34 ° C) to stimulate both the queen to lay eggs and the worker bees to feed and care for larvae.

8. Once the brood rearing has begun, bees must generate whatever heat it takes to maintain the brood nest temperature at approximately 93° F(34° C) until the brood emerges.

9. If the temperature of the outside air rises to 90° F (32° C) or higher, bees normally carry water into the hive, evaporating it by forced air circulation and thus removing the excess heat from the hive (Evaporation of water cools the hive because the specific heat of water is more than 4 times that of air.).

10. Pure hybridization occurs where hot-weather bees (yellow) and cold-weather bees (black/brown) come together naturally by either latitude or altitude with a mean monthly temperature of 75° F (24 ° C).

11. As the inside ambient temperature approaches and/or exceeds both 45° F (7° C) and 106° F (41° C) small black bees approach the breeding condition of thelytoky (Have not been able to accomplish with either yellow-mix or large dark castes.)

12. Humidity in the brood chamber should be about 60% relative humidity, while in the supers where the honey is being ripened it should be 10% relative humidity.

Other basic guidelines for bee breeding

1. Dark (brown/black) cold-weather bees exist naturally below 30° latitude where higher altitudes permit.

2. (Yellow) hot-weather bees exist naturally above 30° latitude where warm thermal areas permit.

3. Small caste races/strains of hot-weather bees exist at the Equator and large caste races/strains of cold-weather bees exist as they approach the poles.

4. As all races/strains of bees advance towards temperature transition-zones at near 30° latitude, hot-weather bees hybridize more, while cold-weather bees hybridize less.

5. Nature breeds constantly and constant when all optimum basic evolutionary needs are met i.e. water, food, shelter, and temperature.

6. Mongrel hybridization is not an evolutionary progression for it separates when artificial stimuli are removed i.e. inappropriate artificial bigger comb size, surrogate geographic areas, and forced climatic breeding.

7. Nature breeds evolutionary change that is progressive, retrogressive, or cloned, when race/strain survivability is at stake.

8. Each race/strain of honeybees has its own separate breeding cycle in Nature providing, an evolution separate from all others, enabling it to exist.

9. Large caste bees on a natural system equate with: 1) fewer bees per brood comb, 2) slower developmental time, and 3) slower mating flight speed.

10. Small caste bees on a natural system equate with: 1) more bees per brood comb, 2) faster developmental time, and 3) faster mating flight speed.

11. Drones take mating flights only on days when bees are able to break cluster and fly outside.

In queen rearing, not the outside air temperature itself is the focal point which beekeepers must consider, but the temperature of the skin surface of the artificially boxed hive where exposed to the sun or the chill-factor of cold winds, which may reach 135° F (57° C) or -100° F (-73° C) (sub zero), or even more depending upon latitude and altitude, and time of the year. This heat or cold passes through the wall/entrance of the hive to its interior, thus increasing or decreasing it to far above or below the outside temperature. Beekeepers seriously breeding bees can help colonies thermoregulate by maintaining tight and painted equipment, and leaving full frames of honey surrounding the brood nests to act as insulation against extremes of cold and heat.

By natural metabolic cluster reactions, honeybees thermodynamically overcome these effects of unfavorable weather conditions within the hive during cold winters and hot summers. However, to bees, the temperature of the skin surface of the artificial box is a trigger mechanism to which they must react, to average the maximum and minimum temperatures of each day. Day in and day out, bees must manipulate natural weather conditions to approach and provide optimum mean temperature conditions for brood rearing and colony survival.

An ambient temperature lower than about 80° F (27° C) inside the colony results in one of two things. Either the brood rearing within the colony decreases and cuts back or, if seasonal conditions cause the bees to react favorably (fresh pollen and/or nectar coming in), the bees will increase their metabolic activity and produce the necessary heat to offset any short-term decrease in temperature, adding a minimum of 12° – 13° F (7° C) of their own body heat to raise brood, if there is a supply of pollen and reserve honey stored. As soon as the brood rearing temperature of 93° F (34° C) is reached, the queen begins to lay eggs and the brood is reared and cared for by the colony.

In spring, when most beekeepers think of rearing queens, they think of progressive breeding techniques, waiting until colonies produce sufficient drones and nurse bees before beginning their queen rearing. Many wrongly believe that hybridization is progressive breeding. It is not! In today's world, hybridization is for the most part mongrel breeding that produces only a short burst of hybrid vigor and then quickly falls apart with each succeeding generation.

For most beekeepers, there should be no breeding from hybrids since it is beyond most beekeepers to control it. The final result is nearly always total mongrelization of local area bee stocks and an uncontrolled mixture of overly aggressive honeybees which makes beekeeping more and more impossible in today's urbanizing world. In a long-term stock improvement program, artificial insemination and various closed-population breeding methods should be avoided, as they lead to severe inbreeding, resulting in poor brood patterns, poor product averages, weak winter cluster carry-over, and colony collapse over a period of 20 – 30 years.

Nature breeds evolutionary changes that are progressive, retrogressive, or cloned, when race/strain survivability is at stake. To accomplish either of the three, beekeepers must remember that all breeding begins with the selection of notable breeding stock of above average overall colony performance. Beekeepers should look for and select honeybee breeder colonies based on a whole-bee theory of field characteristics. To do anything else will, in the long-term, doom the breeding program to problems and necessitate retrogression before being able to proceed further.

Retrogression in a bee hive is not a simple process. We have talked about cell size retrogression and what it involves in physically sizing honeybees back down to natural feral sizing for control of all acarapis mites and their accompanying secondary diseases. This necessary process sets the stage for bee breeding as survivability and variability are achieved. But, just what is progressive breeding, retrogressive breeding (not to be confused with retrogression relative to size), and cloning (thelytoky) as pertains to breeding honeybees?

Progressive breeding

Is the production of uniform progeny within the framework of a fully naturalized breeding program which will true breed and the results of which can only be obtained from uniformly bred colonies. Permanent results can only be achieved by the use of naturally occurring races/strains of honeybees. Since a bee by any other name is still a bee, then beekeepers must use individual or combinations of large or small caste races/strains of hot or cold-weather bees to accomplish this. Artificial hybrids may then be created by mimicking natural hybridization, when two of these races/strains are assimilated. Nature does not produce complex mongrels. Nature transitions in and out from one race/strain to another, with a brief transition-zone between, that is a mixture of each, while always maintaining compatibility to localized geography and climatic thermodynamics.

Retrogressive breeding

Is the reversal of either natural or artificial hybridized combinations of large or small caste races/strains of hot or cold-weather bees, resulting in the production of uniform progeny within the framework of a fully naturalized breeding program, which will then result in each separation achieved, breeding true to their own hot or cold-weather characteristics and large or small caste delineations.

Results can only be achieved by the use of stress-breeding at either the beginning or the end of the selected race/strain breeding cycles where no overlap occurs, one projected breeder-cycle to the other(s). Artificial races/strains can then be created by mimicking natural races/strains where complex mongrelization has taken place, to gain uniformity of characteristics then necessary for the advancement of desirable traits i.e. gentleness and production.

Cloning (Thelytoky)

Is the holding constant of race/strain genetics from one generation to the next naturally or by artificially increasing the propensity of worker bees to lay viable brood, to raise queens as an alternate survival system to supplement normal queen mating in case the virgin queen is lost during the mating flight.

Results can only be achieved by using severe stress-breeding, by using the temperature outside, the beginning or the end of selected race/strain breeding cycles where no overlap occurs, one projected breeder cycle to the other(s).

It is a short-duration phenomenon initiated by extreme stress to allow perpetuation of species, until the first available normal mating can be accomplished, to allow, the colony to permanently requeen itself in the normal manner of mating.

Projecting breeding cycles

Beekeepers should remember when projecting breeding cycles that the color of the exoskeleton is only of significance as a distinguishing character for the purpose of racial analysis where there is the possibility of darker races/strains of honeybees crossing with yellow races/strains of honeybees. In these instances, because the yellow rings of the tergits are so conspicuous, they can be quickly eliminated one from the other. It is because of this that beekeepers from time immemorial have given significance to the coloration of the tergits of the abdomens of honeybees.

Only when more than one race/strain of bees are in a given area do beekeepers need to project breeding cycles to find the best times the drones of their bees have the breeding advantage to maintain racially segregated stock. To project the number of breeding cycle graphs required, beekeepers should first survey colonies in their area that are both domestic and feral (Note – Colonies on oversized artificial brood foundations do not fully correlate with naturally occurring breeding cycles, necessitating that differences be taken into account or excluded from survey). Survey information should include:

1. The number and type of race/strain bees perceived present in the area.

2. Being specific, the approximate dates the worker bees first begin to either raise or eliminate drones from their colonies.

3. Being specific, the weeks/months drones are totally absent from all colonies. (Note – If a few drones are present so note, and under what circumstances i.e. laying worker, extra strong hive, etc.)

To plot breeding cycles, beekeepers need to chart month by month, both the actual "mean monthly temperature" and the "long-term average mean monthly temperature" (see chapter titled "Open-Mating Breeding Charts" Fig. 2). Beekeepers then need to additionally chart month by month the "mean weekly temperature", noting the approximate dates the worker bees first begin to either raise or eliminate drones from their colonies (see chapter titled "Open-Mating Breeding Charts" Fig. 2).

Last, beekeepers need to chart the week(s)/month(s) the drones are totally absent from all colonies. (Note – "Mean" temperatures are used because Nature does not breed by utilization of daily temperature extremes. Honey combs are Nature's regulator for constant breeding transition).

The dominate breeding cycle for the area will be determined by the majority of mean monthly temperature days favoring either right or left of 75° F (24° C) on the "Open-Mating Breeding Chart" see Chapter titled "Open-Mating Breeding Charts" Fig. 1). Beekeepers should then look for either open windows-of-opportunity showing drone breeding advantage, and/or majority-of-temperature dates, to either the cold-weather or hot-weather breeding side of the Open-mating Breeding Chart see Chapter titled "Open-Mating Breeding Charts" Fig. 1). Consequently, for raising dark bees, the closer the breeder can come to the maximum mean monthly temperature, for the warmest month never exceeding 75° F (24° C), the better the results will advantage dark drones. Further, the closer the breeder approaches 57° F (14° C), the darker the results will be. Beekeepers desiring to raise yellow caste bees should follow the same process, only, the closer the breeder approaches 93° F (34° C)) the higher the odds will be for that type of mating.

In areas of complex mongrelization where several races/strains of bees are determined, retrogressive breeding should be a multi-step process. It should start with the separation of yellow races/strains from dark races/strains. Next, beekeepers should separate color by caste size, to be lastly followed by separation of remaining bees by physical characteristics other than size.

By being able to select how to breed bees, either progressive or retrogressive, beekeepers can initiate methods to return beekeeping back to a sound foundation and a future in the 21st century. To go forward, beekeepers must learn they sometimes have to go backwards to rectify breeding and field management problems. Beekeeping in the future can only survive and thrive with uniform, well adapted, peaceable bees in our urbanizing world.

REFERENCES

SECHRIST, E.L., D.F. McFARLAND (1946) - Scientific Beekeeping, Earthmaster Publications, Roscoe, California, U.S.A.

RUTTNER, F. (1988) - Breeding Techniques and Selection for Breeding of the Honeybee, Published by the British Isles Bee Breeders Association by arrangement with Ehrenwirth verlag, Munich

DEGRANDI-HOFFMAN, G., E.H. ERICKSON Jr., D. LUSBY, E. LUSBY (1991) - Thelytoky in a strain of U.S. honey bees, Bee Science, Vol No.3, Pg. 166-171

Dee A. LUSBY, E.W. LUSBY

Suggested Biological Manipulative Field Management for Control of Honey bee Mites – Part 1

Concept & Causes

What is biological field management?

Biological field management of beehives is not new but is seldom practiced anymore. Basically, it is similar to beekeeping the way Grandpa used to do it around the turn of the century.

Because today's conventional drugs and chemicals used in the treatment of bee diseases, pests and parasites are aimed at suppressing disease symptoms by controlling the problem rather than alleviating the problem, they do not have a place in a long-term program of biological field treatment. In the end, chemical controls only add problems for the beekeeper in the form of increased resistance to the very chemicals being used, thus only enhancing the problem.

Colony distress is an important symptom, a signal, which is initiated by the colonies own defense mechanism. Learning to recognize these stress signals is therefore important for early initial natural biological field treatment. To suppress and mask symptoms of honey bee diseases, pests and parasites with chemicals without finding their origin is contrary to the philosophy of long term biological control.

It is of vital importance to realize the various symptoms of honey bee diseases, pests and parasites and should not be viewed as totally negative. Rather, they should be viewed as positive constructive symptoms initiated by the colonies' own healing mechanism, in its effort to restore balance and heal itself.

When this is clearly understood by beekeepers, then time and resources will no longer be wasted on methods that mask symptoms with quick fix remedies and provide only temporary relief, while leaving in their wake the consequences of product and broodnest contamination. Hopefully, the beekeeper will then aim at eliminating and correcting the underlying causative factors of honey bee diseases, pests and parasites, and begin supporting biologically the colonies own recuperative powers of self-correction and healing.

Concept of origin and spread of diseases, pests and parasites

It is a known fact that both honey bees and mites have been on this Earth and have co-existed for many millions of years. Parasites cannot survive if they kill their host. The question then is what has gone wrong? Why do colonies die from Acarapis woodi and Varroa jacobsoni infestations? How do normal healthy beehives change into parasitic mite infested colonies with accompanying secondary stress diseases without cause and effect transpiring?

The well-known colony stress symptoms — unexplainable fatigue, loss of appetite, physical abnormalities, nervous or runny behaviors, lack of housecleaning, poor flight activity –, create increasing degrees of ill health and are considered by many to be consequences of mites.

Since both honey bees and mites have co-existed for many millions of years, it must be assumed that something done artificially to honey bee colonies during their domestication and management by man has created the problem of parasitic mites, that ultimately result in the destruction of the colony population by them and their accompanying secondary diseases. By looking at *cause and effect* we find that beekeepers themselves have wrought cause and effect in several ways. Combined, they have created the situation beekeepers now find themselves in.

First the colonies have to be stressed (the cause) causing the hives to become susceptible to mites and related accompanying stress diseases (the effect). It has been suggested that Acarapis woodi may have evolved very recently, perhaps in Britain, and as recently as 1900 (DEJONG et al., 1982). However, this hypothesis must be treated with caution (see Saga # 16 American mite history background). Nevertheless, the very close similarity of the various species of Acarapis mites (which includes A. woodi, Tracheal mites, & Varroa jacobsoni now renamed V. destructor) does suggest that they evolved symmetrically of Apis mellifera from a common ancestor (Delfinado-Baker and Baker, 1982).

If beekeepers were to study comb size history they would easily perceive that introduction of larger and larger comb cell sizes, used in colonies since the turn of the century, have developed evolutionary changes in honey bees through artificial mutation of body size, therefore making bees more susceptible to parasitic mite attacks.

With today's comb cell foundations now on the market near or exceeding measurements per square decimeter for Apis dorsata, for most of today's European honey bee races, no small wonder there is a parasitic mite problem (see table included). The European honey bees are merely out-of-tune with natural feral races and strains of bees by way of artificially enlarged body and comb sizing.

Based on observations and study of comb cell sizes, it should be hypothesized instead, that honey bees, since the early 1900s, have been artificially enlarged by using bigger and bigger comb sizes, thus causing the parallel evolution of mites as their food source changed.

The causes

1. Artificial oversized brood combs.

Location	Beekeeper	Year	Size
Attica, Greece	Georgandas	1968	733 min, 854 max, 815 avg
Peloponnesus, Greece	Georgandas	1968	846 min, 892 max, 863 avg
Arta, Greece	Georgandas	1968	835 avg
Crete	Georgandas	1968	835 avg
Macedonia	Georgandas	1968	821 avg
---	Collin	1865	854
---	Langstroth	---	838
Italy	House of Frtelli Piana	---	860
Italy	House unnamed	---	813,807,854
---	Baudoux	---	854,807
---	Pincot (for Italian race)	---	764
Burgandy	---	---	798
France	(common black bee)	---	854
France	(degenerated common bee)	---	924
---	Halleux	1890	845
North Africa	Rambaldi	---	940
---	Fremont	1983	825
United States	Grout	1931	857
Netherlands	Swammerdam	1733	857
Italy	Maraldi	1712	789, 954
France	Reaumur	1740	
Germany	Klugel	1700s	
Spain	Castellon	---	763, 828
British Isles	A.D. Betts	1700s	830
India	Rahman & Singh	1946	1013.17 A. indica 2380.61 A. florea 796.10 A. dorsata
United States	A.I. Root	---	825, 850

Since the time of Baudoux in following Huber's experiment in 1791, but by using artificial means instead of drone combs, causing creation of larger worker bees, beekeepers have been artificially mutating the body size of honey bees larger (Grout, 1931). This has placed honey bees with each successive upsizing of comb more out-of-tune with Nature and natural bee flora. Why, because it is difficult to create new honey plants and bees which can be reproduced as such, which have been developed through thousands of years and adjusted to the existing climatic conditions, soil, and especially existing bee flora (Cheshire, 1888; Georgandas, 1968). This then creates and adds to the second cause. Known documented measurements of the dimensions of honey bee brood cells per dm^2 on natural comb.2. Artificial diet causing inadequate nutrition.

Poor nutrition is a serious stress factor of any organism. What happens when key nutrients are present in insufficient quantities for generation after generation? Larger honey bees require richer nutritional diets, yet have access to less in Nature by being out-of-tune through body size to appropriately match natural bee flora.

Colonies can be in a state of inadequate nutrition through either their geographic location placement or placement on artificial enlarged comb foundation creating imbalance with bee flora, and/or fed diets of pollen substitutes and sugars that are inadequate. Because of this, one or more of the key nutrients can be insufficiently represented or entirely lacking in the bee's body adding to immune system deficiency.

Since we believe that a queen reared this way, cannot give to her offspring what she does not have herself, the result is that the queen constitutionally transmits a predisposition for disease and mite attack to her offspring. If honey bees acquire a predisposition for stress diseases due to inadequate nutrition through either their geographic location placement or placement on enlarged brood comb foundation, beekeepers can expect disease and mite infestations in their colonies.

3. Artificial medical treatment by chemicals rather than biological treatment through natural management

...causing neurological disorders (CHANEY, 1988), queen supersedures, brood deaths, resistant mites to chemicals being used for treatment thus enhancing reproduction of same and contaminating internal colony food, leaving the honey bee colony unable to function properly to fight off bee diseases or mites.

Suggested Biological Manipulative Field Management for Control of Honey bee Mites – Part 2

Prevention – A Possibility!

Since a small population of parasitic mites is nondetectable by either chemical or biological examination methods, beekeepers wait for the appearance of a large infestation to tell them that something is wrong. By then it is often too late for the hive. An approach is needed that looks at the situation in reverse. First the honey bee colony drifts into a pathological state, with the final symptom being a severe infestation of parasitic mites.

Logic should compel beekeepers to try to detect the underlying stress signals which are the forerunners of mites, and through biological field treatment manipulations, eliminate the artificial stimulations that result in Acarapis mites attacking colonies at various points of infestation upon bees. This can be accomplished with a long-term biological field manipulative treatment program which can be used to either prevent or wean colonies from parasitic mites and their accompanying secondary diseases. (Note: before we go any further, let's state: *mites have historically been named by point of infestation on the body of the honey bee. in the beginning for Acarapis mites it was head, neck, back, and wing mites; later internal mites, when eventually found in the trachea. nowadays we only talk about tracheal mites and Varroa mites having renamed them just like we have renamed foulbroods several times!*)

There is no denying that methods consisting of heavy medication do wage a battle against parasitic mites and accompanying stress diseases. However, at the same time chemicals only mask the symptoms and perpetuate the problem. In addition, beekeepers run the high risk of chemical contamination and product recall of wax, pollen, and honey crops.

Advanced stages of stress, indicated by symptoms of high Acarapis mite infestation levels at various sites upon the bees' bodies, prevent beekeepers from implementing biological field manipulation treatments easily, because once on chemical dependency treadmills, it is almost impossible to stop treatment without significant loss of colonies.

Stress Symptoms Develop for Several Reasons that Work in Combination.

In the beginning, the honey bee colony is in perfect health without diseases, pests and parasites. Then through the combination of placement on improper artificially enlarged size brood combs for localized geographic regions, and improper nutritional needs brought upon by bees being out-of-balance with natural flora over extended period of time, the colony develops the loss of this healthy condition.

Stress factors weaken the honey bee's natural defense system inherent within the hive. Minor stress symptoms appear in the form of foulbrood and other body diseases. In successive generations, more advanced symptoms appear in the way of various fungal diseases. These diseases, along with accompanying Acarapis mite infestations at various sites upon the bees body can easily gain a foothold within a stressed colony. The end result is that the colony is destroyed from generations of abuse and stress.

The mites and their accompanying diseases are not the problem, they are merely the advanced stages of an artificially caused problem. *The stress resulting from generally accepted beekeeping practices of artificially enlarged brood-combs, nutrition by either being out-of-balance with natural flora or fed artificial diets of pollen substitute and/or sugar/corn syrup, overuse of antibiotics, and chemicals, repeated over many years is the real killer of domesticated honey bee colonies.*

The most important weapon in the fight against the various parasitic Acarapis mites and their accompanying secondary stress diseases is prevention. Beekeepers must be alert to the signs of distress within their colonies from these sources. When stress symptoms are apparent, beekeepers must take action to put their colonies back into a natural biological balance with manipulative field treatments. (Note: *Secondary stress diseases come about because bees ingest bacteria or viral pathogens their systems cannot handle and/or because the mites chew on the bees exoskeleton creating a wound, allowing for bacterial and viral infections to begin through these wounds, that their immune systems cannot handle.*)

This retrogression back to a natural biological balance within the colony can be accomplished through dietary change if an artificial diet is being used, and by replacing the artificially enlarged brood comb with natural sized comb foundation in harmony with the geographic region where the colonies are being maintained. Culling excessive frames of drone combs (more than 10% drone cells drawn on any one comb) will also help.

The downsizing of brood comb by manual shake-down in the field to natural brood comb sizing, before that made for enlarged brood combs at the turn of the 19th century, will realign the bees' body size to again match their native flora. Changing the diet from artificial pollen substitutes and sugar syrups back to pure natural pollens and honey from the colonies own geographic region, will also improve colony vigor.

The removal of stress by beekeepers is, of course beneficial, like the removal of contaminated combs and their replacement with disease free and chemical free (or decontaminated, greatly reduced chemical, processed foundation) combs. But this in itself does not correct the underlying reason the hive came down with the malady in the first place. The whole hive must be restored to full health by retrogressing it back onto a natural biological system, that acts to relieve stress without the use of chemicals, essential oils, antibiotics, or some other crutch that is labor intensive.

If the colonies are still in the early reversible stage of development of stress diseases, the therapeutic administration of natural key nutrients and natural sized brood comb foundation, sized to one's own beekeeping region, will in most cases bring about the restoration of health to the colony, thus naturally controlling mites and their accompanying secondary diseases.

The result is that the bee's own natural defense system and capacity for recovery will again be activated and begin the work of clearing away the problem within the hive. Stress diseases will be eliminated and the mite population will naturally decrease to a level well below economic thresholds for survival of the hive.

Beekeepers must bear in mind that in treating and curing honey bee stress diseases and getting rid of the various accompanying Acarapis mites, that these disturbances to colonies do not possess a capacity for unbridled autonomous growth. Their behavior depends entirely on the state of health of the honey bee colony as a whole harmonious working unit.

The nutritional healing of the colony coupled with replacement back onto natural sized brood comb foundation has a number of important advantages:

1. In a colony that has been restored to health, the natural defense systems of the bees are fully operational again, whereas treatments such as chemotherapy for parasitic mites can have the opposite effect, that of damaging the bees by causing neurological disorders (CHANEY, 1988), as well as probably causing comb and hive product contamination.

2. No secondary infections by foulbroods, chalk broods, etc., can take place because infected brood will be destroyed by the bee's own natural communal defense system.

3. The size of the worker bee returns to normal and again fits the natural flora of the region. This is important because the ratio of worker size honey bees to drone size bees is 20%, a four to five ratio of body size, that remains constant no matter what size the worker bee is and by returning the worker bee to normalcy, you *change the size of the thorax of all bees in the colony*, including the drones.

The automatic downsizing of drone dimensions by the downsizing of worker bees is extremely important for fighting Varroa jacobsoni infestations.

This is important because drones are also periodically thrown-out of hives after each honey gathering season. We further believe that this downsizing of honey bees aids in reducing the Acarapis mite population, no matter what the point-of-infestation is (whether internal and/or external), in important ways:

a. The size of the honey bee is correlated with the capacity of the cell. Small cell, small bee; big cell, big bee (BAUDOUX, 1933). The size remains the same during the whole of the bee's life in perfect ratio one caste to each other.

Since the only place Acarapis woodi (Tracheal mites) mites can get into honey bees is through the first thoracic spiracle (EICKWORT, 1988) on the bees thorax, artificially enlarged cell size is an important artificial mutant that can be rectified by beekeepers through use of naturally sized brood comb foundations. Once placed onto natural sized brood combs the bee's thorax size is reduced (cell size determines the size of the bee's thorax and hence all other body parts in proportion) thus also reducing the hole size of the first thoracic spiracle, and Acarapis mites have lost a very valuable avenue of entry for hive destruction, thus regulating them back to external body mites surrounding the wing region (vagans).

b. In Brazil, cell sizes for Africanized and domestic (European) honey bees when measured averaged 4.5 to 4.8 and 5.0 to 5.1 mm per cell, respectively (MESSAGE and GONCALVES, 1983). They further reported that Varroa infestation rates were 4.8% and 11.5% respectively. CAMAZINE (1988) calculated female Varroa replacement rates for Africanized and domestic (European) honey bees at 1.2 and 1.8 with drones present and 0.8 and 1.5 without drones, respectively. (Note: A female Varroa replacement rate of less than 1.0 indicates that the mite population is declining while a 1.0 rate is indicative of zero population growth.)

Keeping this in mind, it makes perfect sense to downsize (retrogress) artificially enlarged brood combs to natural sized brood combs to take advantage of the 0.8 population replacement of Varroa jacobsoni when drones are seasonally ejected by colonies at the end of each honey gathering season. This could also be accomplished by division of the broodnest for requeening, knowing that following mating with the new queen starting to lay, drones are expelled from colonies also. (Note: equate Africanized with natural feral sizing, which is all it is; and European size with artificially enlarged bees by way of man's interference.)

Additionally, it also makes perfect sense to cull drone combs to less than 10% of all combs in a hive and/or 10% drone cells drawn on any one frame in a hive, to keep Varroa populations down to a minimum.

Thus it may be possible to suppress Varroa populations in domestic colonies by using small natural brood cell foundation to downsize honey bees back to natural racial/strain sizings, as small caste bees on a natural system equate with: 1) more bees per brood comb, 2) faster developmental time, and 3) faster mating flights for breeding advantage over large caste bees, which equate with 1) fewer bees per brood comb, 2) slower developmental time, and 3) slower mating flights.

c. Downsizing also reduces basic food stimuli attractiveness for mites. It has been documented by KULZHINSKAYA in 1956 that worker larvae in enlarged oversized brood cells received 21% more food and 21.4% more protein than worker larvae reared in normal sized cells. He also found that the weight of larvae increased by 12.4% and that of adults reared in enlarged oversized cells by 10.4%.

Since it is common knowledge that mites prefer drone cells, in the case of Varroa jacobsoni, over worker brood cells and Wolfgand Ritter (1988) stated that "Varroa cannot reproduce in the worker brood of Apis cerana, according to Ritter et al, 1980; Koeniger et al, 1981 confirmed this and additionally found Varroa jacobsoni offspring only in drone brood", then logic should dictate that the additional food and protein in enlarged oversized cells does indeed act as a mite attractant.

Hanel (1983) pointed out that one of the reasons for such differential reproductive behaviors of A. cerana bees could be due to their juvenile hormone level. Varroa takes in various amounts of juvenile hormone III during its primary intake of hemolymph (bee blood) when feeding. This induces oviposition (egg laying) in the mite.

In the first 60 hours, the drone larvae of A. cerana and A. mellifera contain more than 5ug/ml JH in their hemolymph (bee blood). Worker larvae of A. mellifera contain 3 – 7 ug/ml and, those of A. cerana contain only 1 ug/ml. The level of juvenile hormone in worker larvae of A. cerana is apparently not sufficient to induce oviposition (egg laying) in the mite.

This has proved to be a selective advantage to the honey bee during the course of its host and parasitic evolution. Only in this manner does the parasite prevent death of its host and thus its own death. F. RUTTER in his paper "Characteristic and variability of Apis cerana" points out that "Contrary to the customary assumption, A. cerana is not generally a small bee when compared with A. mellifera. This frequently-held opinion holds true only when A. cerana is compared with European A. mellifera".

We believe that this is a comparison of a feral sized, naturally occurring type of honey bee, to an artificially over-sized domesticated European sized honey bee that has received more food and protein, thus more juvenile hormone by being reared on artificial enlarged combs. Therefore, retrogressing/down-sizing would have the impact of reducing juvenile hormone levels; and, food and protein contents of the larvae jelly, all of which are mite attractants in oversized cells.

d. Downsizing also compacts the brood nest by density and our observations by inserted temperature probe, show that it raises the brood nest temperature, which we believe helps to speed up the gestation cycle of the brood. Combine this with being able to select for faster developing queens (DEGRANDI-HOFFMAN, LUSBY & LUSBY, and ERICKSON JR, 1989) and it becomes possible to breed for bees with shorter development times as an aid in overcoming Varroa.

Remember in the end, surgical removal of stress (taking diseased frames out by hand) by beekeepers is always possible if the colonies own defense system proves to have been so debilitated as to be incapable of returning to normalcy. If surgery by beekeepers is necessary, a healthy honey bee on a proper nutrient diet will better generate strong recuperative powers once causitory brood combs have been removed and replaced.

Lusby's (LUS bees) Bee Biometrics According to Prof. Koeniger

Prof. Dr. N. Koeniger

INSTITUT FUR BIENENKUNDE

(Polytechnische Gesellschaft)

Fachbereich Biologie der J. W. Goethe-Universitat

Frankfurt am Main

6370 Oberursel 1

Im Rothkopf 5

W.-Germany

May 12, 1986

Dear Mr. and Mrs. Lusby,

Thanks for the letter of March 19th and the samples of bees. We did the biometrics now and it resulted in clear differences of your black bees compared to the usual U.S. mixture. Your bees are quantitatively significant more towards *Apis mellifera carnica* und *Apis mellifera caucasica*. The Italian influence is very limited.

We thank you again for your hospitality. Hope to meet you some day again. Attached you will find the values of your samples (cubital index).

Sincerely

N. Koeniger

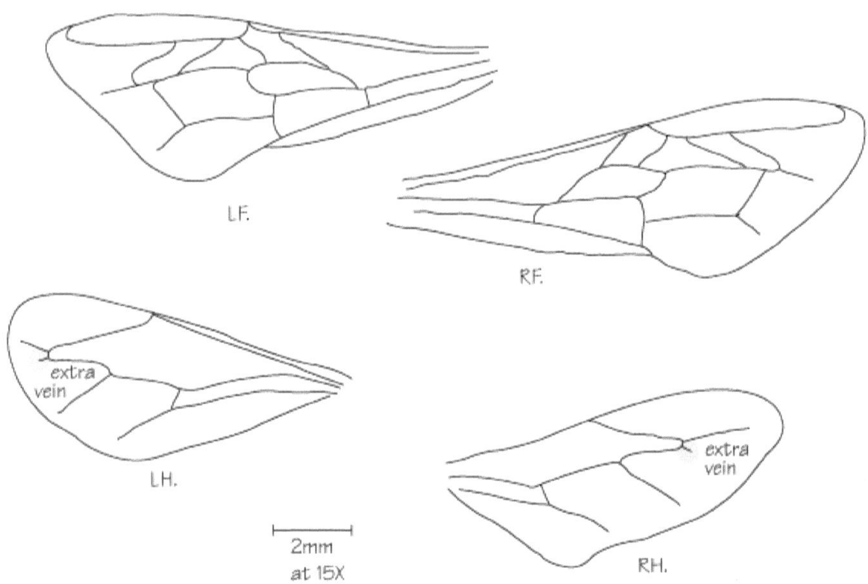

	HD	TH	AB
Length	4	4	15
Width	4	5	5
Depth	2	-	-

Dee's Bees
LA187 8/23/87
#31 Drone DBLFII
DBLHII, DBRFII, DBRHII

Wing vein drawings, by Marilyn Houck, of Lusby's bees.

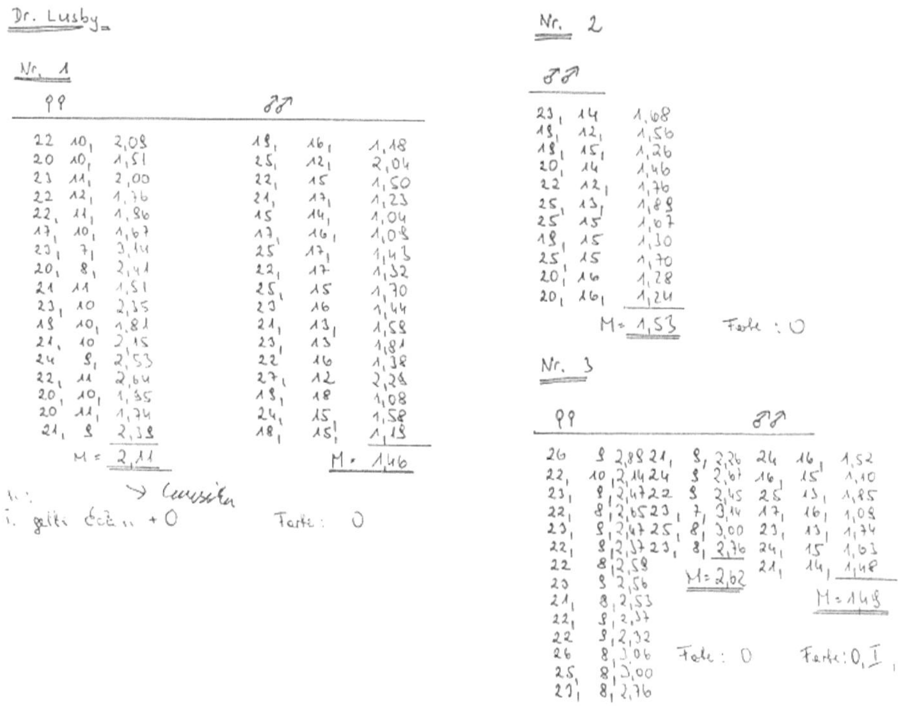

Dr. Lusby

Nr. 1

♀♀			♂♂		
22	10,	2,09	19,	16,	1,18
20	10,	1,51	25,	12,	2,04
23	11,	2,00	22,	15	1,50
22	12,	1,76	21,	17,	1,23
22,	11,	1,86	15	14,	1,04
17,	10,	1,67	17,	16,	1,09
23,	7,	3,14	25	17,	1,43
20,	8,	2,41	22,	17	1,32
21	11	1,51	25,	15	1,70
23,	10	2,35	23	16	1,44
19	10,	1,81	21,	13,	1,59
21,	10	2,15	23,	13	1,81
24	9,	2,53	22	16	1,38
22,	11	2,64	27,	12	2,29
20,	10,	1,95	19,	18	1,08
20	11,	1,74	24,	15,	1,58
21,	9	2,39	18,	15,	1,19
	M =	2,11		M =	1,46

→ [illegible]

Farbe: 0

Nr. 2

♂♂		
23,	14	1,68
19,	12,	1,56
19,	15,	1,26
20,	14	1,46
22	12,	1,76
25,	13,	1,89
25	15	1,07
19,	15	1,30
25,	15	1,70
20,	16	1,28
20,	16,	1,24
	M =	1,53

Farbe: 0

Nr. 3

♀♀						♂♂		
26	9	2,89	21,	9,	2,26	24	16,	1,52
22,	10,	2,14	24	9	2,67	16,	15	1,10
23,	9,	2,47	22	9	2,45	25	13,	1,85
22,	8,	2,65	23,	7,	3,14	17,	16,	1,09
23,	9,	2,47	25,	8,	3,00	23,	13,	1,74
22,	8,	2,37	23,	8,	2,76	24,	15	1,63
22	8,	2,59		M =	2,62	21,	14,	1,48
23	9	2,56					M =	1,49
21,	8,	2,53						
22,	9,	2,37						
22	9,	2,32		Farbe:	0		Farbe:	0, I
26	8,	3,06						
25,	8,	3,00						
23,	8,	2,76						

Biometrics done by Dr. N. Koeniger, of Lusby's bees.

LEGEND for BEE SAMPLES

Dr. Loper ID #	Lusby ID #	Source Location	Color Bees
1	L-5	Tortuga	Yellow
2			
3	L-7	Tortuga	Yellow
4	L-1	Home Breeder Apiary	Black
5			
6	L-8	Tortuga	Yellow
7			
8	L-3	Home Breeder Apiary	Black
9			
10			
11			
12	L-6	Tortuga	Yellow
13	L-2	Home Breeder Apiary	Black
14			
15	L-4	Tortuga	Yellow

Note – The breeder samples used from the home breeder apiary were also taken from one of the three hives used to send samples to the University of Frankfurt, West Germany, for their unbiased quantitative analysis as to what type of bees they were. No Africanization was mentioned in their report back. The USDA/ARS however, highly identified these bees as Africanized.

The breeder hive in question has since been donated to the Carl Hayden Bee Research Laboratory for their research and understanding in basic bee biology as to how this could happen.

For all of the following collections and scores: Last update for discrim coeff for collections of workers was: April 1985

Collection Number: 620

Identity of collection of 5 bees is... Probably European

Score for collection is... 1.353

Probability of being European is... 1.000

Probability of being Africanized is... 0.000

Scores for individual bees

Bees	Score	Pb Eur	Pb Afr	Dev>1.8 SD (expect 5% more than 8, none more than 16
620	1.463	0.996*	0.004	8
621	0.431	0.838*	0.162	12*
622	2.175	1.000*	0.000	9*
623	0.652	0.923*	0.077	7
624	2.295	1.000*	0.000	7

Collection Number: 625

Identity of collection of 10 bees is... Probably Africanized

Score for collection is... -0.227

Probability of being European is... 0.193

Probability of being Africanized is... 0.807

Scores for individual bees

Bees	Score	Pb Eur	Pb Afr	Dev>1.8 SD (expect 5% more than 8, none more than 16
625	-0.738	0.056	0.944*	11*
626	-0.524	0.119	0.881*	8
627	0.796	0.954*	0.046	12*
628	-1.345	0.9006	0.994*	6
629	1.020	0.980*	0.020	12*
630	0.576	0.900*	0.100	12*
631	1.914	0.999*	0.001	9*
632	0.486	0.865*	0.135	8
633	1.194	0.990*	0.010	8
634	0.309	0.765*	0.235	13*

Collection Number: 635

Identity of collection of 5 bees is... Probably European

Score for collection is... 2.212

Probability of being European is... 1.000

Probability of being Africanized is... 0.000

Scores for individual bees

Bees	Score	Pb Eur	Pb Afr	Dev>1.8 SD (expect 5% more than 8, none more than 16
635	0.510	0.875*	0.125	8
636	2.589	1.000*	0.181	8
637	0.396	0.819*	0.181	12*
638	-0.698	0.065	0.935*	6
639	1.446	0.996*	0.004	6

Collection Number: 640

Identity of collection of 10 bees is... Probably Africanized

Score for collection is... -1309

Probability of being European is... 0.000

Probability of being Africanized is... 1.000

Scores for individual bees

Bees	Score	Pb Eur	Pb Afr	Dev>1.8 SD (expect 5% more than 8, none more than 16
640	-1.550	0.003	0.997*	12*
641	2.276	1.000*	0.000	9*
642	-2.132	0.000	1.000*	2
643	-1.720	0.999*	0.001	11*
644	-1947	0.001	0.999*	7
645	0.377	0.808*	0.192	7
646	-1.021	0.020	0.980*	10*
647	0.209	0.310	0.690*	10*
648	-0.662	0.074	0.926*	5
649	-0.285	0.252	0.748*	11

Collection Number: 650

Identity of collection of 10 bees is... Probably European

Score for collection is... 1.325

Probability of being European is... 1.000

Probability of being Africanized is... 0.000

Scores for individual bees

Bees	Score	Pb Eur	Pb Afr	Dev>1.8 SD (expect 5% more than 8, none more than 16
650	1.047	0.982*	0.018	17***
651	0.639	0.920*	0.080	8*
652	2.885	1.000*	0.000	15*
653	2.368	1.000*	0.000	5
654	2.572	1.000*	0.000	8
655	1.701	0.998*	0.002	12*
656	1.821	0.999*	0.001	15*
658	3.242	1.000*	0.000	8
659	2.963	1.000*	0.000	12

Collection Number: 660

Identity of collection of 10 bees is... Probably Africanized

Score for collection is... -1.308

Probability of being European is... 0.000

Probability of being Africanized is... 1.000

Scores for individual bees

Bees	Score	Pb Eur	Pb Afr	Dev>1.8 SD (expect 5% more than 8, none more than 16
660	-0.961	0.25	0.975*	9*
661	-0.714	0.061	0.939*	4
662	2.601	1.000*	0.000	16*
663	0.317	0.770*	0.230	12*
664	0.402	0.823*	0.177	9*
665	-0.653	0.076	0.924*	11*
666	0.212	0.692*	0.308	4
667	-1.476	0.004	0.996*	6
668	-0.176	0.338	0.662*	7
669	-0.869	0.035	0.965*	9*

Collection Number: 670

Identity of collection of 9 bees is... Probably Africanized

Score for collection is... -0.926

Probability of being European is... 0.003

Probability of being Africanized is... 0.997

Scores for individual bees

Bees	Score	Pb Eur	Pb Afr	Dev>1.8 SD (expect 5% more than 8, none more than 16
670	-0.301	0.240	0.760*	15*
671	-3.366	0.199	0.801*	14*
672	0.430	0.838*	0.162	7
673	0.927	0.972*	0.028	7
674	-0.857	0.037	0.963*	15*
675	0.120	0.613*	0.387	6
676	0.026	0.525*	0.475	11*
677	0.605	0.910*	0.090	10*
678	1.802	0.999*	0.001	13*

Collection Number: 679

Identity of collection of 9 bees is... Probably European

Score for collection is... -3.93

Probability of being European is... 0.000

Probability of being Africanized is... 100

Scores for individual bees

Bees	Score	Pb Eur	Pb Afr	Dev>1.8 SD (expect 5% more than 8, none more than 16
679	-2.864	0.000	1.000*	9*
680	-2.398	0.000	1.000*	10*
681	-1.373	0.005	0.995*	7
682	-1.585	0.002	0.998*	9*
683	-2.016	0.000	1.000*	9*
684	-0.106	0.400	0.600*	4
685	-1.803	0.001	0.999*	5
686	-1.929	0.001	0.999*	6
687	-1.681	0.002	0.998*	10*

Collection Number: 688

Identity of collection of 10 bees is... Probably Africanized

Score for collection is... -0.939

Probability of being European is... 0.003

Probability of being Africanized is... 0.997

Scores for individual bees

Bees	Score	Pb Eur	Pb Afr	Dev>1.8 SD (expect 5% more than 8, none more than 16
688	0.607	0.910*	0.090	6
689	0.889	0.968*	0.032	5
690	0.552	0.892*	0.108	8
691	1.689	0.998*	0.002	12*
692	-0.202	0.316	0.684*	15*
693	-0.064	0.440	0.560*	7
694	-0.722	0.060	0.940*	15*
695	-1.370	0.005	0.995*	13*
696	0.122	0.614*	0.386	4
697	1.355	0.994*	0.006	7

Collection Number: 698

Identity of collection of 5 bees is... Probably European

Score for collection is... 1.526

Probability of being European is... 1.000

Probability of being Africanized is... 0.000

Scores for individual bees

Bees	Score	Pb Eur	Pb Afr	Dev>1.8 SD (expect 5% more than 8, none more than 16
698	0.371	0.805*	0.195	7
699	2.023	1.000*	0.000	13*
700	1.723	0.999*	0.001	5
701	2.101	1.000*	0.000	7
702	-1.171	0.011	0.989*	9*

Collection Number: 703

Identity of collection of 5 bees is... Probably European

Score for collection is... 1.526

Probability of being European is... 1.000

Probability of being Africanized is... 0.000

Scores for individual bees

Bees	Score	Pb Eur	Pb Afr	Dev>1.8 SD (expect 5% more than 8, none more than 16
703	1.501	0.997*	0.003	15*
704	0.496	0.896*	0.131	14*
705	-3.089	0.000	1.000*	12*
706	1.252	0.992*	0.008	3
707	1.644	0.998*	0.002	11*

Collection Number: 708

Identity of collection of 5 bees is... Probably Africanized

Score for collection is... -0.922

Probability of being European is... 0.003

Probability of being Africanized is... 0.997

Scores for individual bees

Bees	Score	Pb Eur	Pb Afr	Dev>1.8 SD (expect 5% more than 8, none more than 16
708	-2.815	0.000	1.000*	9*
709	0.211	0.691*	0.309	11*
710	-0.916	0.029	0.971*	9*
711	0.005	0.505*	0.495	4
712	-0.258	0.272	0.728*	6

Collection Number: 713

Identity of collection of bees is... Probably European

Score for collection is... 2.205

Probability of being European is... 1.000

Probability of being Africanized is... 0.000

Scores for individual bees

Bees	Score	Pb Eur	Pb Afr	Dev>1.8 SD (expect 5% more than 8, none more than 16
713	-1.721	0.001	0.999*	3
714	2.481	1.000*	0.000	9*
715	2.226	1.000*	0.000	8
716	2.178	1.000*	0.000	13*
717	1.846	0.999*	0.001	11*

Collection Number: 718

Identity of collection of 5 bees is... Probably European

Score for collection is... 0.222

Probability of being European is... 0.803

Probability of being Africanized is... 0.187

Scores for individual bees

Bees	Score	Pb Eur	Pb Afr	Dev>1.8 SD (expect 5% more than 8, none more than 16
718	1.913	0.999*	0.601	10*
719	-0.742	0.058	0.944*	11*
720	1.074	0.954*	0.016	13*
721	-0.855	0.078	0.824*	8
722	1.846	1.788*	0.213	8

Collection Number: 723

Identity of collection of 5 bees is... Probably European

Score for collection is... 1.721

Probability of being European is... 1.000

Probability of being Africanized is... 0.000

Scores for individual bees

Bees	Score	Pb Eur	Pb Afr	Dev>1.8 SD (expect 5% more than 8, none more than 16
723	1.063	0.983*	0.017	7
724	2.509	1.000*	0.000	12*
725	0.775	0.951*	0.049	11*
726	1.074	0.984*	0.016	14*
727	1.588	0.998*	0.002	11*

Appendix A: Articles and studies with the Lusby's as authors or coauthors

On the Size of Cells – Dee Lusby Preface

February, 1990 – Bee Culture, On the Size of Cells – Dee Lusby Preface

Dee Lusby gives her preface on the article she co-authored.

I do not like this two part series because it was rushed to publication and also because it was published before differences were ironed out on archives. I felt then, and I still feel now, the differences should have been noted along with more of the history on cell size on feral combs. As Ed and I were authors on this article, I also feel that this should have been done, and only proper. A.I. Root made the first mill following original thoughts of the English like Cheshire, Wedmore, etc. and it fit nicely with the chip cut method they had on spiral cutting for the old rhombus method. Just like 5.4 is now 800, so has 5.1 been made 857 cells per dm^2 instead of the old 4.83, which it should be.

Until they go back to the old field sizing that originally set up the system for foundation which was measured parallel wall to parallel wall and then down the diagonals, there will be problems and I just cannot let it go. I felt our author rights were violated then, and the article should have been broader with more pros and cons and if there were two sides to understand to the measurements then both sides should have had the cards laid on the table.

That right was taken away when published without final signoff by all parties concerned, on something we put a lot of input into, gathering samples of combs and digging into history, page by page, out of archives. I have told them since then, many times, they can do what they want with the measurements in the lab for their perfect research for what it is worth, but the field is the field and there you have to do what works and measure the way beekeepers have always measured, namely, simple and fast and parallel then down the diagonals for the feral, always following the bees. Repeat, always following the bees.– Dee Lusby

SPECULATIONS ON FOUNDATION AS A COLONY MANAGEMENT TOOL[1], E. H. EricksonÝ, D. A. Lusby+, G. D. HoffmanÝ and E. W. Lusby

This two-part article is the result of an extraordinary amount of detective work following the twisted routes of many leads. It is about the numerous ways that can be found to complicate an otherwise simple issue. Our purpose is to challenge all in apiculture to question even the most basic assumptions we make when developing sound colony management strategies and interpreting research results.

Domestic honey bee colonies, which beekeepers manage and scientists study, differ in many ways from native or long-established feral (wild) counterparts. These differences are quite similar to those found in other animal species that have undergone domestication. Today, most domestic honey bees exist as artificially selected strains kept in artificial domiciles (box hives). Feral honey bees, on the other hand, exist as naturally selected populations – the colonies are entirely self-sufficient and have adapted to life in naturally occurring cavities. It is imperative that both beekeepers and researchers are aware of these differences when they develop management strategies to solve problems facing the beekeeping industry. Research results from studies using domesticated bees in Langstroth hives are not necessarily applicable to feral bees and vice versa. Periodically, we remind ourselves of this. Yet, in spite of our best intentions, it seems that we (as well as others) often overlook the obvious. So it is with the issue of comb cell size in our bee hives.

Until recently, we gave little thought to the issue of comb cell size. We presumed the subject was adequately researched in the past and all keepers of bees were using similar foundation. However, we have found this is not the case! In fact, beekeepers may be using combs drawn from foundation with differing cell sizes, either in the same apiary or, perhaps in the same hive, particularly if the foundation or combs were purchased from several sources. How can this be, you ask? To answer this question we need to first examine the issue historically.

In the beekeeping literature we found that controversy has followed the issue of optimal cell size for domestic colonies for more than 100 years. Our review starts with the invention of foundation by Mehring in 1857. By the 1880's European beekeepers were using foundation with comparatively small cell impressions. Shortly thereafter, Professor M. Baudoux, through his research at Tervueren, Brussels, Belgium, concluded that this small cell size, 920 cells per square decimeter (= 5.0 mm width per cell), was detrimental to colony development and productivity. He then proceeded to experiment with foundations of increasingly larger cell size. Subsequently, he demonstrated that adult honey bees were larger when reared in comb with larger cells (1). (See "Conversions" next article and footnote for mathematical conversions of some common cell sizes, because some early writers published incorrect conversions.)

Unfortunately, Professor Baudoux was a proponent of the now disproven Lamarkian theory of evolution which proposed that "...environmental changes cause structural changes in animals and plants by inducing new or increased use of organs or body parts..." and that such changes are inherited. This theory would suggest, for example, that the elongated neck of the giraffe is the result of each generation stretching further for the top branches of trees while feeding. Baudoux believed that he could genetically alter the size of honey bees by providing them with larger than normal cells for brood rearing.

Hence, in his research he tested and later advocated the use of oversized cells (as few as 650 cells per dm2 = 6.0 mm per cell). As proof of his theory Baudoux demonstrated, as have others, that bees reared in small cells were significantly smaller than those reared in large cells (4). However, no heritability of size was demonstrated. Neither did he demonstrate that the ability to produce larger cells under these circumstances was genetically determined.

Charles Darwin, in his now widely accepted theory of natural selection, proposed "...that organisms tend to produce offspring varying slightly from their parents and the process of... selection tends to favor the survival of individuals whose peculiarities render them best adapted to their environment..." and that such changes are inherited. Hence, selection for larger queens results in the production of larger daughter queens and worker bees (as well as drones). Larger bees must be reared in larger cells to maintain their size. Conversely, we can select for small bee size and then produce these smaller bees in smaller cells. We now know that ***bee size is a function of BOTH inheritance and cell size.***

It is a curious thing, this conception that bigger is better. Clearly, larger worker bees come from larger, easier to find queens. The workers have longer tongues, larger honey stomachs and store their honey in larger cells (1,2,4). However there is no evidence that a colony made up entirely of larger bees produces a greater honey surplus than a colony of small bees.

Additionally, there are no reports of studies comparing the rate of population growth or peak population size between colonies comprised of small versus large bees. Even so, researchers have shown that colonies of smaller, Africanized bees (AHB) do build up more rapidly than colonies comprised of larger European bees: however, research is needed to determine whether or not cell size is a factor. Finally, and perhaps most importantly, no one has investigated the influence of cell size on the developmental rate of individual bees, susceptibility to disease and parasitic mites, overwintering survival, or other biotic and abiotic stress-inducing factors.

CELL TELL: Documentation of Natural Cell Size.		
Source	Year	Cell Width in mm
Swammerdam	1733	5.1
Maraldi	1712	5.0-5.4
Reaumur	1740	5.3
Klugel	1700s	5.3
Castillion	1700s	5.3-5.5
Latreille	1800's	5.4
Vogt	1800's	5.3-5.5
Collin	1800's	5.2
Langstroth	1800's	5.08
Dadant	1800's	5.08
Root	1868-1917	5.08
Root	1900s	5.26
Chesire	1886	5.06-5.25
Cowan	1898	4.72-5.36
Cook	1904	5.06-5.45
Miller	1910	5.08-5.29
Grout	1937	4.95-5.49
Taber & Owens	1970	4.99-5.45
Dadant	1946	4.99-5.45
Dadant	1975	5.2
Messange & Goncalves	1985	5.07-5.11

The cell size of "natural" worker comb, as measured among the various races of bees, is reported to be variable, ranging from 700 to 950 cells per square decimeter. However, there seems to be a consensus suggesting that, for most races of honey bees, natural worker comb cell size is 857 cells per dm2 (5.1 mm per cell) (5) and ranges from about 830 to 920 cells per dm2 (= 5.0 to 5.3 mm per cell). (Note also that 920 cells per dm^2 was the size which Baudoux argued *against* – See "**Cell Tell**" above.)

How Big: Measurements of from foundation and mills produced from the late 1800s to 1989		
Type	Source	Avg Cell Size in mm
Wax	Africa	4.76
Wax	Africa	4.89
Wax	A.I. Root (circa 1929)	5.05
Aluminum	Unknown	5.10
Mill	A.I Root(circa 1929)	5.12
Wax	Miller	5.18
Wax	Mexico	5.18
Wax 5-3/8"	A.I. Root (1989)	5.18
Wax	Glorybee, OR	5.19
Wax	Tom Industries, CA	5.19
Wax	Honey Acres, WI	5.19
Mill	A.I. Root (1989)	5.20
Wax 8-3/8"	A.I. Root (1989)	5.20
Plastic	Arnaba, HI	5.21
Plastic	Unknown	5.23
Plastic	Unknown	5.28
Wax	W.T. Kelley, KY	5.28
Mill	A.I. Root (circa 1910)	5.29
Wax	Brushy Mountain	5.30
Plastic	Unknown	5.35
Wax 8-5/16"	Dadant (med brood)	5.39
Wax	Honey Acres, WI	5.39
Wax 5-1/2"	Dadant (med brood)	5.39
Duragilt	Dadant	5.40
Wax	Bolivia	5.44
7/11	W.T. Kelley, KY	5.53
Perma-dent	Draper's, NE	5.56
PermaComb	PermaComb	5.05 (allowing for cell walls)

How Big

In the United States, from the late 1800's to the early 1900's, the "standard" cell size for manufactured foundation was 857 cells per dm2. However, in the early 1900's there began a subtle transition to larger cell size by some but apparently not all manufacturers of comb foundation. By 1913 at least 2500 foundation presses with 736 cells/dm^2 (=5.6 mm per cell) were sold in Europe by the Rietsche Co. in West Germany. As a result, the current world industry "standard" for worker cell size is between 725 cells per dm^2 (5.6 mm per cell) and 800 cells per dm^2 (5.4 mm per cell). Most foundation currently manufactured in the United States ranges from 700 to 857 cells per dm^2 (=5.2 – 5.7 mm per cell).

We have examined twenty-five samples of foundation from a number of foundation manufacturers in the United States and around the world. We have also examined three mills, which we were able to obtain for comparison. The cell size of each is summarized elsewhere and is based on 10 measurements each of 10 linear cell impressions (see "**How Big**" above).

Foundation with 700 cells per dm2 has cells 10.7% wider than natural comb cell size. Colonies utilizing the smaller natural cell size (857 cells per dm^2) could produce 22.4% more brood per given area of comb than colonies on 700 cells per dm2. Similarly, such colonies could rear 7.1% more brood than colonies on 800 cells per dm2. Utilization of 857 comb would, almost certainly, require less metabolic energy expended per bee to maintain optimal temperature and humidity for brood rearing. It is possible that developmental time might also be shortened. Both factors would translate into more rapid spring buildup and recovery from bee losses due to parasites, disease or pesticides.

The question that must now be raised is why has the beekeeping industry, in the United States and elsewhere, accepted foundation with 700-800 cells per dm^2 (= 5.7 – 5.4 mm per cell) as a size standard. We may never know, but it seems likely that it has its roots in the mistaken Lamarkian theories which guided the early studies of Baudoux. These studies were followed by those of Gontarski who found that the greatest percentage of bee size change occurs using a cell size of 700 cells per dm2 (5.7 mm per cell) (5). Our investigation suggests that many of the rollers used in mills manufacturing foundation in the United States are made in Europe and the producers of these rollers follow the precepts of Baudoux and Gontarski. At least one of these companies currently making rollers (Rietsche in West Germany) was making flat molds for foundation in 1899. Another explanation might lie in Baudoux's contention that combs with small cells contribute to swarming (1). However, Baudoux also advanced the opposing view that larger bees would produce more body heat leading to increased brood production. Certainly, larger bees resulting from selection and breeding require larger cells for development. There has also been concern that the buildup of larval debris and cocoons in cells reduces cell size. Thus, there is perceived benefit to be gained from starting with a larger cell.

Clearly, reported differences in cell size and in bee size between domestic (European) bees reared in large cells and Africanized honey bees reared in naturally built comb have often been misinterpreted. It is not so much that AHB cells are somehow smaller, but rather the cells built by bees from domestic strains are abnormally large. It is worth noting that the cell size range reported as natural for feral bees has varied little from the 1600's to the present time (see "Cell Tell"). Also noteworthy is the fact that the size range currently cited by various authors as indicative of Africanization (e.g., reported averages = 4.9 – 5.1 mm; range 4.5 – 5.4 mm) significantly overlaps that of natural cells built by European bees (e.g., reported averages = 5.1 – 5.2 mm; range = 4.7 – 5.5 mm) by a wide margin.

The authors wish to thank H. Don who measured all foundation and mill specimens and C. Shipman who helped us assure the accuracy of our mathematics. We also thank all those people who kindly provided us with foundation and mills for examination.

References

1. Baudoux, U. 1933. The influence of cell size. Bee World, Vol. XIV, No. 4, pp. 37-41.

2. Betts, A.D. 1932. The influence of cell size. Bee World, Jan. 1934, pgs. 2-5.

3. Camazine, Scott 1988. Factors affecting the severity of Varroa jacobsoni infestations on European and Africanized honey bees. In Africanized Honey Bees and Bee Mites, Chapter 59, pp. 444-451.

4. Grout, Roy A. 1931. A biometrical study of the influence of size of brood cell upon the size and variability of the honey bee (Apis mellifera L). M.S. Thesis, Iowa State College.

5. Root, A.I. 1978. The ABC and XYZ of bee culture. A.I. Root Company (publs.), Medina, Ohio.

6. Spivak, M., T. Ranker, 0. Taylor, Jr., W. Taylor and L. Davis. 1988. Discrimination of Africanized honey bees using behavior, cell size, morphometrics, and a newly discovered isozyme polymorphism. In: Africanized Honey Bees and Bee Mites. Needham, Glen R. et al. (eds.). Halstead Press, New York, NY.

1 Mention of a trademark, proprietary product, or vendor does not constitute a guarantee or warranty by the USDA and does not imply its approval to the exclusion of other products or vendors that may also be suitable.

U.S. Department of Agriculture, Agricultural Research Service, Carl Hayden Bee Research Center, 2000 E. Allen Road, Tucson, AZ 85719.

Conversions

by Dr. Shipman, USDA

Relative values for cell size using various popular units of measure.

# cells/dm^2 (2 sides)	# cells/inch	# cells/in^2 (2 sides)	cell width in mm^2
650	4.3	41.9	6.0
700	4.4	45.2	5.7
725	4.5	46.8	5.6
750	4.6	48.4	5.5
800	4.7	51.6	5.4
830	4.8	53.5	5.3
850	4.9	54.8	5.2
920	5.1	59.4	5.0
950	5.2	61.3	4.9
1050	5.4	67.7	4.7

Cells per square decimeter (dm^2) is a useful unit of measure for understanding colony population dynamics and for developing management strategies, but, for practical field diagnostics it is easier to measure the width of a row of 10 cells measured side to side. A dm^2 is an area 10 cm by 10 cm; cells on both sides of the comb area are counted. To convert cells per dm^2 to cells per linear dm and then to mm per cell use equation 1:

Equation 1:

cells/dm2 = 2.31XN2 (where N is the number of cells per linear dm)

Example: 850 = 2.31 x (19.18)2

Divide 100 by cells per linear dm to obtain mm per cell

Example: 100 divided 19.2 = 5.2

To convert cells per in^2 to inches per cell:

Equation 2:

cells/in2 = 2.31XN2 (where N = number of cells per inch)

Example: 54.8 = 2.31 x (4.87)2

Equation 3:

cells/dm2 = 15.5 X cells/in^2

Example: 850 = 15.5 x 54.8

The values in this column represent the width of a single cell as measured between centers of opposing cell walls. The actual cell interior width is one cell wall thickness less than this value.

March, 1990 – Bee Culture, SPECULATIONS ON FOUNDATION AS A COLONY MANAGEMENT TOOL1, E. H. EricksonÝ, D. A. Lusby+, G. D. HoffmanÝ and E. W. Lusby+

PART II - *Optimal cell size, which many take for granted, is not as simple a thing as one might expect. The belief that cell size alone is responsible for bee size is not quite accurate. Rather, both cell size and inheritance determine ultimate bee size.*

Clearly, reported differences in cell size and in bee size between domestic (European) bees reared in large cells and Africanized honey bees reared in naturally built comb have often been misinterpreted. AHB cells are not Smaller, *but domestic strains build cells that are* Larger.

How Big

While AHB may produce slightly smaller cells (6), cell size is a poor diagnostic character due to the overlapping size ranges of AHB and natural comb or comb drawn from small commercial foundations (see "**How Big**" chart in "On The Size of Cells, Part 1").

Could it be that the reproductive advantage reported for AHB over domestic bees occurs, at least in part, because AHB have not been subjected to artificial selection pressure for large body size and, as a result, build comb of a smaller size? This is a logical hypothesis if we assume that smaller bees and reduced cell size combine to increase the number of individual bees reared per unit area of comb and shorten the developmental time of the larval and papal stages. Moreover, accelerated spring buildup in smaller cells could lead to early drone production and, hence, a mating advantage of AHB. The logical extension of these hypotheses would suggest that domestic bees would be more competitive with Africanized bees if they were reared in hives with comb of natural cell size and had comparable developmental periods.

Realization of the importance of cell size might also provide a management tool against the Varroa mite. Recently, Message and Goncalves reported that, in Brazil, cell sizes for Africanized and domestic (European) honey bees averaged 4.5 to 4.8 and 5.0 to 51 mm per cell, respectively. They further reported that Varroa infestation rates were 4.8 and 11.5 percent, respectively. Camazine, (3) calculated female Varroa replacement rates for Africanized and domestic honey bees at 1.2 and 1.8 with drones present and 0.8 and 1.5 without drones, respectively. (A female Varroa replacement rate of less than 1.0 indicates that the mite population is declining while a 1.0 rate is indicative of zero population growth). Thus, we think that it may be possible to suppress Varroa populations in domestic colonies by using small strains of bees with shorter development times reared in smaller cells.

Cell size may impact a range of issues from the efficacy of queen excluders, given the variable size of bees that may be produced, to the susceptibility of colonies to disease, parasites and pesticides. Whether or not we can manage Africanized bees and Varroa mites using a combination of smaller brood cells and smaller bees remains to be seen. But, as Cheshire first suggested many years ago, it is possible that departures from the normal size of honey bees may cause or contribute to the severity of problems facing the beekeeping industry. We speculate that this line of reasoning might apply to problems such as tracheal mites (e.g. smaller tracheal openings in smaller bees might confer resistance to tracheal mites), winter mortality and other stress related losses (5). These issues deserve further study.

Some research is now in progress but much remains to be done. For those beekeepers interested in experimenting with cell size on their own, we have, in our early studies, found that packaged bees obtained from a single producer will readily draw out foundation of either 857 or 800 cells per dm2 (see Cell Size below). We simply placed one fully drawn comb in each hive with nine frames of foundation with the same cell size. We assumed that, during the period when the packaged bees were disassociated from comb, the workers would lose their memory of previous cell size and thus ensure that the bees would quickly adapt to the new size.

Cell sizes produced by packaged bees on two sizes of foundation.

Foundation Size	Cell Size
5.12	5.14 ± .09
5.36	5.36 ± .06

Based on 40 colonies per treatment, and 5 measurements per colony.

These data clearly demonstrate the ease with which a beekeeper can effectively reduce comb cell width in colonies. A corresponding reduction in bee size should follow without selection and breeding. However, simple methods are available for further reducing bee size through selective breeding (see our paper entitled "Managing Colony Genetics by Grafting and Selecting for Queens With Shorter Development Times" in the *The American Bee Journal*, November, 1989, pg. 717). Further reduction of cell width appears to follow in such a selection program. We now have samples of comb resulting from the annual insertion of new foundation (5.18 mm per cell) into colonies undergoing continued selection pressure for smaller bee size over a three-year period. Cell width in the most recently drawn comb measures 5.08 mm per cell. However, be cautioned that the extent to which such a program can or should be carried out is yet unclear.

The authors wish to thank H. Don who measured all foundation and mill specimens and C. Shipman who helped us assure the accuracy of our mathematics. We also thank all those people who kindly provided us with foundation and mills for examination.

References

1. Baudoux, U. 1933. The influence of cell size. Bee World, Vol. XIV, No. 4, pp. 37-41.

2. Betts, A.D. 1932. The influence of cell size. Bee World, Jan. 1934, pgs. 2-5.

3. Camazine, Scott 1988. Factors affecting the severity of Varroa jacobsoni infestations on European and Africanized honey bees. In Africanized Honey Bees and Bee Mites, Chapter 59, pp. 444-451.

4. Grout, Roy A. 1931. A biometrical study of the influence of size of brood cell upon the size and variability of the honey bee (Apis mellifera L). M.S. Thesis, Iowa State College.

5. Root, A.I. 1978. The ABC and XYZ of bee culture. A.I. Root Company (publs.), Medina, Ohio.

6. Spivak, M., T. Ranker, 0. Taylor, Jr., W. Taylor and L. Davis. 1988. Discrimination of Africanized honey bees using behavior, cell size, morphometrics, and a newly discovered isozyme polymorphism. In: Africanized Honey Bees and Bee Mites. Needham, Glen R. et al. (eds.). Halstead Press, New York, NY.

For a complete list of references used (33) contact the authors.

1 Mention of a trademark, proprietary product, or vendor does not constitute a guarantee or warranty by the USDA and does not imply its approval to the exclusion of other products or vendors that may also be suitable.

U.S. Department of Agriculture, Agricultural Research Service, Carl Hayden Bee Research Center, 2000 E. Allen Road, Tucson, AZ 85719.

Thelytoky in a Strain of U.S. Honey Bees (Apis Mellifera L.) - Bee Science 1991

May, 1991 – Bee Science, G. DeGrandi-Hoffman, E. H. Erickson Jr., D. Lusby, and E. Lusby, Carl Hayden Bee Research and Biological Control Center – Tucson – Arizona – USA

Introduction

Thelytoky is a type of parthenogenetic reproduction where unfertilized eggs develop into females (Suomalainen 1950). Thelytoky is common in the Cape honey bee (*Apis mellifera capensis* Escholtz), but it occurs with considerably lower frequency in European honey bees (*Apis mellifera* L.) (Onions 1912; Jack 1917; Anderson 1963; Ruttner 1976). In colonies with queens most worker ovaries are suppressed by the pheromone 9-oxo-decenoic acid and other substances produced by the queen (Butler and Fairey 1963), or by the presence of unsealed brood (Kropacova and Haslbachova 1971). However, ovaries can develop and workers can lay eggs after the queen and brood are gone (Perepelova 1929; DeGroot and Voogd 1954; Butler 1957; Butler and Fairey 1963; Jay 1970; Kropacova and Hasibachova 1970, 1971). European workers generally lay unfertilized haploid eggs that develop into males (drones). In rare instances, virgin queens and laying workers produce diploid eggs that develop into females (Mackensen 1943).

Given the high frequency of thelytoky in Cape bees, the relatively rare occurrence in domestic stocks of European bees is unexpected, since populations capable of thelytoky have an advantage over those in which laying worker eggs develop exclusively into drones (Ruttner 1977). Without thelytoky, the survival of a colony rests completely on the successful mating of a single queen which must leave the hive to mate. If this queen does not encounter drones or does not return to the hive, a replacement cannot be produced because female larvae of a suitable age for queen rearing no longer exist, and because the first queen to emerge usually destroys the other queen cells in the colony. However, if brood from laying workers could be raised into queens the colony would have a facultative survival mechanism in case the virgin queen is lost. Thelytoky should occur with greater frequency in populations exposed to conditions that reduce the chances of a queen either taking or returning from a mating flight (Moritz 1984).

A strain of honey bees (hereafter referred to as LUS) has been established from a breeding program in which virgin queens were introduced into broodless colonies (i.e., eggs and larvae did not exist in the colony) from November to March in southern Arizona. The purpose of the breeding program was to select for bees that would rear queens and drones at that time of year. Inclement weather and limited numbers of drones can occur during Arizona winters and prevent queens from successfully mating. Thus, introducing virgin queens at this time of year exerts pressure that could cause the frequency of thelytoky in the population to increase. The purpose of this study was to test for the existence of thelytoky in LUS and determine the frequency of this trait. In addition, observations of worker bees in queenless LUS colonies were made to compare their behavior with that reported to occur in Cape bees.

Methods and Materials

Eighteen queenless four or five frame nucleus colonies of LUS were established using two frames of brood (ranging in age from eggs to pupae) from queenright LUS colonies and two to three frames of honey and pollen. The adult bees covering these frames were included. Different LUS colonies were used to establish each nucleus colony. As controls, three queenless nucleus colonies of a panmictic array of commercial bee lines maintained as a closed population (CP) (Page and Laidlaw 1982; Severson et al. 1986) and six colonies of honey bees carrying the Cordovan (cd) mutant color marker (Laidlaw and Page 1984) were established using the procedure described above. Entrances of the queenless colonies were covered with screen mesh for 24-48 hours after being established to prevent bees from drifting back to their parent colonies. All queenless colonies were examined three to four times weekly while brood from the previous queen was present so that queen cells from the brood could be destroyed. After all the previous queens' brood had emerged, the colonies were examined twice weekly to determine when workers began laying eggs. When eggs from laying workers first appeared in the colonies (i.e., when one or more eggs were seen), 10-20 workers were sampled and dissected to determine the percentage with developed ovaries. The first appearance of eggs was chosen as a means to standardize the time when workers would be sampled, since the percentage of workers with developed ovaries can change over time in queenless colonies (Anderson 1963). Ovaries were considered to be developed if developing eggs were visible in the ovarioles. The brood from laying workers present in either worker or drone cells was sexed while in the pupal stage by removing the cell's cap, and determining gender by the morphology of the head capsule. The presence of queen cells with larvae being actively tended by workers was noted along with whether the cells had a queen emerge or were destroyed by the workers.

Observations of bees on the frames were made during colony inspections. We avoided the use of smoke during these inspections whenever possible to minimize disruption to the workers on the frames. Sometimes during an inspection bees were seen biting each other, or with their abdomens in the cell assuming an egg laying position. We sampled LUS bees being bitten and dissected them to determine if they had ovary development. Whether workers assuming the egg laying position always deposited an egg in the cell also was determined. To conduct more detailed observations of queenless LUS colonies, two frame observation hives were established using one frame of brood and another of pollen and honey along with the adult bees on the frames. The activity of bees on the frames was observed twice daily once in the morning and afternoon, for 30-60 mm. intervals. Observations were begun when all the brood from the previous queen had emerged. The observation hives were not included among the colonies used to test for thelytoky.

Results

Once all the brood emerged in queenless LUS, CP, or cd colonies, worker bees were scattered over the frames giving the colony the distinctive appearance associated with the queenless state. Upon closer examination of bees from the 4-5 frame nucleus colonies and in the observation hives sometimes workers were seen grasping each other with their mandibles. In a LUS observation colony, workers were seen pulling nestmates out of the cells in which they had inserted their abdomens. On other occasions, in the observation hives we saw eggs being eaten by nestmates immediately after the laying worker removed her abdomen from the cell. In the observation hives and the nucleus colonies some bees assumed an egg laying position in a cell, but did not lay an egg. In nucleus and observation colonies we observed bees remaining stationary with their wings spread while nestmates bit them on the dorsal surface of the abdomen and the thoracic area (particularly at the points where the wings articulate). This behavior occurred in LUS, CP, and cd colonies and has been previously described in queenless colonies by Velthuis (1970). LUS from nucleus colonies that were being bitten by other workers were examined for ovary development; 26.7% of these bees had developed ovaries (colonies sampled = 5, total bees examined = 15, SD= 11.4%). We attempted to sample bees being bitten in CP and cd colonies and examine them for ovary development, but sample sizes were too small to obtain meaningful results. Dead bees on the bottom boards of seven LUS colonies were examined and an average of 1.5% of the dead bees per colony had developed ovaries (bees examined = 65, SD = 1.5%). Examination of workers selected at random from the queenless test colonies indicated that an average of 27.1% of the LUS workers had developed ovaries when eggs first appeared in the colony (Table 1). This was a significantly lower percentage than either CP or cd (60.0% and 44.0% respectively).

Table 1: Types of progeny reared from the eggs of laying workers in queenless colonies of U.S. honey bees, Tucson, AZ

Colony type	No. of colonies observed	% workers with developed ovaries ± sd	%colonies rearing			No. of queen emerged
			Drones	Workers	queens	
CP	3	60.0±24.5a	100.0	00.0	00.0	0
Cd	6	44.0±3.3a	100.0	00.0	20.0	0
LUS	18	27.1±15.0b	100.0	55.6	50.0	0

Means followed by the same latter are not significantly different at the 0.05 level as determined by Duncan's [1951] multiple range test.

Of the 18 colonies of LUS tested for thelytoky, 55.6% reared worker brood from the eggs of laying workers, and 50% reared queens. Queens from the brood of laying workers emerged only in the 4-5 frame nucleus colonies, and never in the observation hives. In the nucleus colonies, sometimes a patch of worker brood was produced and the queen cell was constructed within that patch (Fig 1.). A queen cell positioned among worker brood is commonly seen in a colony that is requeening itself in the conventional manner using brood from the previous queen. However, some queen cells from thelytokous LUS colonies were located at the very top of the frame. Neither CP or cd constructed queen cups in this region. A queen produced from laying worker eggs successfully mated and produced worker and drone brood. However, eight of the nine queens produced from workers' brood either did not return to the hive after a mating flight, or were critically injured during artificial insemination.

Queenless CP colonies reared only drones, although queen cells were constructed and eggs from laying workers were placed inside them. The eggs did not hatch, and often were gone the next day. Similarly, cd colonies produced only drones from laying worker eggs, although some colonies reared larvae in queen cells. These queen cells were larger and longer than those produced by LUS or commonly seen in colonies rearing queens from a mated queen's brood. During colony inspections the cd workers were observed crawling over the capped queen cells just as the LUS bees did in their colonies. However, within 3-5 days in the cd colonies the queen cells were torn down by the workers.

Discussion

LUS were selected from commercial European honey bee stock, indicating that thelytoky may exist as part of the overall *Apis mellifera* gene pool. However, reports indicate that in managed colonies thelytoky is expressed at a very low frequency (Mackensen 1943). This may be because beekeeping practices inadvertently select against thelytoky. For example, swarming and supersedure can be minimized through various management techniques, and thus the possibility of a colony becoming queenless due to the loss of a virgin queen can be reduced. If colonies lose their queens and do not have brood to produce replacements, the queens often are replaced with new ones by beekeepers. Hence, there is no selective pressure for thelytoky in colonies managed in this manner. Conversely, the conditions under which the LUS strain was derived may have inadvertently selected for thelytoky. Virgin queens introduced into broodless colonies during the winter may not have been accepted by the workers in some cases, while in others the queens may not have mated or were lost on mating flights. Some of the colonies that survived may have done so because they requeened themselves with brood from laying workers. The winter requeening procedure was repeated annually using queens produced from brood of colonies that survived the previous year's winter requeening. If thelytoky was at a low frequency in the LUS strain at the beginning of the breeding program, the frequency possibly was increased because of continued selection followed by the production of new queens from brood of the survivors.

Unfortunately, all but one of the queens produced from laying worker brood were lost before they could begin egg laying. Still, queens reared from the brood of LUS laying workers apparently have the potential to mate and produce worker and drone brood. We stopped finding eggs in the colonies once the queen emerged and was present in the hive. The colony whose queen successfully mated, behaved like any other colony with a new queen. After mating the queen began laying worker brood which was cared for by the adult workers in the colony. The colonies that reared queens but lost them did not rear others. The colonies subsequently dwindled and died or were robbed by workers from other colonies thus causing the LUS workers to abandon the hive. Colonies composed of 4-5 frames of workers and brood apparently have only one chance at rearing a queen from laying worker brood. If the queen is lost, the workers will not produce another perhaps because the workers are too old, the colony is too weak, or some combination of both. Whether a colony that had a larger population at the time of queen removal would have enough bees of the appropriate age to rear another queen from laying worker brood if they lost the first one needs to be tested.

When queenless nucleus colonies were inspected, the use of smoke was minimized to limit the disruption of the bees. Still, opening a colony is disruptive because it changes colony temperature and perhaps odor and pheromone levels within the hive's environment. We cannot be sure of the repercussions of opening colonies on the workers' behaviors we observed on the frames. Observation hives enabled us to make more detailed behavioral observations of LUS workers in queenless colonies without having to open the colony. However, how well the results from the observation hives mirror the behaviors of bees in the nucleus colonies is not known. Workers in observation hives reared fewer larvae into adults compared to the nucleus colonies, and never reared queens. Perhaps the populations were too small or temperatures could not be properly maintained in the observation hives for brood rearing to approach the levels seen in the nucleus colonies.

There are both similarities and differences between laying workers of Cape bees and LUS. Cape bees can have workers with developed ovaries while brood is present (Anderson 1963). We have not found this to occur in LUS (DeGrandi-Hoffman unpubl. data). Internal fighting among nestmates following the removal of a queen and a subsequent increase in the number of dead bees on the bottom board occurs in Cape, LUS, CP, and cd bees. As in Cape bees, most of the dead LUS bees did not have ovary development. In Cape bees an average of 28% of the workers have developed ovaries 13 days after queen removal, and in LUS the average is 27% when eggs from laying workers are first seen (Anderson 1963). Significantly fewer workers in queenless LUS colonies have developed ovaries compared to CP or cd, suggesting that worker ovaries might be more effectively suppressed by the presence of laying workers in LUS (Velthuis 1970). Cape bee workers lay unfertilized diploid eggs because during ana-phase II the egg pronucleus and the central descendent of the first polar body fuse to form a diploid zygote nucleus (Verma and Ruttner 1983). Whether a similar cytological mechanism exists in LUS is yet to be determined.

A honey bee colony's ability to requeen itself with the eggs of laying workers requires not only that some workers can lay diploid eggs, but that the workers can foster the cooperation from nestmates needed to construct a queen cell and rear the egg into a queen. When laying workers developed in CP or cd colonies, often queen cells were constructed and sometimes eggs were deposited inside them. However, the eggs were either cannibalized by other workers or left unattended and did not hatch. Other than in LUS, the greatest cooperation among individuals to rear a queen from laying worker eggs was in cd bees where workers actively cared for the larvae in the cells. Queen cells were capped in some instances, but were destroyed soon afterwards. Our study indicates that attempts at requeening occur in non-thelytokous lines of honey bees, but apparently these bees lack some of the physiological and behavioral attributes needed to rear a viable queen.

Acknowledgments

The authors would like to thank H. H. Laidlaw, R. E. Page, and A. Cohen for reviewing earlier versions of this manuscript.

Literature Cited

ANDERSON, R. H. 1963. The laying worker in the Cape bee, *Apis mellifera capensis. J. Apic. Res.* 2: 85-92

BUTLER, C. G. 1957. The control of ovary development in worker honey bees *(Apis mellifera). Experientia* 13: 256-257

BUTLER, C. G. and E. M. FAIREY 1963. The role of the queen in preventing oogenesis in worker honey bees. *J. Apic. Res.* 2:14-18

DEGROOT, A. P. AND S. VOOGD 1954. On the ovary development in queenless worker bees (*Apis mellifera* L.). *Experientia* 10: 384-385

JACK, R. W. 1917. Parthenogenesis amongst the workers of the Cape bee. Mr. G. W. Onions experiments. *Trans. Entomol. Soc. London* 64: 396-403

JAY, S. C. 1970. The effects of various combinations of immature queen and worker bees on the ovary development of worker honey bees in colonies. *Can. J. Zool.* 48:169-173

KROPACOVA, S. AND H. HASLBACHAVA 1970. The development of ovaries in worker honey bees in queenright colonies before and after swarming. *J. Apic. Res.* 9: 65-70

KROPACOVA, S. AND H. HASLBACHAVA 1971. The influence of queenlessness and unsealed brood on the development of ovaries in worker honey bees. *J. Apic. Res.* 10: 57-61

LAIDLAW, H. H. AND R. E. PAGE JR 1986. Mating designs. In *Bee Genetics and Breeding*, T.E. Rinderer Editor. Academic Press Inc. Orlando FL. pgs. 323-344

MACKENSEN, O. 1943. The occurrence of parthenogenetic females in some strains of honey bees. *J. Econ. Entomol.* 36: 465-467.

MORITZ, R. F. A. 1984. Equilibrium of thelytokous and arrhenotokous parthenogenesis in populations of the honey bee (*Apis mellifera*). In *Advances in Invertebrate*

Reproduction, W. Engels, Editor. Elsevier, Amersterdam, New York, Oxford. pg. 615.
ONIONS, G. W. 1912. South African fertile worker bees. *Agric. J. Union of South Africa 7*: 4446.
PAGE, R. E. JR., AND H. H. LAIDLAW 1982. Closed population honey bee breeding. 1. Population genetics of sex determination. *J. Apic. Res.* 21: 30-37
PEREPELOVA, L. 1929. Laying workers, the ovipositing of the queens, and swarming. *Bee World* 10: 69-71
RUTTNER, F. 1976. The Cape bee - A biological curiosity? In *African Bees: Taxonomy, Biology, and Economic Use*. D.J.C. Fletcher, Editor. Proceedings of the Apimondia Interntl Sym. Pretoria, S. Africa.
________.1977. The problem of the Cape bee (*Apis mellifera capensis* Escholtz): Parthenogenesis - size of population - evolution. *Apidologie* 8: 281-294.
SEVERSON, D. W., R. E. PAGE JR., and E. H. ERICKSON JR. 1986. Closed population breeding in honey bees: A report on its practical application. *Amer. Bee J.* 126: 93-94.
SUOMALAINEN, E. 1950. Parthenogenesis in animals. *Advances in Genetics* 3:193-253.
VELTHUIS, H. H. W. 1970. Ovarian development in *Apis mellifera* worker bees. *Entomol. Exp. and Appl.* 13: 377-394.
VERMA, S. and F. RUTTNER 1983. Cytological analysis of the thelytokous parthenogenesis in the Cape honey bee (*Apis mellifera capensis* Escholtz). *Apidologie* 14: 41.57

Abstract

A strain of U.S. domestic honey bees (*Apis mellifera* L.) with the ability to rear workers and queens using the eggs of laying workers has been isolated. Previously, thelytoky was assumed to occur rarely in honey bees with the exception of the South African Cape bee (*A. mellifera capensis*). Our thelytokous line, hereafter referred to as LUS, was developed from commercial stocks of European honey bees. Comparisons of worker behavior and ovarian development were made among queenless colonies of LUS and two arrhenotokous lines hereafter referred to as CP and cd. LUS had a significantly lower percentage of from laying workers first appeared in cells than either CP or cd. All three lines constructed queen cells and deposited laying worker eggs in them, but viable queens emerged only from LUS. The CP line did not rear larvae in the queen cells but in some instances the cd line did. However, the cd bees destroyed the queen cells either prior to or soon after capping them. Comparisons between behaviors of queenless LUS colonies and those reported to occur in queenless Cape bee colonies also are discussed.

Contact Address: Drs. DeGrandi-Hoffman and Erickson: Carl Hayden Bee Research and Biological Control Center, U.S.D.A. – A.R.S., 2000 East Allen Road, Tucson, AZ 85719; Mr. and Mrs. Lusby: Rangeland Honey, Tucson, AZ

Dr. Gloria DeGrandi-Hoffman is a Research Entomologist at the U.S.D.A. – A.R.S. Carl Hayden Bee Research Center in Tucson, AZ. Her research efforts have been devoted to the construction and validation of computer simulation models of biological systems. Dr. Eric Erickson Jr. is the Research Leader and Center Director of the Carl Hayden Bee Research Center. Dr. Erickson's research has focused on crop pollination and honey bee behavior and morphology. Delores and Edward Lusby are commercial beekeepers in Tucson, AZ.

Accepted for publication on 29 January 1991

Key Words: parthenogenesis, Cape honey bees, laying workers

The Chemical Treadmill

It has been well known since Rachel Carson first wrote "Silent Spring" in the early 1960s that agriculture has been getting deeper into trouble with continued pesticide use.

Today within our beekeeping industry, this has never been more apparent. Yet, strangely, many beekeepers are uneducated in the field about what pesticide resistance is and how it escalates into a pesticide highway to hell and eventual colony destruction. In the past in the 1960s and 1970s forward, it was often described as a pesticide treadmill, because once you are on it with your agricultural field management, it is virtually impossible to wean your self off. Herein therefore is the danger, namely a growing resistance, creating a growing chemical dependency, requiring stronger and stronger treatments of various dopes.

Someone sent me a perfect example, when they wrote, "I have the gut feeling this is how it is as when you have treated with acid, (referring to chemical resistance we had corresponded about with honeybees and parasitic mites), you get a feeling the colony is like a magnet on mites or starts producing more. I have an example from a beekeeper, he treated and treated different kinds, different acids, culling dronecomb, Apistan, everything, and saved all mites and counted them. Though he treated like a maniac, he managed to produce 10,000 mites from that colony in a year. And it was in a miserable state the next spring, but still alive, which surprised me. After all these chemicals. Amazing."

Well, this could be described as a perfect example of a chemical treadmill in action going full steam with a high degree of resistance to chemicals by the parasitic mites, now totally – out of control!

For understanding, Pesticide resistance or "pest" resistance should be simply stated, – that some insects within each given species are naturally more resistant to certain chemicals. You never can kill 100% of the little trouble makers with any given dope treatment. There is always an exception. Has to be or evolution would stop!

Within our beekeeping community as those treating with various dopes try to kill parasitic mites infesting their colonies, most are not aware that as the dopes do their job and make the susceptible parasitic mites die, the survivors multiply, passing their resistance onto the next generation for that particular level of doping (strength of chemical used).

When the pesticides are perceived to "no longer control" (beekeeper sees a growing body of prolific mites) at normally recommended rates (strength of chemical used), a pest resurgence occurs, when the parasitic mites killed by the dopes return in larger numbers.

Basically, what this is, is the reproduction of those little trouble makers, the exceptions, the dopes could not kill in the first place, now reproducing without chemical effect, their next generations. We say the pesticides no longer control at normally recommended rates, instead of saying we need more killing power to now go back and refight the exceptions that were stronger in the beginning and really needed more dope to finish the job, but now the insect, somehow knows what to expect; and it also requires a higher dosage. Our problem is figuring out how much and of course, here we go again trying to get another 100% kill, which we know is technically impossible. Consequently a circle of treatment or a pesticide treadmill is created.

Now during this treatment, some of the parasitic mites will develop what is called cross-resistance. This is basically where resistance to one chemical means resistance to a second chemical with a similar mode of action (method of killing) as the first. Multiple resistance also is known to occur, where there is resistance to several classes (different chemical groups). This is now currently accomplished, by beekeepers all using different types of treatments within a given area and then as the honeybees co-mingle in the field or drift from colony to colony, the mites transfer rides on the backs of the bees themselves and cross mate, passing each others resistance on to the others linage.

Now to add insult, beekeepers need to understand that depending upon the type of resistance (type of dope used) and the species of parasitic mite (tracheal or Varroa or ?), resistance tends to last in the absence of the dopes when control measures are stopped. What this means, is that the breeding accomplished through survival (surviving the various dopes thrown upon their bodies) is now considered inherent within the linage of the parasitic mites.

Now, as beekeepers use stronger and stronger dopes other problems begin to set in to complicate the already growing bad situation created. Many times other problems created involve not only pest (mite) resurgence, but also the creation of secondary pests/insects now becoming serious primary pests not known to effect colonies i.e.

beetles, ants, earwigs, moths, and also secondary diseases.

This happens because when we attempt to kill what we consider our primary pests, namely parasitic mites, we also inadvertently also kill their natural enemies that would help to keep them in balance around our hives. Two things normally happen here. Either the natural enemies of our pests are killed or they leave the area since their food source is no longer available. This leaves opportunity for the treated surviving parasitic mites to reproduce before their natural enemies return (other insects, or birds, etc.).

Secondary pests become serious primary pests when their natural predators are killed. The whole class of mites is very widely studied for this vary reason for the havoc caused when dopes kill their natural predators. This adds to the pesticide treadmill. Then the corresponding treatment of the secondary pests i.e. beetles, moths, ants, etc., add to the problem of our mites and our ability to contain their damage. Look at beekeepers here in the USA currently treating for parasitic mites, now also being forced to treat for beetles with chemicals, now used not only within the colonies, but now all around the colonies on the ground. This is nothing more than a serious speeding up of the treadmill to disaster because you have increased the scope and width of the playing field for chemical management.

Add to this now, besides external dopes, internal dopes like oxytetracycline that blow out the bees internal gut for beneficial digestive bacteria and you now create not only eating disorders, but internal susceptibility to various secondary diseases and shorten the life of our poor honeybees at the same time.

Yet to stop the dopes both external and internal is pure hell. For now there is no natural backup help to be found, the bees are in a weakened state not able to digest natural food, they, for the most part can no longer defend themselves, and then we wonder why colonies collapse.

What are we left with? I would say a lot of empty equipment growing in quantity each and every year as the treadmill, now worldwide accelerates, persistent residues in soil and living tissue for those stupid enough to eat this wholesome food we still call honey, and more and more decreased pollination service sure to effect our global food supply in the future. Yet, if I may be so bold to state - very few are willing to do what it takes to go back to full biological to correct the problem. But in the end I really feel in my heart they may be forced to, and what a price we are all going to pay as a worldwide global community.

- Dee Lusby

More on Small Cell Foundation for Mite Control

Lusby, ABJ – June 1997

Since you printed my letter to the editor in your November 1996 issue of ABJ, my husband and I have again been swamped with letters and phone calls seeking more information. Consequently, I feel I should write more as other readers may also be interested in what the information desired has centered on. Namely:

Telling beekeepers more on how to measure combs with diagrams.

Explaining how our honey bees chew out Varroa and what beekeepers should look for.

Converting to naturally smaller brood combs for one's own area.

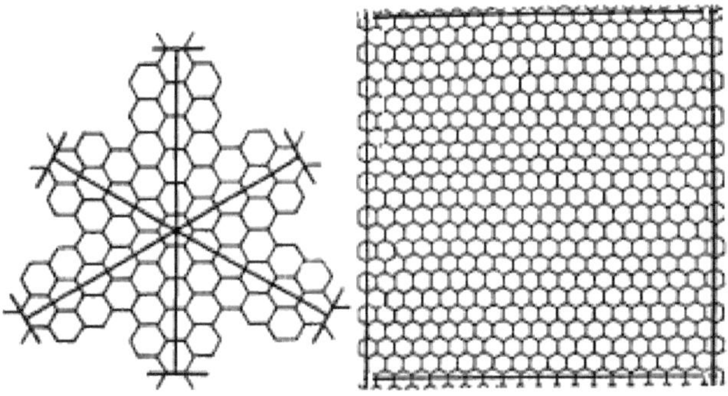

Dia. 1- (Spivak, et al., 1988 #39, Africanized Honey Bees and Bee Mites. Dia. 2 - Counting a square

decimeter for number of cells using square measurement.

Concerning honeycombs - Honey bee comb cells are measured parallel wall to parallel wall in three directions. They are not measured point to point, nor measured with a mixture of both, one way in each direction. They can, however, be counted either parallel wall to parallel wall or with a mixture of both, one way in each direction. (See diagrams 1, 2 and 3). What is so interesting is that each is correct when properly applied, with the exception of exclusively measuring point to point.

Beekeepers in the field measure parallel wall to parallel wall. Normally, the mean diameter of the worker cells is recorded by measuring the distance by 10 to 20 linear cells with a ruler. In all cases, the three diagonals of the rows of hexagonal cells are measured to average out any cell irregularities.

No measurements should be made of drone comb, comb filled with nectar or honey, or comb towards the edges of the brood nest, as these cells tend to be larger. (See diagrams 1 and 3). I know of no beekeepers who measure the size of the cells in the field by measuring parallel wall to parallel wall and then straight down the frame. (See diagram 2). This measurement is normally used for counting the number of cells in a square decimeter along with the rhombus method. (See diagram 3). Counting the number of cells by each method, you would find that with the rhombus count you would have 800 cells total. This is because to arrive at the square decimeter total, you must use the cell count from both sides of the frame. In counting the number of cells by the square measurement method (Diagram 2), you would find that you would have 920 cells total.

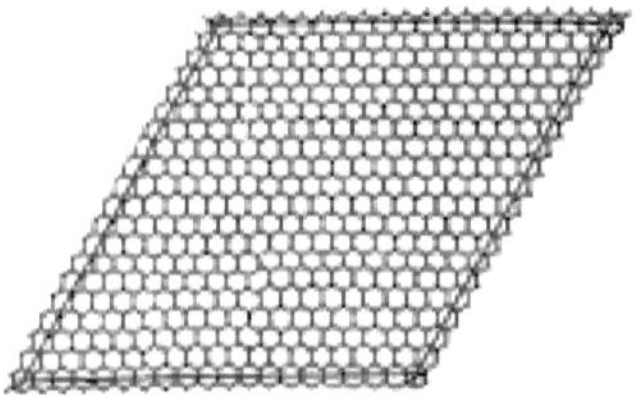

Dia. 3 - Counting a square decimeter for number of cells using rhombus measurement.

Again this is because to arrive at the square decimeter total, you must use the cell count from both sides of the frame and factor in the half cells. You gain three extra rows of cells to count, because you measure point to point in one direction.

This is important to understand – The cell count from using the square measurement method for a square decimeter is only good in the laboratory, not in the field. The cell count from using the rhombus measurement method for a square decimeter has direct correlation to the field, as the size of the cell between the parallel walls of the cell regulates the size of the honey bee worker, and then by ratio the corresponding sizes of the drones and lastly the queen.

If you as beekeepers want your honeybees on natural 800 size foundation relative to the field in which they fly, you must learn to measure your cells for the field. Five cells (5) to the inch is 20 cells in four inches and equals 800 cells by rhombus measurement method for a square decimeter. Interestingly, 19 cells in four inches times 19 cells by the rhombus method, equals 361 cells on one side of the frame and multiplied by 2 for a square decimeter total, equals 722. This would tend to indicate a 700 series with 22 cells left over, which when divided by (2) equals (11). So you would have a 700 series with 11 extra cells on each side of the comb. Anyone keeping bees knows the significance of 7/11 comb foundation?

Concerning chewing out Varroa: We have had several beekeepers want to know how and what to look for, to see if their worker bees are chewing out and/or removing Varroa mites from infested larvae cells. This is what we have told them to look for. Here in Arizona, you will see this chewing out of Varroa mites on the downside of the honey flow. It will start slowly as the queens stop raising drones, pick-up speed as the drones are expelled from the hive, then taper-off just prior to brood nest cleansing time. By the time the brood nest is re-situated and cleaned by the workers, with the pulling out of old larvae cocoons and reshellacked, you will find Varroa mites down to a non-detectable level in most cases; and under control by the workers. In Arizona, we see it happening approximately twice a year with the primary chewing out season in the Fall. Other times you will see it occurring in spurts and will be right after requeening, when the hive workers are throwing out drones and getting ready to roll again. You will see it mostly around the edge of the brood nest of sealed worker cells, although it can occur as a buck-shot brood pattern in weaker hives or in a strong hive where large numbers of mites are transferring from drone to workers.

Look for uncapped worker brood with the pupae exposed and in many cases cannibalized. If there was only one Varroa and it was located on the head between the eyes, many times the pupae will be unharmed, as the worker bees have only to remove the mite to rectify the situation. If the Varroa is on the back of the head between the thorax, the worker bees will eat the head off to get to the Varroa. If the Varroa and/or another is on the thorax, they will eat down to that also. If the Varroa and/or more are located on the abdomen, lodged with the tergits, the bees will continue eating down. You will notice that when the worker bees are doing this and working only with removing Varroa mites from healthy bees, the pupae will be a healthy white color, which shows that the worker bees are not removing diseased or infected larvae/pupae. When the Varroa is removed from the top of the head and the pupae left unharmed, you will usually notice that the pupae are at a stage of purple darkening eyes. The bees seem to chew out the Varroa when other chores of the hive are not pressing i.e. honey gathering and major brood rearing. Until then, the Varroa mainly infest drone larvae and pupae. Thus the drones, although they do no work physically in the hive, do act as the best attractant by body mass and therefore a better basil food target, to pull disease and parasites to themselves, so workers can survive throughout the active season by raising vital brood and gathering stores or honey and pollen. Then as the season winds down the drones are thrown out, the worker brood acts as a living liver in the hive purging the overpopulation of Varroa mites to bring it into a balanced parasitic mite host relationship similar to Apis cerana in Southeast Asia. Each new brood rearing season, the cycle starts again. Check of sealed worker brood, not uncapped by workers, have revealed non-infested pupae by Varroa. When you see this, you know that your bees are doing what they should to handle the problem. Caution: Do not confuse this phenomena with starving bees that need pollen and or honey or both. These hives were not starving and had plenty of stores in them. Beekeepers must learn to see with their eyes and understand the difference. If you look close, you will see which types of queens and characteristics to recognize, to know by body color and conformity. that your bees can handle mites.

Concerning converting to small naturally sized brood combs: First rule of thumb here is to *follow the bees not popularity themes of how you should do it!* There are fast ways to convert brood combs and bees if you want to and can hustle (work). If you cannot find small natural foundation for your area, you can use swarm-catching hinged frames and using cutouts, take advantage of the free comb available in Nature. Then, to get more foundation, take empty frames and wire vertically with (7) to (8) vertical wires and crimp the wires. Place this between two frames of good laid up worker brood and you will find out the bees will draw wax fine. However, do not place between drone combs or badly spotted droned worker comb. This is also a good way to get stock living in Nature that more readily adapts to handling the Varroa problem. After a while you will learn to spot feral queens by body and characteristics that can handle the problem. Learn again to open your eyes and look while not getting panic fever and run for the fast gimmick treatment in the bottle.

Remember above all, learn to follow the bees, in the end all will have to live with what Nature says will live and size down to natural bees through the three to four tiered retrogression periods it takes, in queening. You will find only old stick frames and/or crimp-wired frames will work to allow bees to drawn-out their own naturally sized combs that suit them, or you will have to put in smaller 800 size or 850 sized brood comb, preferably wax based without plastic, so the bees can modify it to their local regional needs. In the end, all else will in the long term not work.

Dee A. Lusby

Small Cell Size Foundation for Mite Control - ABJ July 1996

ABJ, November, 1996 – Page 758-760, Dee Lusby

Since you were kind enough to publish my letter to the editor in your July 1996 issue of your magazine on small natural cell size foundation for mite control, my husband and I have received numerous letters and phone calls seeking more information. Some of the letters have come all the way from Europe. Consequently, I feel I should write again as other readers may also be interested in some of it's history and what field experience we have gained these past 10 years plus in working with small natural cell diameter foundation without the use of chemicals.

To begin with, my husband and I consider ourselves "Naturalist Beekeepers". To us this means keeping bees in accordance, as much as possible, on a natural system of breeding and management without the use of drugs, chemicals in the field, and therefore hopefully in tune with the laws of God and Nature. We do not believe in insemination and its inbreeding practices. We take our honey crop with a brush, frame by frame, the old way, without Bee-Go. Yet, we are considered commercial beekeepers trying to run about 1,000 hives. We believe that every time you work a hive you should examine the brood nest and that hives should be worked approximately every 6-8 weeks in rotation throughout the beekeeping year, which to us is a twelve-month program.

We have had our share of honey bee losses due to mites over the past 10 plus years, with some apiaries in elevations over 4400 feet experiencing up to 80% losses more than one year in a row. We know what it is to requeen three times a year, Spring, late Summer, and Fall and still not outrun the mites because queens played-out all too soon! We know what it is to lose half of all our stock in one winter and say enough is enough, there must be a better way. In the Mid-1980s we made a decision to go back to naturalist beekeeping and never look back. Chemicals are too expensive and in the end who can afford to replace combs every three to four years to outrun contamination problems that might end in product recall and/or poison the brood nest area because the propolis in the brood cells has become contaminated, thus killing and harming the brood besides contaminating the wax.

What we did was tighten our belts and dig in for the long-haul and decide to see if Nature's way, therefore God's way, was the solution and make the journey back into a natural system of beekeeping, keeping as much modern technology as possible so we wouldn't break our backs. We never dreamed that so much cause and effect had been overlooked that had the combined potential of creating this huge mite mess.

Since it is a known fact that both honey bees and mites have been on this Earth many millions of years together and survived quite nicely, the question then is – WHAT HAS GONE WRONG? We decided in reading to try to go back to the beginning and check for cause and effect. This is what we have been telling the beekeepers who have written and called us, as we try to tell them where to find the information they need so they too can make their own management changes that fit their own needs and style of beekeeping.

Everything we have learned so far seems to revolve around of all things, our comb foundation sizes and the size of our honey bees since the invention of artificial wax foundations in 1857 by a Mr. J. Mehring of Frankinthal, Germany. The early wax sheets made in England and Germany back then were simple and had no sidewalls, just simply indentations. A person by the name of Wagner twenty years later made improvements adding sidewalls, making the sheets much more like natural comb. Up until 1875 all foundation was made with a pair of plates. Then a Mr. A. Washburn made a machine that would roll out a continuous sheet of embossed wax for foundation making. Following this a Mr. Chas. Olm of Fond du Lac, Wisc., invented an automatic machine which cut with a set of knives the embossed surfaces of the roller mills, as Mr. Washburn's were hand stamped and labor intensive. This brought foundation mills affordable for widespread industry use.

We have found out through reading, that mills have been made in many sizes over the years, all the way up to 3 1/2 cells to the inch. This is where our industry has gotten into so much trouble. Cell sizes, their size and how to measure them. Most beekeepers universally agree that five cells to the inch is worker size and four cells to the inch is drone size in the feral population, but the domestic size our bees today have ended up on is quite different and has wrought havoc, causing much disease and parasitic mite attacks as artificial combs have gotten bigger. The stress upon our honey bees caused by being out-of-balance with natural flora has opened Pandora's box to foulbrood diseases, chalk, and viral infections. The stress upon our honey bees caused by being too big by way of artificial mutation through use of oversized combs, has resulted in parasitic mite infestations as our now pseudo-drones (workerbees) are perceived as a new food source by Varroa and Tracheal mites. With all this damage being done, we find no one teaching the history of use of artificial comb foundation sizes in the United States, so our new upcoming beekeepers can make rational decisions concerning proper usage by size. Just like our woodenware, our various sizes were originally designed with meaning, now forgotten.

Let's get something straight right now, no one person or company is to blame either in the United states or elsewhere in our world for artificially increasing honey bees so big as to cause disease and parasitic mite problems to overtake our bees by placing them in a situation out-of-tune with the laws of God and Nature. Nothing was hidden and everything was written and published out in the open. The only thing that has happened, is that through the passing of each successive generation over the past 100 years or so, information was not passed on from father to son and/or teacher to pupil in our institutions of higher learning. Ours is a world looking always for bigger and better, faster and cheaper, simpler and not labor intensive, and circumstances have finally caught up with us and we now have to sort past-written information out, so we can correct and go forward.

We have forgotten as an industry how to measure the comb foundations our industry was built upon. We have forgotten as an industry how to do the simple field management inspections that keep bees healthy, instead relying upon quick fix patties. We have forgotten as an industry how to basically breed bees, many now preferring to buy quick requeening/restocking solutions, not considering how they fit into one's own regional individual environments. When we as an industry forgot the basics and stopped teaching it, we got into trouble. We cannot rely upon quick fix gimmicks to solve our problems, for they have created them. The information is there to solve our disease and parasitic mite problems and its about time it was gone over again, for the new generation of beekeepers to review and use, to make constructive decisions with, so they can manage their bees against today's diseases and parasitic mites. My husband and I really don't believe it's all really that hard. The basic principles are simple. It's really believing in a natural biologically controlled system to correct the situation – that's hard, for to believe in and put your trust in Nature, you must also believe in Nature's system and therefore God. The only catch is, going back to naturalist beekeeping involved work – hard field work and tightening the belt for a few years to come through the tunnel to survive.

Basic Principles to Review for Management Changes

1. Honey bee comb cells are measured parallel wall to parallel wall in three directions. They are not measured point to point, nor measured with a mixture of both, one way in each direction differently.

Metal mill rollers were originally made by making the bottom of the cells out of three chip-out little lozenge shaped plates, that when put together formed the bottom of the cell. This was done so that the bees could beautifully build what is called a "Rhombic Dodecahedron". Beekeepers know this figure as a common bee cell. When beekeepers measure comb foundations, they should measure the combs using the dimensions inside that of a rhombus, because in doing so they measure parallel wall to parallel wall and can arrive at an accurate figure that corresponds to that used by the mill maker in creating the mold that duplicates Nature. When beekeepers measure comb foundations today, many make the mistake of measuring parallel wall to parallel wall across the first row and then down straight to make a cell count determination.

Combs are measured in what is called a "Square Decimeter", but a square decimeter can be measured one of two ways. It can be measured either with a perfect square or by a rhombus method. By changing to a perfect square measurement, we have gotten into deep trouble because. . .*the numbers arrived at in the totals are vastly different. it is this vast difference that has wrought down upon us our parasitic mite problems as many of us try to use what we think is the proper size foundation our honey bees should be using, but in actuality it is not*.

By trying to approximate the old U.S. Standard of 856 and the old World Standard of 800 cell sizes to the square decimeter many beekeepers have used foundation bases geared to a square decimeter using square measurements rather than a square decimeter using rhombus measurements. The error is proving fatal to say the least.

2. Beekeepers should be actively culling their drone combs in their hives.

It has been previously demonstrated that Varroa mites prefer drone brood to worker brood for reproduction in the feral population of honey bees. Generally, about 40% pf drone cells are infested, while for workers, the average is close to 10%. (For Tracheal mites the feral average is also about 10% for workers for infestation levels). It has been demonstrated that the larvae food is the stimulant in the bigger cells for attracting Varroa infestation. For many years it was taught to cull drone combs as much as possible, but since the advent of Varroa, this practice has been reversed to the detriment of our hives. Beekeepers should go back to the old way of thinking, as there will always be plenty of drones reared in corners of the frames or in cells that become enlarged by accident. It should always be remembered that the drones do no work physically in the hive, but they do act as the best attractant to pull disease and parasites to themselves so workers can survive throughout the active season. Then, when the honey is in and new queens mated, their jobs done, they are cast out to cleanse the hive of its disease and parasite problems. On a natural system, the few phoretic mites that remain are quickly filtered out through the brood nest by the workers chewing out and/or removing mites from infected larvae cells. This happens during each transition period between summer and winter bees, short or long-lived bees, happening twice each year here in the Arizona desert Southwest. By us culling drone brood frames which are excessive (more than 10%) we therefore limit our infestation and reduce it down using the 40% vs 10% infestation level difference to our own hive management advantage.

Further, by changing out oversized artificial combs in our brood nests (some on the market are as much as 40% oversized) we reduce the attraction for Varroa to enter pseudo-drone cells (worker cells artificially enlarged with more larvae food for mites) and reproduce at higher than natural 10% infestation levels also.

3. Beekeepers should be actively out-breeding their colonies and not practice inbreeding which lowers hive productivity and life.

It is a known fact that the life span of honey bees becomes shorter and bee colonies develop less as inbreeding degree is increased. It has been noted that F4 out-bred queens can produce 300% more brood then F4 inbred queens. Bees were not meant to be artificially inseminated using inbreeding methods when in the wild under the laws of God and Nature they naturally outbreed.

4. Beekeepers should not actively feed artificial pollen substitutes nor sugar for feed for long periods of time for they lower hive productivity and life.

Most often artificial food substitutes fed to honey bees are to be used for short-term duration of 6-8 weeks. Feeding honey bees on sugar in the Fall has a negative effect on the physiological processes in their preparing for wintering. Bees by having to process sugar syrup all Winter and into the Spring cause the hypopharyngeal glands fat body and ovaries to deteriorate and this can result in low brood rearing in the spring when fast build-up is needed to out-run mites. Colonies fed on honey and real pollen result in larger emerging bees and more vigorous bees.

5. Beekeepers should not be practicing 100% pre-medication preventative care of their honey bees. It's like treating a person with chemotherapy before he has cancer and wondering what is making him sick.

If beekeepers medicate only what is sick in their hives, on a regular 6-8 week full hive inspection program their colonies would be better off. Why, because they would be catching most problems in the beginning, rather than when finding them out of control, with just early Spring or late Fall inspections. We do not buy artificial drugs to treat our colonies anymore. Instead we rely upon 2000 plus years of Bible history and use "Propolis" the old "Balm of Gilead" to heal and medicate our hives. We have found it quite successful in treating bacterial, fungal, and viral diseases. Furthermore, it is the only universal antibiotic manufactured within our own beehives that is used by man and we see no reason why the bees, therefore, should not be medicated with it themselves when necessary, on a case-by-case basis only.

My husband and I have told beekeepers calling and writing, that if man should ever seek to change honey bees so that they no longer relate to Nature's and God's law, they would likely intervene in such a way as to preserve the necessary balance originally created. For there is some reason to believe that in the plan of Nature, the honey bee was not only created to conform to the necessity of its mission as a pollinating agent, but that the plants and their bloom may have been fashioned to conform to the convenience of the bee also in one large masterful plan. There is a barrier we have crossed as an industry worldwide we need to retreat from, that seems to have been deliberately placed there by God and Nature to prevent any wide deviation of the honey bee in size and action from what they designed that it should be, this being accomplished by limiting the size of the bee to that of the cell in which it is developed, as set down in the feral bee, beyond which it cannot go far without being forced back. Diseases and parasitic mites are forcing us back now into balance with native regional floras. My husband and I believe beekeepers should pay heed or they'll go out of business. Time is short now for our industry as our inbreeding mistakes catch up with us. Time is short now for our industry as our chemical treatment mistakes catch up with us. Time is short now for our industry as our oversized artificial combs catch up with us. We as an industry want to run fast and cheap and it won't work.

For those who want to go back to naturalist beekeeping, it is not hard, but it does take field management time and a belief in Nature and biological controls. We have checked comb sizes all over this country. Very few manufacturers have natural sizes available, for you have to be an old company with deep roots in industry to have it available. Dadant has small natural foundation, five (5) cells to the inch. They call it 900 series (use the Square measurement method) rather than the 800 Rhombus it really is (use the Rhombus measurement method). Beekeepers reading this should really call them up and bug them hard to get them to advertise it to the industry again out in the open. My husband and I fully believe that our industry will have to change back their brood combs to natural feral size or die. Many who can't change back will go out of business for lack of time and resources at this late date, but small natural comb should be made readily available to the industry for those who wish to try a biological natural solution and are willing to work. The bottom is a 2/3 bee loss to adjust their colonies which is a 3-4 year acclimatizing process (like going through alcohol withdrawal). But, those who can graft and make bees from survivors can work with a base of 35-40%. Beekeepers cannot work with 90% blowouts and no plan of survival, with the only recourse. buy more bees and throw in more chemicals and blowout again. Where's the plan and the end of this insanity.

It's about time we as an industry got in tune with history again and Nature's system. Call Dadant and bug them to advertise small natural comb again and dust off their old mills. Remember their 900 is 800 and the heck with trade secrets of what it really is. We need small natural comb foundation sold again and advertised on a regular basis.

Dee A. Lusby

Arizona Beekeeper Believes Smaller Size Cell Diameter is the Answer to Mite Problems - ABJ 1997

ABJ, December, 1997 – Page 837-838, Dee Lusby

On 11 September Dr. Eric H. Erickson, the director of the Carl Hayden Bee Research Facility in Tucson, Arizona, went with us to two bee locations, in unisolated areas, to test for both tracheal mites and Varroa mites. Samples taken in the center of the brood nest also contained drones where possible. We choose unisolated locations because we wanted to show him, to beat the problem, one must be able to accomplish business as normal in doing bee management within the field. Please note that beekeepers around us have severely lost bees, as we ourselves have, to both mites over the years. When taken, several adjacent yards within 2 miles were being treated, crashing, or being fed to keep them alive. Our bees were building; and at the Carmen yard were very fast drawing new foundation.

We began putting the 4.9 cm cell size in hives in May. We did a second round the end of June and did a third round ending Labor day. The Carmen yard we took samples from was worked Labor Day along with the Knight location. The Carmen yard had been drawing wax and averaged 4-8 or more frames per colony drawn. A few colonies had a full box (10 frames) drawn. The Knight location had less than 3 frames drawn on average and most brood laying was on 5.0 cm comb. Both yards still had 2-3 (3-Carmen 2-Knights) one super hives (nucs) still laying on the larger Duragilt that refused to change. Note these one super hives are now dead, not having survived through to mid-October. So much for Duragilt (5.44 cm).

With smaller 4.9 cm comb which is still bigger than the 4.83 cm comb this country was founded on in Southern latitudes, (Northern latitudes were founded on 4.9 cm to

5.0 cm sizes), we are now getting our Varroa populations down to field tolerant coexistent levels so we can mimic natural environment living conditions. Tracheal mite levels are down there also, having regulated the mite back to external Acarapis vagans status, as was the norm condition around 1917 in our country, before we artificially mutated the bee's thorax and breathing tube bigger on the thorax to create a parasite problem. At 0-6% tracheal mites, bees have no problem coexisting. At 10-11%, Varroa mites are on the cuff for trouble. In Southern latitudes in times of plenty they do fine; in times of dearth the bees do poorly and both require constant management to control secondary diseases. This is on 5.0 cm size comb. At 0-7% Varroa mites, changing to 4.9 cm comb sizing, bees draw wax well and hives no longer require constant management to control secondary diseases. Business is back to normal for management in the field. We hope to cut percentages again this coming year 1998 as brood nests are continued with 4.9 cm comb and all frames are converted in our broodnests.

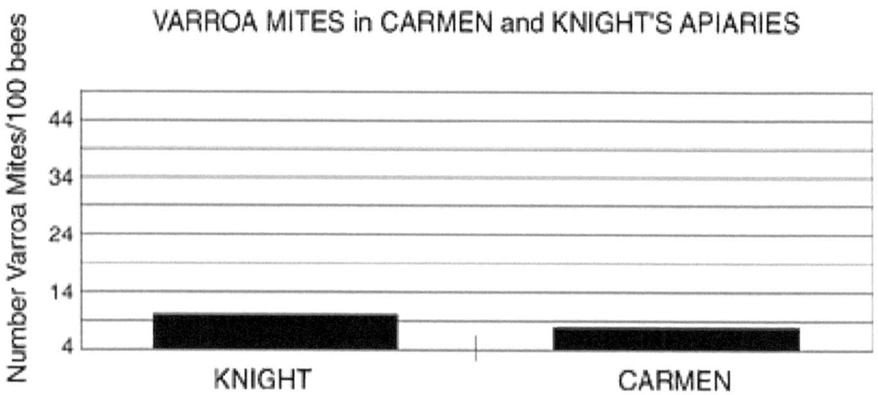

This shows breeding is not all the solution. We figure comb is 1/3, diet is 1/3 and breeding is 1/3. Comb must be put in by half (5) to full boxes to work.

Dee Lusby

Note from author regarding article:

ABJ, December, 1997 - Page 837-838

NOTE:

Re: ARIZONA BEEKEEPER BELIEVES SMALLER SIZE CELL DIAMETER IS THE ANSWER TO MITE PROBLEMS

Even though we have asked for it several times, no subsequent retesting has been done by the USDA to see if mite counts have gone lower as predicted. We find this odd since so many are trying to find a solution to the problem of parasitic mites.

- Dee Lusby

HONEY BEE PARASITES FROM CARMEN YARD

VARROA MITES			
Colony #	# Bees	# Varroa	# Varroa/100 Bees
A	175	34	19.43
B	186	30	16.13
C	161	39	24.22
D	186	5	2.69
E	157	7	4.46
F	183	13	6.99
G	169	13	7.70
H	148	5	3.38
I	187	2	1.07
J	149	6	4.03
K	185	5	2.70
L	164	7	4.27
M	188	7	3.72
N	156	5	3.21
P	163	8	4.91
Q	179	17	9.50

TRACHEAL MITES IN 30 BEES		
Colony #	# Tracheal Mites	% Tracheal Mites
A	0	0.00
B	0	0.00
C	0	0.00
D	0	0.00
E	1	3.33
F	0	0.00
G	1	3.33
H	0	0.00
I	1	3.33
J	7	23.33
K	0	0.00
L	1	3.33
M	1	3.33
N	1	3.33
P	0	0.00
Q	2	6.67

HONEY BEE PARASITES FROM KNIGHT YARD

VARROA MITES			
Colony #	# Bees	# Varroa	# Varroa/100 Bees
A	165	1	0.61
B	186	15	8.06
C	142	13	9.15
D	177	18	10.17
E	168	21	12.50
F	184	23	12.50
G	171	26	15.20
H	186	9	4.84
I	181	53	29.28
J	200	8	4.00
K	189	19	10.05
L	182	4	2.20
M	175	23	13.14

TRACHEAL MITES IN 30 BEES		
Colony #	# Tracheal Mites	% Tracheal Mites
A	2	6.67
B	3	10.00
C	0	0.00
D	0	0.00
E	0	0.00
F	0	0.00
G	5	16.67
H	8	26.67
I	0	0.00
J	4	13.33
K	2	6.67
L	1	3.33
M	1	3.33

Note: Cell sizes are recorded in general for zones, and where altitude and higher latitude occur, use Humbolt's Law for deviations.

Special Note: After studying correlations, we believe brood 4.9 mm cell size could be the upper limit for cell size control for mites worldwide.

MONTHLY TEMPERATURE MEANS	GENERAL CELL SIZES	COLOR ZONE
-0F - 60F / 80F	5.0 mm - 5.2 mm	
0 / 40F - 60F / 80F	4.9 mm - 5.1 mm	
0 / 40F - above 80F	4.8 mm - 4.9 mm	
40F / 60F - 40F / 60F	4.9 mm	
40F / 60F - 60F / 80F	4.8 mm - 5.0 mm	
40F / 60F - above 80F	4.7 mm - 4.9 mm	
60F / 80F - 60F / 80F	4.6 mm - 4.8 mm	
60F / 80F - above 80F	4.7 mm - 4.9 mm	
above 80F - above 80F	4.5 mm - 4.7 mm	

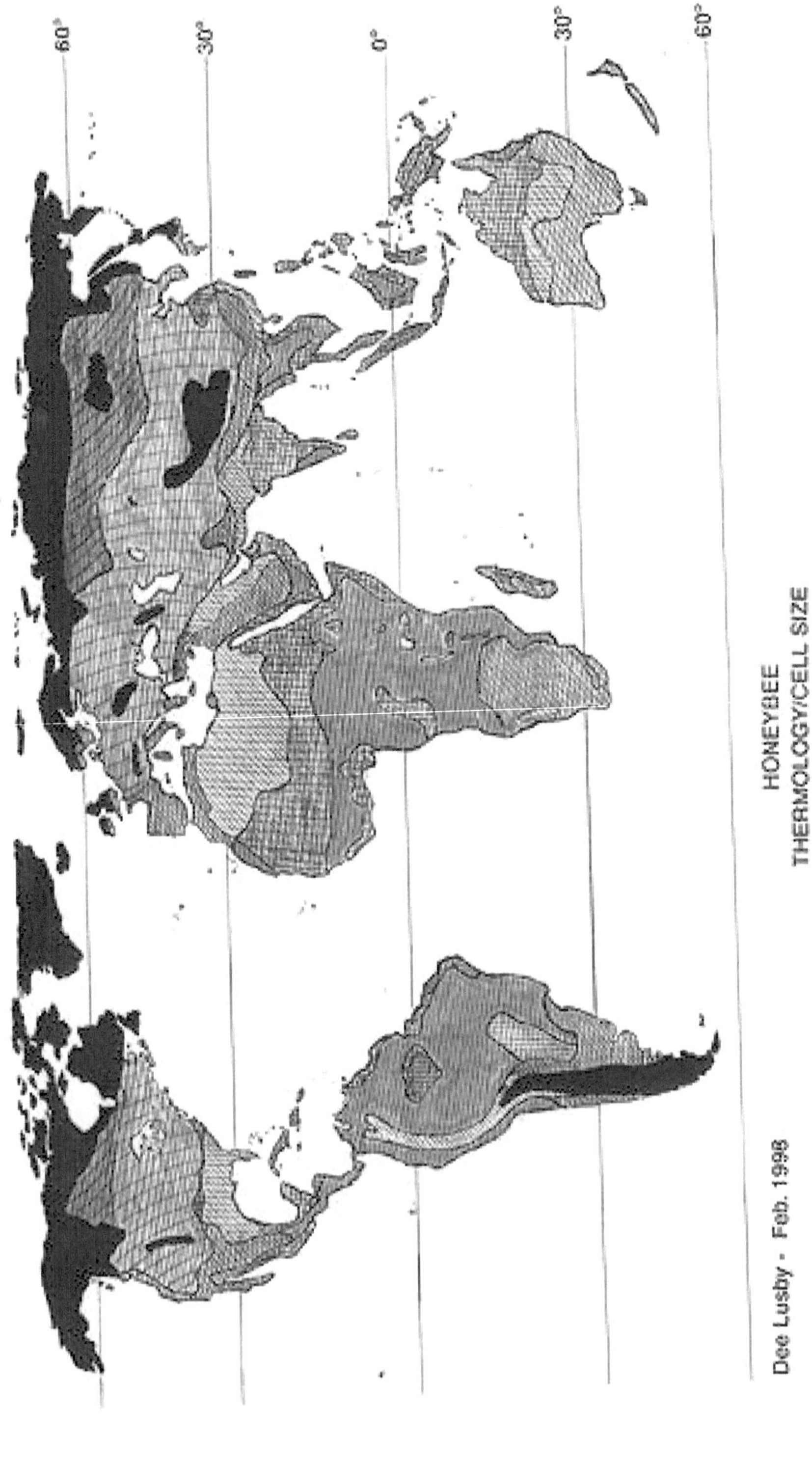
60°
30°
0°
30°
60°
HONEYBEE
THERMOLOGY/CELL SIZE
ZONES
Dee Lusby - Feb. 1998

Is Smaller Better? - Bee Culture 1998

Lusby, Bee Culture – June 1998

The cell size debate may soon be over.

We have been trying to figure out the best brood comb cell-size for some years for our area (Tucson, AZ). It started locally, but then beekeepers in other parts of the United States, and even overseas wanted to know what we thought their best natural brood comb cell size should be also. Eventually we constructed a world map showing thermal zones and corresponding cell size. (see previous chapter for map)

To create the map we combined many pieces of information relative to history and world environment, such as: 1) What was the actual recorded cell size prior to the use of artificially enlarged foundation and how was it measured? 2) Where were honey bees capable of living when looking at natural zones of heat and cold during a full year? 3) Were there naturally occurring variables that could change cell size or bee habitat? 4) Would it all fit together – that is, recorded cell sizes compared to the environment they were in?

This map based on atlas composites of hot and cold land area maps accurately reflects the history of recorded cell sizes in published records prior to the use of artificially enlarged foundation. Climate and rainfall were variables allowing habitat transition into and out of recorded zones. Cell sizes are recorded in general for the zones and, where altitude dictates Humboldt's law, should be used in higher mountainous areas and areas of higher latitude.

Every thermal/cell size zone has a small, medium, and large range to allow for bees to transition into and out of habitat areas as vegetation and rainfall vary throughout the yearly cycle. With smaller cell sizes you gain variability, and with larger cell sizes you have less. We found this published many times.

It seems logical that as you go up the hill bees get bigger to match the colder, higher altitudes, but also, they encounter less habitat to live and breed in. Therefore, when they reach the top of the mountain they have to go down again to regain variability or suffer extinction.

Some of the last of the artificially-enlarged cell size research was done round 1941 with cells of 5.4mm diameter, with the experiments lasting a good four years. Between 1957-1963 other artificially-enlarged cell size research project was undertaken with 5.65 mm diameter cells. The research was considered successful, but the experiments were done in the Romanian mountains. In that area the research showed cell size of natural honeycomb presented great variability, depending upon the altitude. In fact, when all the work was done it was recommended that bee colonies should have at their disposal honeycombs with cells as nearly as possible to the size cell which the bees themselves naturally build.

This made us wonder why we were using "high altitude honey-comb" in "low altitude areas" and then compounding the problem further by placing it into the broodnest. Could this have an impact on the disease and parasite problem? We assumed that since honey bees, mites and other pressures have coexisted for many years, it could be further assumed that something artificial – like oversized cell size – may have disrupted that co-evolved condition, rendering the bees more susceptible to parasites and diseases. So, all we are doing is placing our bees back onto the natural cell size for our area and letting Nature take her course.

We have indeed taken on a puzzle. However, we think we have now found many of the pieces. First, we put our bees on 5.0 mm cell size foundation by making over 40,000 sheets for our brood nests. A long drought forced us to reconsider this cell size though. We have now placed into the field over 4,000 frames of 4.9 mm cell size foundation. That was done in 1997 during the drought and before going into Winter with our bees. We culled heavily to get the job done, shaking down our hives into only 1-2 deep brood boxes to over-Winter.

We spent this past Winter preparing over 10,000 frames with 4.9 mm cell size foundation to put onto our bees between March and the end of July this year. We'll put on as much extra as we can until frost in November using an additional 10,000 frames of 4.9 mm foundation that will be ready by mid-Summer. That will give us a total of 24,000 frames of smaller, more natural 4.9mm cell size to use once and for all if going back to natural sizing works well for disease and mite control.

The experiments performed in 1941 and 1957-1963 that pointed to the "bigger is better" theory were done with an average of 5-6 colonies. Those experiments changed the world's beekeeping industry. It became the belief of the day that bigger was better – for more honey, for less swarming, for ease of extracting and spinning out honey. But in the end, what did we gain?

We don't experiment with only 10 frames or even a few hives as a basis for saying whether or not something works. We're considered small commercial beekeepers, and averaged 900-1100 hives up until the mites came. We believe we'll never manage over 700-800 hives with a more intense, smaller, natural bee management system, but we also believe it will be profitable. With El Nino's help this past year producing lots of plants in the southwest desert, we'll see how much 4.9 mm foundation our bees can draw out.

We can only come back now as fast as we can draw comb. We've been sitting under a mesquite tree now for nearly three years waiting for the chance to draw a large amount of comb. We're ready. It will be a challenge to see if we can retool all our combs, with the goal of running 700-800 hives, by the end of 1999. all in 4-5 deeps, but if we make it so can everyone else.

Look at the thermal/cell size zone map (see right). So far everything is pretty much matching up. The trouble is, now the questions start. What have we lost by having too-big bees relative to crop pollination? Is there an upper cell size limit for controlling mites relative to altitude and cooler latitudes? What's the relationship of cell size to disease, chemical contamination, pesticide sprays, inbreeding, pollen and propolis for human cures? Are there limits to taking bees out of one zone and placing them into other zones? If so, is it beneficial, or detrimental?

Ed and Dee Lusby

Managing Colony Genetics by Grafting and Selecting for Queens with Shorter Development Times

ABJ, November, 1989 – Page 717-719 *by GLORIA DeGRANDI-HOFFMAN*, DELORES A. LUSBY**, and ERIC H. ERICKSON, JR.**

**Carl Hayden Honey Bee Biology and Insect Biological Control Center U.S.D.A.-A.R.S., 2000 East Allen Road, Tucson, Arizona 85719*

***Rangeland Honey, Tucson, Arizona*

Two serious problems facing the beekeeping industry are the migration of Africanized honey bees into the U. S. and the spread of Varroa mites. Now more than ever beekeepers must manage the genetics of the bees in their colonies if they hope to deal with these problems.

The strongest tool that a beekeeper has for controlling colony genetics is the grafting needle. Colony characteristics that are favorable to a particular beekeeping operation or are adapted for a specific geographic area can be increased by grafting queens from colonies that possess the desired traits. By grafting their own queens, beekeepers can create lines of bees tailored for the conditions of their apiary sites and beekeeping practices.

A trait that may be an important component in solving Africanized bee and Varroa problems is queen development time. The first queen to emerge destroys the remaining queen cells and becomes the matriarch of the colony. The colony's behavior and attributes will reflect the genetic composition of the queen and the drones with whom she has mated. Queen development time could be partially responsible for Africanized traits being expressed by bees in geographical areas that previously were inhabited by European strains, if the development period for Africanized queens is shorter than that of European queens. In Africa, queens of Apis mellifera scutellata develop in 14-15 days while European queens require 14-17 days (Anderson, Buys, and Johansmeier 1973). If Africanized queens emerge first the colonies will express many traits associated with that line of bees. Queen development time apparently is an inherited trait. A line of honey bees (hereafter referred to as Lusby bees (LUS) that has been selected for shorter queen development time now has queens with an average development period of 14.1 days (with a range of 12.4-15.8 days). We conducted an experiment to determine the variability in queen development time using a closed population (CP) line of bees composed of stocks that can be purchased from commercial package and queen breeding operations throughout the U.S. (Page and Laidlaw 1982). Larvae from LUS bees were also grafted for comparison. Three CP colonies and two LUS colonies were used for grafting. The resulting queens will hereafter be referred to as CP 1, 2, or 3 or LUS 1 or 2 queens. In this experiment, only 12-24 hour old larvae were grafted (age was determined by size of the larvae). The grafting technique was similar to the procedure outlined by Laidlaw (1981) in which larvae were placed in a drop of royal jelly at the bottom of queen cups. The grafted larvae were then placed in starter-finisher hives containing the same line of bees from which the larvae were grafted. Five days after grafting, the capped cells were placed in individual plastic vials and put in an incubator set at 94° F (34.50° C) and 78% relative humidity (Fig. 1). The incubator was checked every 4-5 hours for newly emerged queens.

The emergence times for CP and LUS queens are shown in Table 1. LUS queens emerged 9.5-10.6 days after grafting (13.5-14.6 days total development time), while CP queens emerged 10.4-11.0 days after grafting (14.4-15.0 days total development time). LUS 2 queens had the shortest average development time. The average development time of LUS 1 queens was not significantly different from those of any of the CP queens.

Table 1. Total development times (egg to adult) of grafted queens from two different strains of honey bees.

Strain	Colony Number	No. of queens	Queen emergence times (days) after grafting
Lusby	1	17	14.6ac
	2	63	13.5b
Closed Population	1	28	15.0a
	2	19	14.4c
	3	23	14.8ac

Means followed by the same letter are not significantly different at the 0.05 level as determined by Scheffe's S test.

Differences among colonies concerning queen development times are revealed in greater detail by examining the percentage of queens from each colony emerging over time (Fig. 2). Almost 20% of LUS 2 queens had 12-13 day total development times and emerged before other LUS or CP queens. CP 2 had some queens with 13-14 day development times as did both LUS colonies. Most of CP and LUS 1 queens had 14-15 day development times. A relatively small percentage of LUS 1 and CP 2 queens had 15-16 day development times, while almost 60% of CP 1 queens and 20% of CP 3 queens emerged at this time.

Grafting queens and documenting their development time using an incubator is a simple procedure that can be done by any beekeeper. The first step is to determine the range of queen development times in existing stocks, particularly those with other attributes that need to be perpetuated. To do this, graft larvae of the same age. When the cells are sealed, place them in individual plastic or glass vials, and transfer them to an incubator. We use a plastic foam Little Giant poultry incubator, Miller Mfg. Co. Inc., St. Paul, Minn., that costs about $30.00. Check the incubator every 4-6 hours to determine emergence times. Label the vials with the time that the queen emerged, and estimate the total development time. To apply selective pressure for shorter queen development time, introduce only those queens which emerge 9-10 days after grafting. By repeating this process with the offspring of these queens, the frequency of shorter queen development time can be increased in the next generation. Once this trait is established in a colony, it will be retained even if the colony re-queens itself (assuming that larvae of nearly the same age are selected by the bees to be reared into queens).

Additional studies are currently being conducted using the grafting and selection techniques described here to determine whether queens with shorter development times produce workers with this trait. We are testing factors that could influence development rates. One such factor is temperature which in many insect species strongly influences development rates. In honey bee colonies workers control temperature particularly in the brood nest, and thus may be influencing development rates through temperature regulation.

Shorter development times may be associated with smaller body size. We are examining the size and weight of queens (and possibly workers) with the shortest and longest development times to determine if they differ. If, indeed, queens with shorter development times produce offspring with this trait, they may show resistance to Varroa mite infestations since fewer female Varroa mites will have the opportunity to develop before the adult worker or drone emerges (Camazine 1988).

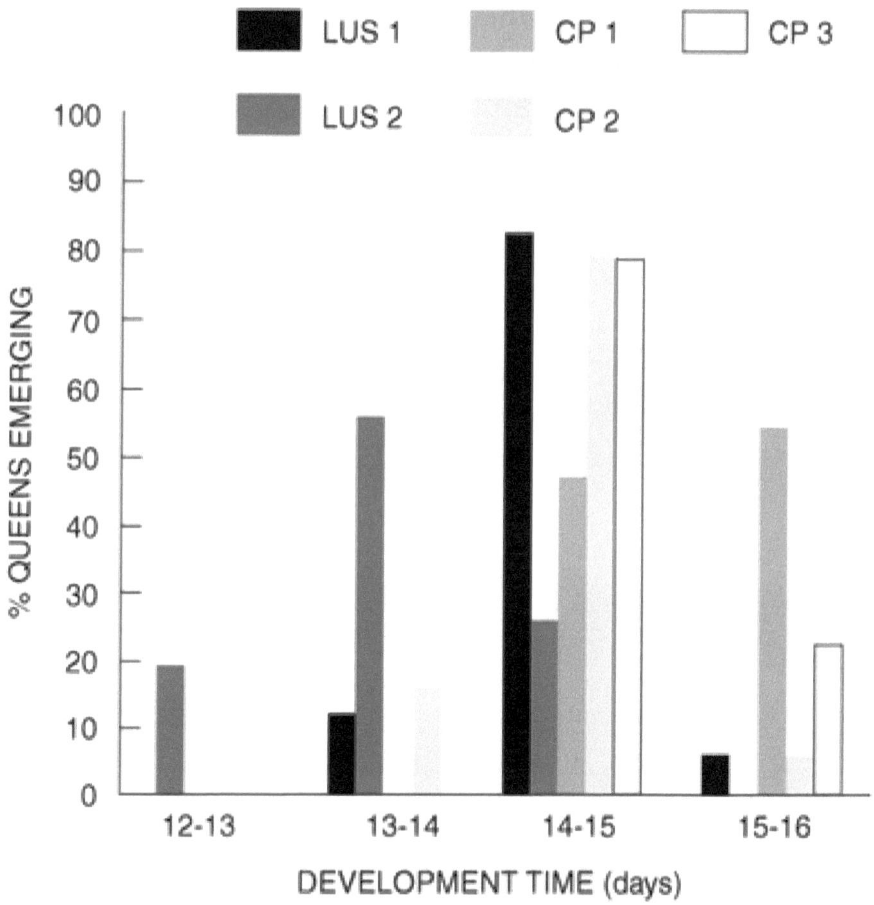

Figure 2. The percentage of queens emerging from each test colony over time.

LITERATURE CITED

Anderson, R. H., B. Buys, and M. F. Johannsmeier. 1973. Beekeeping in South Africa, Dep. of Agric. Technical Services Bull. No. 394.

Camazine, S. 1988. Factors affecting the severity of Varroa jacobsonii infestations on European and Africanized honey bees. In: Africanized honey bees and bee mites. G. R. Needham, R. E. Page Jr., M. Delfinado-Baker, and C. E. Bowman, eds. Ellis Horwood Limited, Chichester, West Suffix, England.

Laidlaw, H. H. Jr., 1981. Contemporary queen rearing. Dadant and Sons, Hamilton, IL.

Page, R. E. Jr., and H. H. Laidlaw. 1982. Closed population honeybee breeding. 1. Population geneties of sex determination. J. Apic. Res. 21: 30-37.

Suggested Biological Manipulative Treatment for Control of Honey bee Mites

Apiacta XXVII, 109-117 (1992), Dee A. LUSBY, E.W. LUSBY

What is biological management?

Biological management of bee-hives is not new but is seldom practiced anymore. Basically, it is similar to beekeeping the way Grandpa used to do it around the turn of the century.

Because today's conventional drugs and chemicals used in the treatment of bee diseases, pests and parasites are aimed at suppressing disease symptoms, they do not have a place in a long-term program of biological treatment and control. In the end chemical controls only add problems for the beekeeper. Colony distress is an important symptom, a signal, which is initiated by the colonies own defense mechanism. Learning to recognize these stress signals is therefore important for early initial biological treatment. To suppress and mask symptoms of bee diseases, pests and parasites with chemicals without finding their origin is contrary to the philosophy of long-term biological control.

It is of vital importance to realize that the various symptoms of bee diseases, pests and parasites should not be viewed totally negatively, rather they should be viewed as positive constructive symptoms initiated by the colonies' own healing mechanism, in its effort to restore balance and heal itself. When this is clearly understood by the beekeeper, then time and resources will no longer be wasted on methods that mask symptoms with quick fix remedies and provide only temporary relief. The beekeeper will then aim at eliminating and correcting the underlying causative factors of bee diseases, pests and parasites, and begin supporting the colonies own recuperative powers.

Concept of origin and spread of diseases, pests and parasites

It is a known fact that both honeybees and mites have been on this Earth and have co-existed for many millions of years. Parasites cannot survive if they kill their host. The question then is what has gone wrong? Why do colonies die from Acarapis woodi and Varroa jacobsoni infestations? How do normal healthy beehives change into parasitic mite infested colonies with secondary stress diseases without cause and effect transpiring?

The well-known colony stress symptoms – unexplainable fatigue, loss of appetite, physical abnormalities, nervous or runny behavior, lack of housecleaning, poor flight activity -, create increasing degrees of ill health and would have to be considered consequences of mites. Since both honeybees and mites have co-existed for many millions of years, it must be assumed that something done artificially to honeybee colonies during their domestication and management by man has created the problem of parasitic mites that ultimately result in the destruction of the colony population by them and their secondary diseases. By looking at cause and effect we find that beekeepers themselves have wrought cause and effect in several ways. Combined, they have created the situation they now find themselves in.

First the colonies have to be stressed (the cause) causing the hives to become susceptible to mites and related stress diseases (the effect). It has been suggested that Acarapis woodi may have evolved very recently, perhaps in Britain and as recently as 1900 (DEJONG et al., 1982). However, this hypothesis must be treated with caution. Nevertheless, the very close similarity of the various species of Acarapis mites does suggest that they evolved symmetrically of Apis mellifera from a common ancestor (DELFINADO-BAKER and BAKER, 1982). If beekeepers were to study comb size history they would easily perceive that introduction of larger and larger comb cell sizes used in colonies since the turn of the century have developed evolutionary changes in honeybees through artificial mutation of body size, therefore making bees more susceptible to parasitic mite attacks. With today's comb cell foundations now on the market near or exceeding measurements per square decimeter for Apis dorsata for most of today's European honeybee races, no small wonder there is a parasitic mite problem (see table below). The European honeybees are merely out-of-tune with natural feral races and strains of bees by way of body and comb sizing. Based on observations and study of comb cell sizes, it should be hypothesized instead that honeybees have since the early 1900s been artificially mutated larger by beekeepers using bigger and bigger comb sizes, thus causing the parallel evolution of mites as their food source changed.

Location	Beekeeper	Year	Size
Attica, Greece	Georgandas	1968	733 minimum
			854 maximum
			815 average
Peloponnesus, Greece	Georgandas	1968	846 minimum
			892maximum
			863 average
Arta, Greece	Georgandas	1968	836 average
Crete	Georgandas	1968	835 average
Macedonia	Georgandas	1968	821 average
- - -	Collin	1865	854
- - -	Langstroth	- - -	838
Italy	House of Fratelli Piana	- - -	860
Italy, House (unnamed)	- - -	- - -	813, 807, 854
- - -	Baudoux	- - -	854, 807
- - -	Pincot (for Italian race)	- - -	764
Burgundy	unk	- - -	798
France (common black bee)	- - -	- - -	854
France (degenerated common bee)	- - -	- - -	924
- - -	Halleux	1890	845
North Africa	Rambaldi	- - -	940
- - -	Fremont	1893	825
United States	Grout	1931	857
Netherlands	Swammerdam	1733	870
Italy	Maraldi	1712	789, 954
France	Reaumur	1740	832
Germany	Klugel	1700s	832
Spain	Castellon	- - -	763, 828
British Isles (200 years ago)	A.D.Betts	- - -	830
India	Rahman & Singh	1946	1013.17 A.indica, 2380.61 A.florea, 796.10 A.dorsata
United States	A.I.Root	- - -	825, 850

The causes

1. Artificial oversized brood combs. Since the time of Baudoux in following Huber's experiment in 1791, but by using artificial means instead of drone combs, causing creation of larger worker bees, beekeepers have been artificially mutating the body size of honeybees larger (GROUT, 1931). This has placed honeybees with each successive upsizing of comb more out-of-tune with Nature and natural bee flora. Why, because it is difficult to create new honey plants and bees which can be reproduced as such, which have been developed through thousands of years and adjusted to the existing climatic conditions, soil, and especially existing bee flora (CHESHIRE, 1888; GEORGANDAS, 1968). This then creates and adds to the second cause.

2. Artificial diet causing inadequate nutrition. Poor nutrition is a serious stress factor of any organism. What happens when key nutrients are present in insufficient quantities for generation after generation? Larger honeybees require richer nutritional diets, yet have access to less in Nature by being out-of-tune through body size to appropriately match natural bee flora. Colonies can be in a state of inadequate nutrition through either their geographic location placement or placement on artificial enlarged comb foundation creating imbalance with bee flora, or fed diets of pollen substitutes and sugars that are inadequate. One or more of the key nutrients can be insufficiently represented or entirely lacking in the bee's body. Since we believe that a queen reared this way, cannot give to her offspring what she does not have herself, the result is that the queen constitutionally transmits a predisposition for disease and mite attack to her off-spring. If honeybees acquire a predisposition for stress diseases due to inadequate nutrition, beekeepers can expect disease and mite infestations in their colonies.

3. Artificial medical treatment by chemicals rather then biological treatment through natural management, causing neurological disorders (CHANEY, 1988), queen supersedures and brood deaths, leaving the honeybee colony unable to function properly to fight off bee diseases or mites.

Mite prevention – a possibility

Since a small population of parasitic mites is nondetectable by either chemical or biological examination methods, beekeepers wait for the appearance of a large infestation to tell them that something is wrong. By then it is often too late for the hive. An approach is needed that looks at the situation in reverse. First the honeybee colony drifts into a pathological state, with the final symptom being a severe infestation of parasitic mites. Logic should compel beekeepers to try to detect the underlying stress signals which are the forerunners of mites, and through biological treatment manipulations eliminate the artificial stimulations that result in mites attacking colonies. This can be accomplished with a long-term biological manipulative treatment program which can be used to either prevent or wean colonies from parasitic mites (LUSBY and LUSBY, 1992).

There is no denying that methods consisting of heavy medication do wage a battle against parasitic mites and stress diseases. However, at the same time chemicals only mask the symptoms and perpetuate the problem. In addition, beekeepers run the high risk of chemical contamination and product recall of wax, pollen, and honey crops. Advanced stages of stress, indicated by symptoms of high parasitic mite populations, prevent

beekeepers from implementing biological manipulation treatments easily, because once on chemical dependency treadmills, it is almost impossible to stop treatment without loss of colonies.

Stress symptoms develop for several reasons that work in combination

In the beginning, the honeybee colony is in perfect health without diseases, pests and parasites. Then through the combination of placement on improper sized brood combs for localized geographic regions, and improper nutritional needs over extended periods of time, the colony develops the loss of this healthy condition. Stress factors weaken the honeybee's natural defense system inherent within the hive. Minor stress symptoms appear in the form of foul-brood and other body diseases. In successive generations, more advanced symptoms appear in the way of various fungal diseases. Both diseases, along with mite infestations can easily gain a foot-hold in a stressed colony. The colony is destroyed from generations of abuse and stress. The mites and diseases are not the problem, they are merely the advanced stages of an artificially caused problem. The stress resulting from generally accepted beekeeping practices of artificial enlarged combs, nutrition, and chemicals repeated over many years, is the real killer of domesticated honeybee colonies.

The most important weapon in the fight against parasitic mites and their secondary stress diseases is prevention. Beekeepers must be alert to the signs of distress within their colonies. When stress symptoms are apparent, beekeepers must take action to put their colonies back into biological balance with manipulative treatments. This can be accomplished through dietary change if an artificial diet is being used, and by replacing the brood comb with natural sized comb foundation in harmony with the geographic region where the colonies are being maintained. Culling excessive drone combs will also help. The down sizing of the brood comb foundation will realign the bees' body size to again match their native flora. Changing the diet from artificial pollen substitutes and sugar syrups back to pure natural pollens and honey from the colonies own geographic region will also improve colony vigor. The removal of stress by beekeepers is, of course beneficial, like removal of contaminated combs and their replacement with disease free combs. But this in itself does not correct the underlying reason the hive came down with the malady. The whole hive must be restored to full health by placing it back onto a natural system that acts to relieve stress.

If the colony is still in the early reversible stage of development of stress diseases, the therapeutic administration of natural key nutrients and natural sized brood comb foundation, sized to ones own beekeeping region, will in most cases bring about the restoration of health to the colony. The result is that the bee's own natural defense system and capacity for recovery will again be activated and begin the workof clearing away the problem within the hive. Stress diseases will be eliminated and the mite population will naturally decrease to a level well below economic thresholds for survival of the hive.

Beekeepers must bear in mind that in treating and curing honeybee stress diseases and getting rid of parasitic mites, that these disturbances to colonies do not possess a capacity for unbridled autonomous growth. Their behavior depends entirely of the state of health of the honeybee colony as a whole harmonious working unit. The nutritional healing of the colony coupled with replacement back onto natural sized brood comb foundation has a number of important advantages:

1. In a colony that has been restored to health, the natural defense systems of bees are fully operational again, whereas treatments such as chemotherapy for

parasitic mites can have the opposite effect, that of damaging the bees by causing neurological disorders (CHANEY, 1988), as well as probably causing comb and hive product contamination.

2. No secondary infections by foulbroods, chalk broods, etc., can take place because infected brood will be destroyed by the bee's own natural communal defense system.

3. The size of the worker bee returns to normal and again fits the natural flora of the region. This is important because the ratio of worker size honeybees to drone size bees is 20%, a four to five ratio of body size, that remains constant no matter what size the worker is and by returning the worker bee to normalcy, you change the size of the thorax of all bees in the colony, including the drones. The automatic downsizing of drone dimensions by the downsizing of worker bees is extremely important for fighting Varroa jacobsoni infestations. This is important because drones are also periodically thrown out of hives after each honey gathering season. We believe that this downsizing of honeybees aids in reducing the parasitic mite population in important ways:

a. The size of the honeybee is correlated with the capacity of the cell. Small cell, small bee; big cell, big bee (BAUDOUX, 1933). The size remains the same during the whole of the bee's life in perfect ratio one caste to each other. Since the only place Acarapis woodi mites can get into honeybees is through the first thoracic spiracle (EICKWORT, 1988), cell size is an important artificial mutant that can be rectified by beekeepers through use of natural sized brood comb foundations. Once placed onto natural sized brood combs the bee's thorax size is reduced, and Acarapis mites have lost a very valuable avenue of entry for hive destruction.

b. In Brazil, cell sizes for Africanized and domestic (European) honeybees when measured averaged 4.5 to 4.8 and 5.0 to 5.1 mm per cell, respectively (MESSAGE and GONCALVES, 1983). They further reported that Varroa infestation rates were 4.8% and 11.5% respectively. CAMAZINE (1988) calculated female Varroa replacement rates for Africanized and domestic (European) honeybees at 1.2 and 1.8 with drones present and 0.8 and 1.5 without drones, respectively. (A female Varroa replacement rate of less than 1.0 indicates that the mite population is declining while a 1.0 rate is indicative of zero population growth.) Keeping this in mind, it makes perfect sense to downsize artificially enlarged brood combs to take advantage of the 0.8 population replacement of Varroa jacobsoni when drones are seasonally ejected by colonies at the end of each honey gathering season. It also makes perfect sense to cull drone combs to less than 10% of all combs in a hive to keep Varroa populations down to a minimum. Thus it may be possible to suppress Varroa populations in domestic colonies by using small strains of bees with shorter development times reared in smaller cells (ERICKSON et al, 1990). Both these points appear now proven and have been incorporated into a biological manipulative treatment program for long-term control of parasitic mites by 1) queen rearing techniques (DEGRANDI-HOFFMAN et al, 1989), and 2) biological field manipulative techniques (LUSBY and LUSBY, 1992).

c. Downsizing also reduces basic food stimuli attractiveness for mites. It has been documented by KULZHINSKAYA in 1956 that worker larvae in enlarged oversized cells received 21% more food and 21.4% more protein than worker larvae reared in normal sized cells. He also found that the weight of larvae increased by 12.4% and that of adults reared in oversized cells by 10.4%. Since it is common knowledge that mites prefer drone cells, in the case of Varroa jacobsoni, over worker cells and Wolfgang RITTER (1988) stated that "Varroa cannot reproduce in the worker brood of Apis cerana, according to RITTER et al, 1980; KOENIGER et al, 1981 confirmed this and additionally found Varroa jacobsoni off-spring only in drone brood", then logic should dictate that the additional food and protein in enlarged oversized cells does indeed act as a mite attractant.

HANEL (1983) points out that one of the reasons for such differential reproductive behavior of A. cerana bees could be due to their juvenile hormone level. Varroa takes in various amounts of juvenile hormone III during its primary intake of hemolymph when feeding. This induces oviposition in the mite. In the first 60 hours, the drone larvae of A. cerana and A. mellifera contain more than 5 ug/ml JH in their hemolymph. Worker larvae of A. mellifera contain 3-7 ug/ml and, those of A. cerana contain only 1 ug/ml. The level of juvenile hormone in worker larvae of A. cerana is apparently not sufficient to induce oviposition in the mite. This has proved to be a selective advantage to the bee during the course of its host and parasitic evolution. Only in this manner does the parasite prevent death of its host and thus its own death. F. RUTTNER in his paper "Characteristics and variability of Apis Cerana" points out that "Contrary to the customary assumption, A. cerana is not generally a small bee when compared with A. mellifera. This frequently-held opinion holds true only when A. cerana is compared with European A. mellifera". We believe that this is a comparison of a feral sized naturally occurring type of honeybee to an artificialized over-sized domesticated European sized honeybee that has received more food and protein, thus more juvenile hormone by being reared on artificial combs. Therefore, downsizing would have the impact of reducing juvenile hormone levels, food and protein contents of the larvae jelly, all of which are mite attractants in oversized cells.

d. Downsizing also compacts the brood nest by density and our observations by inserted temperature probe, show that it raises the brood nest temperature, which we believe helps to speed up the gestation cycle of the brood. Combine with being able to select for faster developing queens (DEGRANDI-HOFFMAN et al, 1989) and it becomes possible to breed for bees with shorter development times as in aid in overcoming Varroa. Remember in the end, surgical removal of stress by beekeepers is always possible if the colonies own defense system proves to have been so debilitated as to be incapable of returning to normalcy. If surgery by beekeepers is necessary, a healthy honeybee on a proper nutrient diet will better generate strong recuperative powers once causitory brood combs have been removed and replaced.

REFERENCES

DE JONG, D.; R.A. MORSE; G.C. EICKWORT (1982) – Ann. Rev. Entomol. 27 pp. 229-252

DELFINADO-BAKER M.; E.W. BAKER (1982) –Internat J. Acarol., 8 pp. 211-226

GROUT, R.A. (1931) – A biometrical study of the influence of size of brood cell upon the size and variability of the honeybee (Apis mellifera L.). M.S. Thesis, Iowa State College

CHESHIRE (1888) – Bees and beekeeping, pp. 317-318

CHANEY, W.E. (1988) – The effect of synthetic pyrethroid insecticides on honey bees in Indiana; laboratory studies and a survey of beekeepers and pesticide applicators. PHD Thesis, Purdue University

BAUDOUX, U. (1933) – The influence of cell size. Bee World, Vol XIV, No.4, pp. 37-41

MESSAGE, D.; L.S. GONCALVES (1983) – The effect of the size of honey bee cells on the rate of infestation by Varroa jacobsoni. 29 International Congress of Apiculture, pp. 250; Apiacta 1984, pp.62

GEORGANDAS, D. (1968) – Natural comb of Greek bees and comb foundation. American Bee Journal, Jan 1968, pp. 14-15

KULZHINSKAYA, K.P. (1956) – Apicultural Abstracts, 37, pp. 177

ERICKSON, LUSBY, HOFFMAN, LUSBY (1990) – On the size of cells. Gleanings in Bee Culture, February 1990, pp. 98-101, Part 1 and March 1990 pp. 173-174

DEGRANDI-HOFFMAN, G.; D.A. LUSBY; E.H. ERICKSON Jr.; E.W. LUSBY (1989) – Managing colony genetics by grafting and selecting for queens with shorter development times. American Bee Journal, Vol 129 (II) 717-719

CAMAZINE, S. (1988) – Factors affecting the severity of Varroa jacobsoni infestations on European and Africanized honeybees. In Africanized Honey Bees and Bee Mites, Chapter 59, pp. 444-451

EICKWORT, G.C. (1988) – The origins of mites associated with honeybees. In Africanized Honey Bees and Bee Mites, Chapter 40, pp. 332-333

RITTER, W. (1988) – Varroa jacobsoni in Europe, the tropics, and subtropics. In Africanized Honey Bees and Bee Mites, Chapter 42, pp. 349-351

HANEL, H. (1983) – Apidologie, 14, pp. 137-142

KOENIGER, N.; G. KOENIGER; H.P. WIJAYAGUNASEKARAN (1981) – Apidologie 12(1), pp. 37-40

RITTER, W.; T. SAKAI; K. TAKEUCHI (1980) – Apimondia Symposium, Bad Homburg, pp, 69-71

LUSBY, D.; E. LUSBY (1992) – Suggested biological management program for control of parasitic mites. (Unpublished)

RUTTNER, F. – Characteristics and variability of Apis cerana (Fabr.), pp. 130-133

BETTS, A.D. (1932) – The influence of cell size, Bee World, Jan 1934, pp. 2-5

SWAMMERDAM – Bee World (1937) pp.43

ROOT, A.I. (1978) – The ABC and XYZ of bee culture, A.I. Root Company (publs.) Medina, Ohio

RAHMAN, K.A.; S. SINGH (1947) Bee World September 1947

Dee A. LUSBY

Edward W. LUSBY

"Housel Positioning" – How I View Its Importance To Beekeeping!

By Dee A. Lusby, 21-22 Sep 02

Just a few weeks before this meeting, in discussion with Michael Housel, of Orlando, Florida, I received information concerning proper positioning of wild feral combs built by honeybees he had been monitoring and observing in his local area hanging on limbs of trees.

Intrigued by, and recognizing the value of the information concerning the positioning of the wild feral combs, my husband and I immediately started incorporating the information into our field management program, by resequencing close to 35,000 frames in our colonies, to match their positioning.

So just what is this proper positioning of feral combs Michael Housel told me about?

It concerns understanding the "Y" formation of the pyramids formed at the base of the wild combs, and in manufactured beeswax foundation at the base of the cell imprints, that beekeepers place into their colonies, to help domesticated honeybees replicate wild feral combs.

Foundation used by beekeepers is basic to field management. It is used to stimulate domesticated honey bees to build both brood and honey combs, using beeswax secreted from glands on the workerbee's body. It was originally copied from wild combs in the 1800s.

The "Y" formation has been there since the beginning in the making of beeswax foundations. It's in understanding it, and it's proper positioning and placement that Michael Housel has recognized, and we just resequenced our colonies to duplicate, that I hope others here today listening and learning about it, will want to duplicate also, in their own beekeeping operations.

If you copy something exactly to use, which is the purpose of our foundations, and then you don't use it as originally designed and placed by the bees themselves, how can beekeepers blame bees for building and doing things wrong within a beehive? For then in actually, it's man's improper alignments and positioning of manufactured foundations, contrary to original natural design, that could then be causing much of today's bee's internal problems relative to working and drawing combs.

How can scientists do research even, with improper positioning of foundations, not relative to actual positions in the wild? Is science, science, if based upon an artificial world of enlargeness, and improperly positioned combs at the same time, that matches nothing in a real world? How do you know if the research you are doing is good or bad for what it is supposed to relate to, if the combs in the domesticated colonies being reviewed do not match the positioning of wild combs?

The "Y" formation

A "Y" is formed where lozenge-shaped rhombic plates come together to form a Y impression at the bottoms of cells on beeswax foundation. The formation of the "Y" is also seen in wild combs at their cell bases.

There is a right and left side to each foundation and comb when viewed, whether in a man-made colony, or hanging down from a limb.

The right and left sides for facing foundation and drawn combs in a beekeepers hive are determined by the top or bottom positioning of the "Y" formation.

This changes by either being right or left of an imaginary center line in domesticated hives. In the wild there is one special center comb hanging down from a limb. In our man-made hives which we call colonies this does not occur, and so an imaginary line must be drawn and used, for positioning right or left of center, and up or down, of the "Y" formation.

Beekeepers can easily turn a wild comb and see this. Likewise beekeepers can turn a man-made frame or piece of foundation and see this formation also.

When wild combs are cut down, should not they be positioned in alignment like those obtained from the wild colony, to aid the now domesticated bees placed into a man-made hive, to continue to grow and properly expand?

If you have not seen or noticed this before, take a sheet of foundation and put it in front of you on a flat spot to look at.

Then with the rectangle sheet of foundation with long-ways on top and bottom, and short ways on sides, carefully look at it.

There are two ways to rotate a sheet or comb (in frame) when looking at it to observe the "Y" formed at the bottom of the cells.

Most beekeepers are taught early on to carefully rotate a sheet or frame with bees, from top to bottom (vertically up and down), with a twist of the fingers and wrist, so as to disturb the bees on the comb as little as possible, to observe the broodnest for conditions relative to disease, mites, egg laying, and larva size, applicable for grafting.

When beekeepers rotate a frame this way, no change to the eye takes place, though you rotate to see both inside the top and bottom of the cells. Beekeepers are taught this motion to observe bees for various fouls, and mite fecal for evidence of Varroa present.

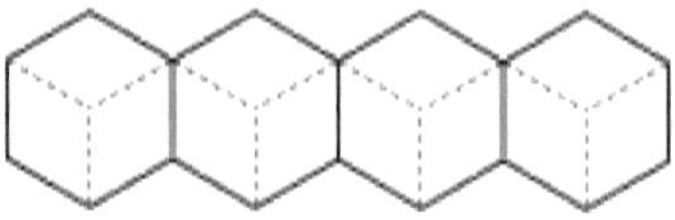

Next, with the sheet of foundation in front of you, turn the sheet *not vertically, but instead from left to right horizontally!*

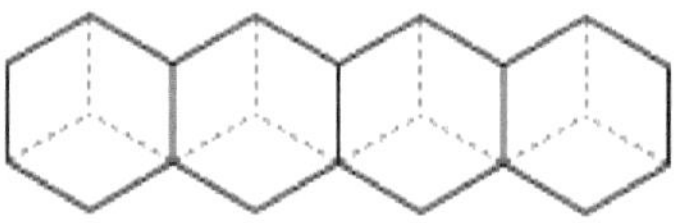

Now, when you look at the cell bottoms with the "Y" formation it should change from top to bottom, every time you turn the sheet over.

Explaining "Housel Positioning"

In the wild, there is one center frame that is first drawn when honey bees swarm onto a limb. In spring or following normal swarming the first comb built is worker (exception being more towards fall, following the summer solstice and longest day, when bees swarming can sometimes want to build drone/honey comb first to obtain stores for winter and then once a certain amount is drawn and realized, they then start workercombs).

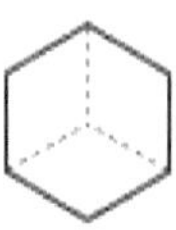

Now this comb is built with the "Y" inverted and upside down on both sides of the comb. So I now type "^I^" to show the inverted "Y" on both sides of the comb. There is only one of these combs made.

For hives that normally swarm, wanting worker larva for continuation of species, bees need optimum cells for workerbrood immediately, especially in areas of short flows. Hence, this specially drawn first comb.

This starts the wild nest with a center comb expressly designed for maximum production of worker bees, that are needed immediately for continued rearing of new brood and collection of stores, as the field force dies off.

Each comb then, on each side of the center comb follows position, for continued maximum rearing of brood, and then collection of stores of pollen and honey, as comb building progresses and expands the nest.

From here, the "Y" formation stays inverted first, facing center with the "^" down. This continues formation of a slanting ledge the larva rest on, allowing for maximum field bees to be used for gathering stores of nectar needed for comb production, with lesser numbers of nurse bees required.

I now type "^" to show the inverted "Y" for side facing center comb (or center of imaginary line in center of man-made colony) with slanted ledge.

On the other side of the comb the "Y" formation faces up, and helps to form a slanted roof, to help once the bees manage to build enough comb, to protect larva and stores gathered from sun, rain, etc. I now type "Y" to show the "Y" right side up with roof, for side facing away from the center comb.

What beekeepers end up with then, is all foundation or combs in colony with the "^" down formation facing towards center, and all foundation or combs in colony with "Y" up formation, facing towards the sides of the boxes/supers, away from an imaginary center line. I now type ^IY to show this.

Now, the combs in the center on frames are the smallest and are worker cells, and only at the periphery of the worker cell broodnest change into drone cells.

This can be done two ways.

On either side of a good drawn workercomb you can have periphery drone cells, including the bottoms.

Once an average of four or so worker combs are drawn on each side of the center worker comb, beekeepers will find the next combs built a combination of drone/honey combs. So what you are looking at in broodboxes/supers then is:

YI^,YI^,YI^,YI^,YI^,^IY,^IY,^IY,^IY,^IY

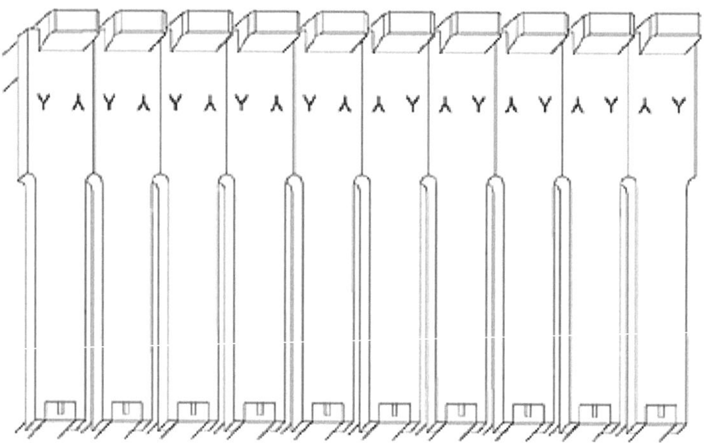

What you are looking at in wild combs hanging is:

YI^,YI^,YI^,YI^,^I^,^IY,^IY,^IY,^IY

This transition to larger starts slow but gets more pronounced the closer to the outside of the broodnest you go across the first workercell combs built from the center main comb or imaginary line.

On good flows, beyond this, especially in wild colonies, you can get combs drawn with cells even bigger then drone cells, but rarely seen except in exceptional years.

Now, the placement of these bigger combs/dronecombs on the outside periphery, is to protect the worker combs from damage. Animals attacking a feral hanging nest will pull off the outside larger combs for food and many times go on after eating their fill. Wind if strong, along with rain will knock or blow/rip down these outside combs. They are weaker combs with less wax cell walls, and thus more easily tear loose. But, they serve to protect the inside combs, by their side alignment and positioning, from both the elements and animals. This then leaves the smaller worker combs safe, which can and often do, contain honey besides pollen, as the active year progresses and brooding cuts back, and are the strongest combs with maximum wax for strength.

The positioning of the combs in man's domestic hives should follow the above for drone/honey cell positioning relative to worker/pollen/honey cell positioning.

All good drawn-out worker combs should be placed to center, then frames/combs with peripheries of drone cells (not more then 10% kept), then lastly badly drawn-out transition combs. This way, beekeepers end up with 4 good worker combs in the center of broodboxes, and the three on each side for combinations of combs containing worker/drone, pollen/honey storage, and only the immediate outside frame position, for absolute hodge-podged transition cull comb, until the beekeeper can work it up and out during routine field work, for taking back to the honey house for extracting and recycling by melting down.

Importance of "Housel Positioning to Field Beekeeping Management

As I said earlier, intrigued by, and recognizing the value of the "Housel Positioning" relative to wild feral combs, we have resequenced close to 35,000 frames in our colonies and will do more as we continue to work our bees. By resequencing our combs to match wild comb positioning, final internal colony problems relative to our honeybees drawing-out of foundation and how the bees work the combs, appear to be lessening or stopping altogether. Much stress seems to have been eliminated.

My husband and I manage our hives using 4.9mm small cell beeswax foundation, with unlimited broodnest management of 2-3 deep boxes, with 1-2 deep supers for honey production, with an overall average colony size of 4-5 deeps. We see no problems in using 4.9mm foundation in conjunction with "Housel Positioning", as all this does, is copy wild naturally small honeybee comb positioning found hanging from a limb on a tree. This way, we end up with a field management program that is biologically harmonious to wild honeybees, in both comb size and positioning, but under man's control for production.

At the same time, by not having to use various treatments of chemicals, drugs, essential oils, FGMO and acids for parasitic mite control, accompanying secondary diseases and miscellaneous bee pests, we also gain clean products of the hive to sell, and bees harmonious with Nature again that live.

Final internal colony problems lessening or stopped by proper "Housel Positioning" following resequencing of combs have been:

1. Queens not laying in inserted drawn combs placed into the broodnest. Many times beekeepers, as a part of field management throughout the active beekeeping year, insert drawn combs into the broodnest for their queens to lay in, as a means of producing more honeybees for production of products they sell.

These combs can be dry combs or extracted wet combs. But on subsequent hive checks, that can be days and even weeks later, the beekeeper comes back to find the comb not used, but the combs on either side being utilized and laid in. Loss to buildup of workerbees, necessary for production, is then the loss of brood that could have been generated, for each 21 day brood cycle of workerbees, not laid by the queen.

2. Excessively bulged/drawn-out honey combs with the next frame either burred or hardly drawn. It is not uncommon for beekeepers to find bulged/drawn-out honey combs with newly drawn-out comb 2-3 inches thick in supers with new foundations, while the adjoining new frame of foundation next to it is hardly touched or is burred in pattern.

Transporting such honey combs home can be trying as bumps are driven over, that cause the frames to knock and rub together, causing the honey to run out the bottoms of stacks of supers, before reaching the honey house and creating messes that then need to be cleaned up.

Through observation, we now know that the foundation/frame positioning in the super was wrong, and that the frame that was either burred or hardly touched, next to the bulged overdrawn-out honeycomb, was backwards in position to other combs in the honey super relative to positioning of wild combs.

3. Bees refuse to move up into next higher box/super of either drawn frames or new foundation. While this does not happen too often on good honey flows, on average to poor honey flows this can be a problem with bees showing reluctance to expand up into the next higher box/super, to either fill empty combs there, or draw-out foundation. This found happening in a few hives can lessen workerbrood raised and honey stores gathered. Once frames are repositioned according to the way the "Y" formation is facing, the bees move up and continue to expand and work.

4. Odd frames of foundation not drawn and/or bees sidewinding. From time to time beekeepers place a new frame of foundation into a broodbox or super of drawn combs only to have their bees ignore it. Or they may have 2-3 frames of either new foundation or drawn empty combs or combination of these, the bees seem to ignore in a broodbox/super. Through observation, we now know the "Y" positioning of the new frame or frames was probably faced wrong, causing the bees to go around the improper sequencing and positioning relative to wild combs.

5. Burred foundation or overlaid foundation. From time to time beekeepers find frames of new foundation that has been overlaid with sections of either bigger or smaller combs drawn out. We have seen bigger drone/honey combs overlaid on frames positioned with the "Y" formation inserted backwards. We have also seen worker/pollen combs overlaid on frames positioned with the "Y" formation inserted backwards. When looking at the overlaid comb, interesting to note, is the fact that the bees in overlaying the pattern, seem to be reworking the facing of the "Y" formation.

Many places of overlay face the same way as the foundation is placed, yet in other areas on the overlaid face, the bees it seems, are actually trying to reverse it's positioning to that of the foundation which was improperly positioned. Each burr overlaid formation tells it's own little story of the bees working it, trying to adapt the "Y" formation. This leads to much transition comb if these frames are allowed to be continued. Our combs are more evenly smaller now, because our bees are more uniformly maintained and bred, so we mainly see our bees trying to determine which way to face the "Y" formation now. Various sizes of differing transitional burr combs are not so prevalent with cells sizes strikingly different to the eyes.

6. Transitional combs containing various cell sizes are built. Similar to overlaid combs built upon new sheets of foundation, beekeepers can find transitional combs being built by honeybees containing numerous cell sizes. These cells are normally built by colonies upon foundations with "Y" formations positioned wrong and can range up to .2mm to .3mm bigger on average.

7. Queens are suddenly raised at wrong times of the active year causing swarming problems. Beekeepers in adding empty drawn combs or freshly extracted wet combs into the broodnest sometimes go back and find hives requeening at odd times of the active year. Beekeepers can also add odd frames of new foundation into the broodnest to be drawn-out and end up with a few queens being raised along with worker larva. They can also have changed nothing from the previous year in the broodnest, but all of a sudden requeening starts even though they know the queen they have is young and this should not be happening. This can be especially frustrating when a honey flow is coming on or in progress, or they actively follow breeding programs trying to requeen their colonies yearly to avoid this. Why would colonies want to requeen more than once throughout the active beekeeping year?

From what we have seen in our colonies, it is a comb positioning problem with the frames in backwards. With the comb positioned backwards and thus out of alignment with other combs in proper sequence, beekeepers can trigger spontaneous requeening in colonies by failing to note which way the "Y" formation is facing. Beekeepers must take note and remember one way the formation of the "Y" faces is inverted and down "^", creating a ledge for larva to lay upon that honeybees use for fast build-up following swarming, etc.

On the other side of the comb and/or foundation, the "Y" formation faces up and helps to form a slanted roof, to protect larva and stores gathered from sun, rain, etc. But, the slanted roof of the "Y" formation facing up has another purpose in a colony! For it is only on the side where the "Y" formation faces up, and helps to form this slanted roof, that honeybees raise "queen cells" that face downward for requeening.

Therefore, beekeepers not positioning foundation and drawn combs properly can spontaneously trigger superceding, and thus swarming in their colonies. With hives under stress already from disease, pests (beetles), and predators (mites), besides often on programs of various treatments for same, improper positioning then takes less effort to trigger problems, one of which can be spontaneous requeening.

Whose fault is it then! The bees or the beekeepers, for not following proper "Housel Positioning" for sequencing of managed colony combs, relative to proper positioning of wild combs?

One last note, in going back to colonies that were resequenced with proper "Housel Positioning" of frames, the disposition of the bees was noticed to be gentler then before.

Appendix B: Articles about the Lusbys

Lusbys Receive W.A.S. Memorial Award

ABJ – December, 1997

The Western Apicultural Society is proud to announce the selection of Ed and Dee Lusby, Tucson, AZ, as recipients of the A. I. Root Roy Thurber Memorial Award for 1997. The award, which has been presented intermittently since 1986, specifically recognizes 'Inventiveness in Beekeeping'.

Ed and Dee are full time commercial beekeepers who rely on native desert plants rather than cultivated crops for their honey and pollen production. Ed is a fourth generation beekeeper. His great grandmother, Ada Duncan, of Newsite, Alabama began keeping bees in the mid 1800's. After marriage to his great grand-father, Joseph, they continued to keep bees on the frontier plains near Amarillo, Texas and later on the Sandstone Creek north of Elk City, Oklahoma Territory. To assure a family livelihood following a severe decline in the cattle business around 1888, the family turned more to beekeeping. In 1927, his grandfather Bill moved to Tucson where the family business has thrived ever since. Dee grew up in Montrose in Westchester County, New York, but spent many summers on her uncles upstate farm. Her uncle kept a few colonies of bees as a hobby and it was here that Dee was first exposed to the world of bees. After a stint in the United States Air Force, she married Ed in 1984 and began beekeeping in earnest.

Dee and Ed work side by side in all phases of their operation. In addition to the conventional activities of beekeepers such as maintenance of equipment, colony management and the harvesting of honey and pollen, they mill their own woodenware and wax foundation, select and maintain a stock of slightly smaller bees highly adapted to their area, produce their own queens, and market a portion of the bee products they harvest. Their non-chemical 'back to basics' approach to beekeeping leads them to spend much of their spare time in libraries where they search for obscure bits of information which, when assembled in logical order, yield insights into old problems such as bee kills due to the use of pesticides, and new problems like parasitic mites. Such has been their pursuit of an understanding of the importance of comb cell diameter, an issue emanating out of their bee breeding activities and search for non-chemical methods of resolving disease and mite problems.

The Lusbys found that comb cell diameter differs among the various sources of foundation manufactured in the United States and around the world. Following publication of this discovery in 1990, they undertook an all out effort to resolve the question of optimal natural cell diameter and its potential impact on colony vigor. Having identified, to their own satisfaction, optimal cell diameter for their geographic area (Southern Arizona), they have nearly completed converting their entire operation to a natural system incorporating their concept of smaller cells. They have widely reported to beekeepers that their use of optimal natural cell diameter has significantly reduced disease and mite infestation in their colonies while simultaneously increasing brood viability and colony productivity. Convinced, a number of beekeepers have embraced the Lusby's management strategies. Ed and Dee have now turned their attention to developing a world map that will identify, for beekeepers, optimal natural cell diameter by latitude. Publication of this map will be forthcoming.

The Lusbys have worked tirelessly over many years to mitigate the impact of agricultural pesticides on honey bee colonies, even though their own apiaries are seldom if ever impacted by these chemicals. They have established an informal bee/pesticide information network which beekeepers, particularly those in southern Arizona, often access. They have taken the time to learn the rules and regulations governing the use of pesticides, to assure that state and federal authorities are adequately informed regarding the hazards of pesticides to honey bees, and to encourage these authorities to ascertain when pesticides are improperly applied and take appropriate corrective action.

Dee and Ed hold or have held elective offices in the Southern Arizona Beekeepers Association and the Arizona Beekeepers Association. They have contributed to or written several publications. They share a passion for improving the business of beekeeping, not only for themselves, but for the welfare of fellow beekeepers around the world. To this end they have selflessly contributed their time, including the hundreds of hours they have spent in libraries and on the phone gathering facts, as well as their own financial resources. They are known to generously share their time, ideas, and knowledge with fellow beekeepers around the world. Beekeepers in Third World countries are particularly interested in their 'do-it-your-self' approach to producing one's own hives, frames and foundation. Least well known is the quiet generosity that Ed and Dee have shown in assisting beekeepers less fortunate than themselves, some of whom they have been known to help financially on occasion.

Ed and Dee are to be applauded for their selfless dedication to improving the art and science of beekeeping for beekeepers everywhere. Their contributions reflect the true spirit of the A. I. Root Roy Thurber Memorial Award.

Apiculture News- July/August 1997

Dr. Eric Erickson was our final presenter at the meeting. He was a little disappointed that his AHBs were so tame at the time of our visit. So he showed us a video of what the same colonies were like earlier in the season. I'm glad they were "off" when we were there.

Eric reported that local bees, living in combs with smaller cell sizes, have survived the onslaught of Varroa and have less than 10% infestation rates. Tracheal mites are very low, too. They are hoping to breed from those queens for mite resistance. [Editor's Note: The AHBs in the USDA yard were living in previously drawn EHB combs. Their mite level in uncapped drone brood was about 50%.]

If you have knowledge of one or more colonies of bees that seems to have survived for years without treatment for either mite, Dr. Erickson would like some samples of the workers from those colonies.

The WAS Conference wrapped up with the annual awards banquet. Dee and Ed Lusby, Tucson area beekeepers, received the Thurber Award for Inventiveness in Beekeeping. The Lusby's have spent many years and personal resources on breeding a bee specific for desert beekeeping. They also have experimented with reduced sized comb foundation and cell sizes. When adequately small, the cell size seems to positively affect disease and mite control in a colony.

Sincerely,Eric Mussen

Beekeeping on the Fringe, with Ed & Dee Lusby

Bee Culture – January, 1998 by Kim Flottum

Dee and Ed Lusby are fringe people. They live in Tucson, Arizona, but they keep bees in places most people wouldn't; their management style works, but requires a lot of labor and attention; and they've taken on a biological puzzle that may or may not provide answers to a bunch of honey bee problems.

The Lusbys run a small commercial operation with about 250 colonies at the moment. . . . That rather low number is the result of strong culling, colony loss from mites, and a several-year significant drought.

"We used to run over 600 colonies, and if the weather breaks and our new management scheme (see below) works the way we think it will, we'll be back to over 600 or 700 by the Fall of '98," Dee told me while we were bouncing down a washboard road heading to an isolated beeyard in the Kit Peak area outside Tucson. We'd gone down back roads, then back trails, through gates locked to keep out people not belonging on the King Ranch land we were on, and finally on wagon tracks and over riverbeds that only coyotes, lizards and honey bees could find.The Lusbys produce honey, raise queens, and make splits in this by turns incredibly hostile and wonderfully abundant land. The extremes make or break anyone who takes on the natural order. The Lusbys, as stated above, live on this fringe.

Keeping bees in the Desert Southwest requires using a calendar different from the one followed by most beekeepers in the United States.

Early January is the beginning of the active time. By the end of the month, brooding up is in full swing, and comb whitening is taking place by the middle of February. When this starts, supers are added – the second or third brood chamber. For colonies with three supers, when the third is 50% full of brood, a piggyback split is made. So by the end of March, these strong colonies have been split, and by April, they have been requeened and separated to their own stand.

After April, all colonies are supered as needed. Beeyards are visited about once every three weeks, and supers are added all Summer. Colonies are evened out as needed during the Summer.

About the first of May, full honey supers are pulled to extract, but the flow is over by about mid-July. This dearth lasts until about mid-August in a good year, or early October in a bad year.

Supering starts again with the honey flow and continues until about Thanksgiving – supering, harvesting, and supering again. When the Fall flow is over, Fall divides are made, piggyback style, from the strongest colonies, and queens are raised from mid-October to about the first of December. From then until January, colonies are only maintained, and then the process starts all over.

Honey production in the area is modest by some standards, with, on average, 10 colonies producing a barrel of honey in a year. Drought years can mean that over 20 colonies are needed to produce that 600 plus pounds.

A piggyback split is made by separating the top box of a three-story colony (or a very strong two-story colony) with an excluder or division screen. Then the top is requeened, or the old queen is moved up and the bottom is requeened. Since a growing colony 'tends' to have a queen moving up, the queen is often in the top box – but not always. These splits are made about the middle of October and the middle of March, depending on the season, colony strength and honey flows.

Unlike the Hines operation (see article on page 33), Africanized Honey Bees seem not to have caused the management problems common to the region. Opening colonies in a couple of yards showed no apparent excessive defensiveness. Most were worked without veils. The Lusbys had no explanation for this, even considering the fact that the first yard we visited was alongside a mountain range that feral swarms followed north from Mexico into Arizona.

The Lusbys raise their own queens, and follow a fairly typical technique, though their selection criteria are unique to their operation and needs. Basically, they select breeder queens for honey production, no brace comb production, gentleness and 'health.'

Once a breeder is selected, frames of day-old larvae are pulled and grafted into Kelley wax cell cups. "We use Kelleys because they tend to be smaller, about 5/16-inch rather than the larger 3/8-inch," Dee said. "This ties in with our bee-size (see box) operation," she added.

Cups are primed with royal jelly before grafting because of the arid climate, and grafts are done every three days. Bars containing 60 to 120 cells are moved into cell builders at the home yard right in town, then moved after four days into one of three incubators, which are in constant use during the season, to finish.

Finished queens emerge, and the virgins are captured in a bottle, the cell cup acting as the bottle top. Honey is added for food, and bottles are changed when soiled.

Queen selection begins by appearance only, even before introduction into a colony. In the first 12 hours after emergence, her color and banding appear. Size, too, is beginning to be obvious. The Lusbys select immediately for small, black queens, with long red legs and wings longer than the abdomen. "Average-sized queens tend to be more variable than the ones we select," Ed pointed out, "and our experience has shown that those we choose at first tend to work best."

Virgin queens are placed in mating nucs and open-mate with local drones – source pretty much unknown – which, it seems, has worked so far relative to progeny. But Tucson is pretty much Africanized, so this may change.

Once mated, new-queen nucs are evaluated for brood patterns before being introduced to a new colony. Half-moon patterns are not acceptable since the Lusbys want a wall-to-wall pattern, which produces more brood in a colony. "We need fast, fast buildup to react to a rapidly changing environment," Dee says, "and the more brood the faster, the better."

Once a queen is selected, she's introduced to a colony. Introduction works like this: A colony, recently made queenless, or a split without a queen, is smoked at the bottom until smoke comes out the top. The queen is then direct-released into the center of the colony. Supers are lifted, and she's popped out of her cage. Or, she's released right on top. This technique results in an 80% to 90% take, which is enough, so far, for the operation.

Until mites came on the scene, this management program worked for the Lusbys. But mites, and the problems associated with them, changed all that. And, even though the techniques haven't changed, the fundamentals of traditional management have been questioned by Dee and Ed, and found wanting.

The basic change in their philosophy has been the assumption made on the cell size of traditionally sold foundation. That, plus their 'natural' treatment extender patties (propolis, sugar and vegetable shortening), have set them on a path of their own choosing. Varroa, some disease and tracheal mites have taken a heavy toll on their stock. But those left are thriving, and queens selected from these are thriving, too.

Traditional management schedules and non-traditional equipment seem to be making a difference in this operation. No chemicals what-soever and selective breeding, along with some good luck with the weather, promise a bright future for the Lusbys and their on-the-fringe operation in the Desert Southwest.

NATURAL COMB

The Lusbys have, for several years, been investigating the ramifications of the cell size honey bees use. Their extensive research has turned up some interesting, and intriguing information.

Historically, man-made foundation started the same size as the size bees naturally produced. However, the cell size bees naturally produce is to some degree dependent on where in the world they are. Like many animals, those closer to the equator tend to be smaller than those closer to the poles. That is, honey bees in the southern U.S. naturally build cells a tiny bit smaller than bees in

Canada. This discovery has complicated what is 'natural,' but not the fact that natural is still, well, natural.

Years ago beekeepers believed that larger bees would be better able to take advantage of flowers with the nectar deep within, normally out of reach of the bees' tongues. Long tongues were selected for, and some advantages were gained. However, larger bees were deemed the answer to even longer tongues, and to produce larger bees foundation with slightly larger cell base size, hence slightly larger (eventual) cells. It was believed bigger was better. Well, maybe, maybe not.

The Lusby's theorized that this larger cell size, and larger bee, produced an environmental stress on both individual bees, and the entire colony. Generally, colonies handle this subtle but persistent pressure with indiscernible outward signs. It is, however, difficult to measure because essentially all comb produced now is artificially large, at least to some degree.

Measuring cells produced by feral colonies in their part of Arizona, coupled with the results of their research led the Lusbys to initially produce foundation with a smaller-sized cell base on an experimental basis. Their first attempt was a cell base with a parallel-side-to-parallel-side measurement of 5.0 mm. (see diagram).

Since most foundation produced now is in the 5.22 mm to 5.55 mm range, reducing cell base size to 5.0, (a 4.3 – 10% reduction) seemed significant. But after a few trials the differences seemed minimal. Their time spent, however, continued to uncover more information supporting their belief that 'natural' cell size was better for bee stress reduction.

The 5.0 mm cell size did show promise, however. The Lusbys noted some reduction in parasite infestation and less incidence of disease. But not enough to be commercially economical, and colony losses continued.

It should be noted here that along with the inclusion of smaller cell-sized foundation in their management scheme, the Lusbys discontinued the use of all drugs, medications and acaricides, except a propolis, sugar and vegetable shortening combination in a patty. The early results were predictable – colony losses mounted, but not as rapidly as other, untreated colonies. Something was going on.

Further research indicated that, at their latitude, a cell size of 5.0 mm may have been 0.1 mm too large, and they found a cell size of 4.9 mm may be better. Precise measurements of feral comb supported the 4.9 mm size, so they began to search for a foundation mill to produce this size cell base. This wasn't an easy task. Not only were current manufacturers not using mills that small, most were reluctant to make a switch without some hard evidence the cost would be worthwhile.

One did, however. Tom Industries, in El Cajon, CA agreed to make a few small-sized mills, for a price, to see if they worked.

Lusby's make their foundation the old fashioned way, one sheet at a time. They dip a board in melted wax, let the wax cool and peel it off. One dip is enough. Then they run this through the hand powered mill. Result – eight sheets to the pound.

So far they've found that colonies on their new natural comb seem to swarm less (There is more space for brood – about 1250 more cells in a two-story chamber than using Duragilt.) and have fewer mites.

The few-mites thing, along with less disease incidence, has been aided by continual selection for tolerant colonies. But the two have worked. Independent sampling by USDA researchers have confirmed that, indeed, fewer mites than normal are present in these small-cell colonies.

So far the Lusbys have changed over most of their colonies. Their early observations indicate faster build up, healthier colonies and more honey production. But these are early results.

They are passionate in their belief that this management scheme is the answer to the stresses of desert beekeeping, both mites and whatever diseases their bees encounter. Others have changed, too, in their belief, but for the most part the jury is still out. Two, or better three seasons of continued success, without medications, will tell.

But for now, the Lusbys are busy switching from their old, too-big combs to the new, just-right foundation.

Natural Size Foundation is the Best - ABJ 1996

ABJ, November, 1996 – Page 757-758

I'm a Belgian beekeeper and I read with interest the article "Natural suppression of honey bee tracheal mites in North Dakota: a five years study" (A.B.J. May 1996). I read with even more interest the answer by Mrs. Dee A. Lusby (Letters To The Editor, July 1996). I fully agree with the remarks and the questions raised in this letter. I was interested because it was in 1893, more than 100 years ago, that Baudoux stated that "Bigger would make more honey" (*Le Progres apicole*, June 1893). Since that day till this moment, there have been discussions and quarrels about these questions.

Baudoux, his disciples and his followers asserted that it was possible to enlarge the honey bees by at least 40% in the length of the tongue and the capacity of the honey stock by 60%. His statements were contradicted many times: Dr. Miller, USA; Dr. Goub, USA; Roy A. Grout, USA; Perret-Maisonneuve in his book "apiculture intensive", Zappi - Italy; Descourt - France; Baldensperger - France; Michailov - Russia, etc.

These researchers said that the enlargement was only 2%. The discussions were often vehement - and that is only about the possibility to enlarge the bees.

Another question is: if it is possible, is it also profitable? "Yes!", said Baudoux, "Bigger bees make more honey because the bees are healthier, stronger and live longer and to prove that, he compared too small (950) cells/dm2) with large bees (700 or 640 cells/dm2). It is clear that in this case the smaller bees are at a disadvantage – the more so as the bees brought on the small cells came from large cells and had to adapt.

As far as I know, Baudoux never compared the natural cell size and comb distance with his own invention.

I, myself, together with other beekeepers, often compared colonies on 750 cell foundation at a comb distance of 37 mm with colonies on natural built comb at a distance of 34 mm. The "natural" colonies were always stronger, developed faster and had less winter loss. As a result, they gave more honey (an average of 20% in my apiary of 30 beehives). The bees were also healthier: This past year there was an outbreak of chalkbrood, only the "natural colonies" had no trace. If foundation with the natural number of cells (±850) was available, I would fit all my hives with it.

To do the tests, I used small strips of foundation or specially designed top bars and after 10 years, I can tell you, it's not easy to make the bees build nice, regular combs. I don't think there is a method, something that works well one year, doesn't the next.

My opinion?

I think the natural size of brood cells and the natural distance of the combs is the most profitable for our honey bees. I do know there are small bees (*Apis cerana* and even *florea*) that are adapted to build small nests and to forage at a short distance. I also know *Apis dorsata* that build great colonies and many colonies together and forage at a great distance.

Our bees became what they are during an evolution period of millions of years and they are adapted to forage at the distance they do and to build the nest the way they do. In fact, they are still wild. I think we can't intervene in the life of the bees the way we did with other animals. We bred chickens that laid more eggs, pigs that produced more meat in a shorter time, cows that gave more milk. To do that we have to provide these animals with balanced diets, artificial light and warmth (not to mention hormones and other additives). We can't do that with our bees. We can't feed them sugar from which to make honey.

During their evolution period, it was not the colonies that swarmed most that survived, not the colonies that put every drop of honey that was gathered in the brood, but the colonies that gathered the most honey and that saved that honey carefully – that's our goal.

If the goal was a beautiful body, I suppose we could breed bees with blue eyes and red wings. If the goal was to cook them for dinner, we could breed fat, large bees as we did with the cows and the pigs. But, as long as honey and pollination are the goal, we can do no better than did nature by natural selection. The only thing we can do is to try and repair the damage caused by mismanagement and errors of the past (and at present).

Marcel Arnst

Essen, Belgium

Appendix C: Historical Data on the Influence of Cell Size

December, 1887 – Bee Culture

It is common to speak of combs as containing 25 worker-cells to the square inch, and 16 drone-cells. Ought we to speak so loosely? My attention was first called to it in Frank Cheshire's book, where he gives 28-13/16 as the number of worker-cells to the square inch, and 18-178/375 drone-cells. If the cells were square, 25 and 16 would be correct; but they are hexagons.

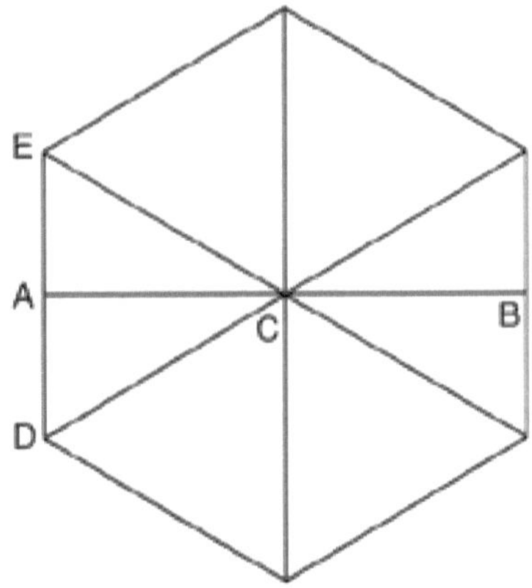

Diagram to show exactly the number of worker-cells per square inch.

Any one curious in the matter can easily solve it by algebra, or even by arithmetic. In the figure of a hexagon here given, having the diameter AB we have 6 equal equilateral triangles. Bisecting one of them gives us the right-angled triangle ACD, with the hypotenuse CD. If AD is 1, CD is 2, and AC is found by taking the square root of the difference of the squares of AD and CD. The squares of AD and CD are 1 and 4, their difference 3. The square root of 3 is 1.73205, the measure of AC. But in a worker-cell the diameter is 1/6 of an inch; and half the diameter, or the line AC, .1 of an inch. To reduce AD to the same scale, we find by proportion that 1.73205 is to .1 as 1 is to .057735, the true measure of AD in a worker-cell. To get the contents of the triangle CDE, multiply half the base, or .057735, into the altitude AC, .1, and we have .0057735. There are six such triangles in the hexagon, and 6 times .0057735 is .034641, the contents of a worker-cell represented in the decimal of a square inch. This is contained in one inch 28.8676 times, a trifling shade more than 28-13/15.

Counting 25 cells to the square inch gives us 3600 to the square foot, against 4157 by the true measurement, a difference of 557 in a square foot. Will it not be better, ordinarily, to say 29 worker-cells to the square inch?

C.C. Miller, Marengo, Ill.

Friend M., I am glad to see you have had proof that the quality of the honey is sometimes, if not always, greatly improved by being ripened in the hive; that you have also learned that this thoroughly ripened honey will sometimes, if not always, remain clear, without candying. I have, like yourself, seen honey, so poor as to be pronounced almost unfit for use, become beautiful honey in the course of time, by being simply ripened in the hive thoroughly. – Regarding the number of cells to the square inch, I have long been aware of the point you make; but if you measure several square inches of ordinary comb as you find it – that is, comb manufactured by the bees –

you will find it runs less than 25 cells to the square inch oftener than it does more. This matter was discussed a good deal years ago. Since most of the combs in our hive are now, however, made of cells built on foundation, the case will probably be different. You are basing your conclusions on the statement that five worker-cells side by side measure just an inch across. This is not true however, if I am correct, with little if any of the foundation we have in the market. Years ago we settled down on a size of the worker-cells, so that 24 equal 5 inches; and I believe that most manufactures of foundation-rolls have followed us in the matter.

Raising and Introduction of Queens - Cheshire 1888

Bees and Beekeeping – Cheshire, 1888

Page 176

The most puzzling of all variations remains to be noticed, for no observer has discovered even the key to the gauging of the dimensions of the cell by the wax worker. It cannot be put in evidence that the size of her body or head, or reach of her jaws, determines it, for, under certain conditions which are perfectly uniform, she discards the 1/5in. diameter, and starts constructing cells 1/4 of an inch between the parallel sides, and these are used for the storing of honey or the raising of drones, and so are commonly called drone cells. The statement, many times made, that twenty-five and sixteen of these respectively cover a square inch is erroneous, as the outline is not square, the correct numbers being as below:

	Diameter	Length	# on sq. in. one side of comb	# on sq. ft. one side of comb	# on sq. ft. both sides of comb
Drone	1/5 in.	15/32 in.	28 13/15	4157	8314
Worker	1/4 in.	9/16 in.	18 178/375	2660	5320

Pages 317-318

Creatures grow by transfusion of material in their living bodies, and the more solidity their tissues have, the more slowly does this transfusion occur. Some flesh-flies, in the earlier part of their larval state, will increase in weight two or three hundred times in twenty four hours - a rate of development absolutely forbidden, by physiological and chemical laws, to creatures of larger proportions; for, other things being equal, as the size increases the rate of development must decrease. The inconceivably minute monad, weighing a fraction of a billionth of a grain, by absorbing nutrition doubles its weight and divides every four minutes. If food abound, and the fluid surrounding the creature be free of enemies and not circumscribed, it, in the course of three or four hours may produce in its descendants an amount of living, moving material exceeding the weight of the largest elephant; while the latter animal, with its digestive and assimilative powers stimulated to the uttermost, could only, in the same time, add a few ounces to the weight of its body.

The economics of the question must not be overlooked. In gathering from clover, it has been shown that about 1/350th grain is secured at each visit. Let us imagine that our bee is enlarged twice, by which its weight has grown eight-fold. As it flies, carrying its large body from clover-bloom to clover-bloom, an amount of wear and tear is involved which is eight times as great as that accompanying similar movements in the normal bee. This wear and tear is replaced by food - of course, proportionately augmented, and which has to be deducted from the 1/350th grain secured. The net increase to the stock is, therefore, less at each visit, in the case of the larger bee, than in that of the normal one. The former, however, has the advantage of being able to decrease its return visits to the hive to unload, because its honey-sac is larger; but this is the only gain, and it is much more than counter-balanced by the fact that, with normal bees, eight independent gatherers would be at work simultaneously for only the same wear and tear that would permit of the efforts of one if the bulk were increased as supposed. Selection has gone on for ages regulating the proportions of the wondrous insect between those extremes in which the loss by excessively frequent returns to the colony, and the loss through excessive bodily weight, balance each other, and has thus given us a bee whose size yields the best possible results.

The botanical reason for desiring no alteration was expounded in Vol. 1. Flowers and bees have been constantly interacting. The build of every floret is adapted to that of its fertiliser, and, could we suddenly increase the dimensions of our hive bees, we should throw them out of harmony with the floral world around them, decrease their utility, by reducing the number of plants they could fertilise, and diminish equally their value as honey-gatherers. Mechanics, physiology, economics, and botany alike, show any craving after mere size to be an ill-considered and unscientific fancy, for which it would be even difficult to find an excuse.

ABC and XYZ of BEEKEEPING, A.I. Root – 1891 – Pages 172-178

Everybody knows that the cells of the honey-comb are 6-sided, and I presume most people know why they are 6-sided. If they were square, the young bee would have a much more uncomfortable cradle in which to grow up, and it would take a much greater space to accommodate a given number of bees. This last would, of itself, be a fatal objection; for to have the greatest benefit of the accumulated animal heat of the brood, they must be closely packed together. This is not only the case with the unhatched bees, but with the bees of a whole colony in winter; when each bee is snugly ensconced in a cell, they occupy less room than they could by any other arrangement.

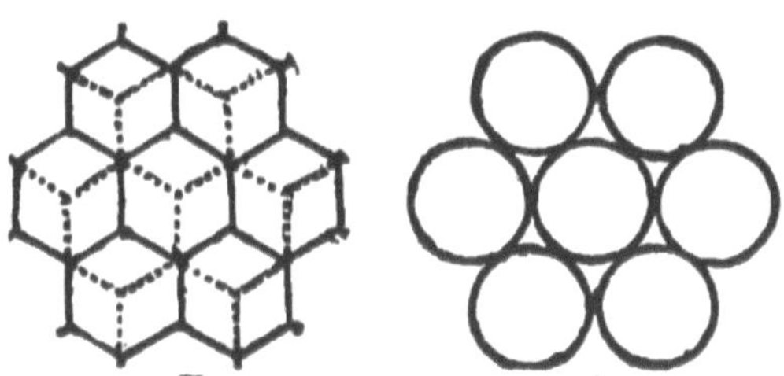

B A

Why the cells of the honey comb are made 6 sided

If the cells were round, they could be grouped together much in the same way as they are now; viz., one in the center, and 6 all around it, equally distant from the central one, and from each other, like the cut, in the figure A; but even then, the circles will leave much waste room in the corners, that the bees would have to fill with wax.At B, we see the cells are nearly as comfortable for the young bee as a round one would be of course. I mean from our point of view, for it is quite likely that the bees know just what they need a great deal better than we do and, at the same time, they come together in such a way that no space is left to be filled up at all. The bees, therefore, can make the walls of their cells so thin that they are little more than a silky covering, as it were, that separates each one from its neighbor. It must also be remembered that a bee, when in his cell, is squeezed up, if we may so term it, so as to occupy much less space than he otherwise would; and this is why the combined animal heat of the cluster is so much better economized in winter, when the bees have a small circle of empty cells to cluster in, with sealed stores all around them.

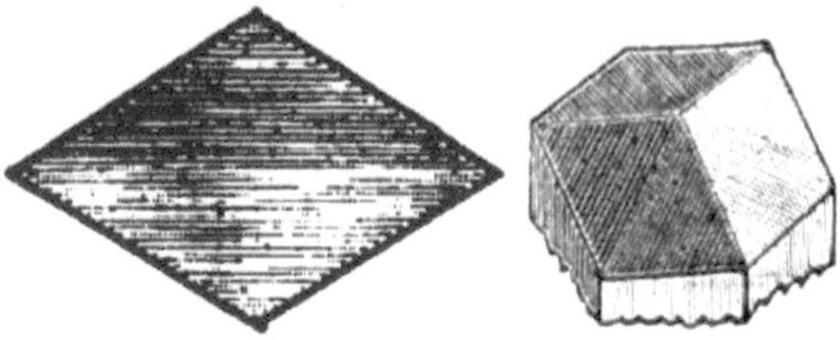

How the bottom of the cell is made

But, my friends, this is not half of the ingenuity displayed about the cell of the bee. These hexagonal cells must have some kind of a wall or partition between the inmates of one series of cells, and those in the cells on the opposite side. If we had a plain partition running across the cell at right angles with the sides, the cells would have flat bottoms which would not fit the rounded body of the bee, besides leaving useless corners, just as there would have been if the cells had been made round or square. Well, this problem was solved in much the same way, by making the bottom of the cell of three little lozenge-shaped plates. In the figure below we give one of these little plates, and also show the manner in which three of them are put together to form the bottom of the cell.

If the cells were round, they could be grouped together much in the same way as they are now; viz., one in the center, and 6 all around it, equally distant from the central one, and from each other, like the cut, in the figure A; but even then, the circles will leave much waste room in the corners, that the bees would have to fill with wax.

Now, if the little lozenge plates were square, we should have much the same arrangement, but the bottom would be too sharp-pointed, as it were, to use wax with the best economy, or to best accommodate the body of the infantile bee. Should we, on the contrary, make the lozenge a little longer, we should have the bottom of the cell too nearly flat, to use wax with most economy, or for the comfort of the young bee. Either extreme is bad, and there is an exact point, or rather a precise proportion that the width of this lozenge should bear to the length. This proportion has been long ago decided to be such that, if the width of the lozenge is equal to the side of a square, the length should be exactly equal to the diagonal of this same square. This has been proven by quite an intricate geometrical problem; but a short time ago, while getting out our machine for making the foundation, I discovered a much shorter way of working this beautiful problem.

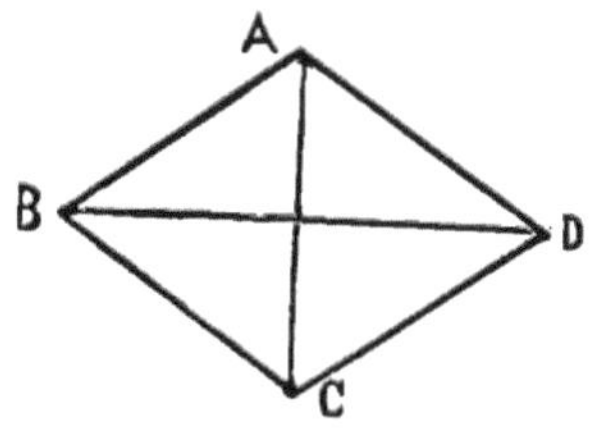

In the figure above, let A B C D represent the lozenge at the bottom of the cell, and A C, the width, while B D is the length of said lozenge. Now, the point I wish to prove is, that A C bears the same proportion to B D that the side of a square does to the diagonal of the same square.

THE MATHEMATICS OF THE HONEY-COMB.

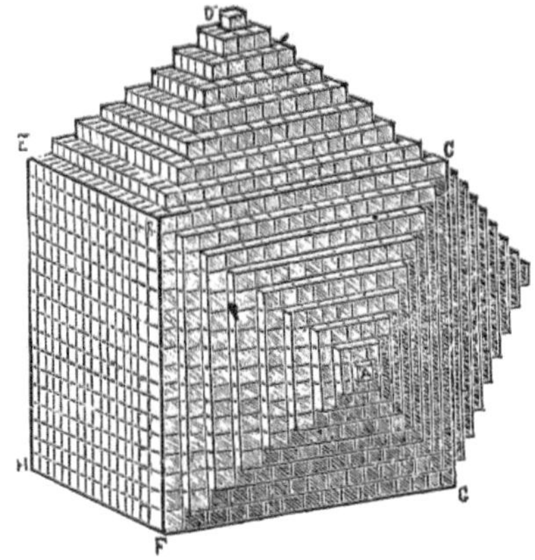

Suppose we have a cubical block, E B C G F, and that we pile small blocks on its sides as shown, so as to raise pyramids of such an inclination that a line from any apex to the next, as from A to D, will just touch the edge of the cube, B C. Now A C D B is the geometric lozenge we are seeking. Its width, B C, is equal to one side of the square, E B F H, for it is one side of the cube. Now, to prove that A D is equal to the diagonal E F, we will use the diagram below.

Let E B F H represent the cube, and the dotted lines the pyramids. If the pyramids are so made that the line A D is a straight, continuous one, it is evident, by a little reflection, that the angles A and D will be right angles. If this is so, A D is exactly equal to E F, the point we were to prove. Now, referring to the former figure, if we should go on building these pyramids on all sides of the cube, we will have the beautiful geometrical figure called the rhombic dodecahedron; it is so called, because it is a solid figure having 12 equal sides, and each side is a rhomb, or lozenge, such as we have described. Where the obtuse angles of three of these rhombs meet, as at C, we shall have the exact figure of the bottom of a honey-comb cell. A picture of the geometrical solid we have mentioned is given below.

RHOMBIC DODECAHEDRON.

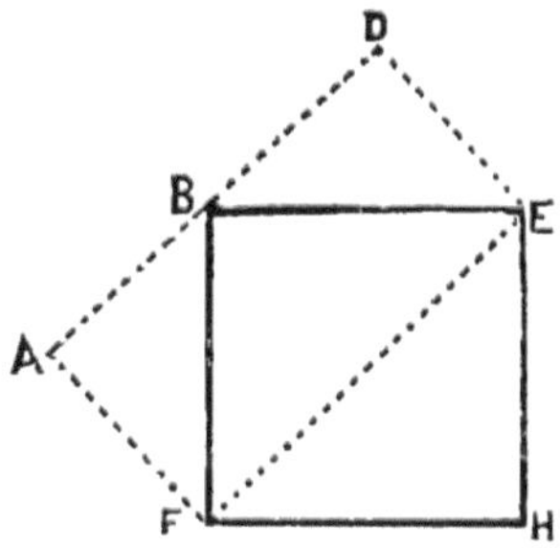

How does it come that the bees have solved so exactly this intricate problem, and know in just what form and shape their precious wax can be used, so as to hold the most honey, with the very least expenditure of labor and material? Some are content with saying that they do it by instinct, and let it drop there; but I believe God has given us something farther to do than to invent names for things, and then let them drop. By carefully studying the different hives in a large apiary, we see that not all of them build comb precisely alike, and not all colonies are equally skilled in working wax down to this wonderful thinness. Some bees will waste their precious moment and wax in making great, awkward lumps of wax; coarse, irregular cells; crooked, uneven comb, etc., with very bad economy either for the production of brood or for the storing of honey; while others will have all their work so even and true, and so little wax will be wasted, that it is wonderful to contemplate the regularity and system with which the little fellows have labored. Now, it does not require any great amount of wisdom to predict that the latter would, in a state of nature, stand a far better chance of wintering than the ones that were wasteful and irregular in their ways of doing things. If this be the case, those queens whose progeny were best laborers, most skillful wax-workers, as well as most energetic honey-gatherers, would be most sure to perpetuate themselves, while the others would, sooner or later, become extinct. I have found more of a tendency in bees to sport, or to show queer peculiarities, than in any other department of the animal or vegetable kingdom. They vary in color, in shape, in size, in disposition, in energy; and almost every colony, if studied closely, will be found to have some little fashion or way of doing things, different from all the rest in the apiary. Now, when we take into account the fact that many generations can be reared in a single summer, we see how rapidly, by fostering and encouraging any desirable trait or disposition, the bees may be molded to our will. The egg that is laid by a queen to-day may, by proper care, be made to produce a queen laying eggs of the same kind herself, in the short time of only 25 days, as I have explained heretofore. Well, if we should pick out a queen whose progeny made the thinnest comb, and rear others from her, doing the same thing for several generations, we should probably get bees whose combs would break down by the weight of the honey. In a state of nature this extreme would correct itself, as well as the other; but the point I wish you to see is right here: Geometrical accuracy in the shape of the cells can never be overdone, and can be reached only by absolute perfection; and this absolute perfection, the bees have been constantly aiming at through endless ages. Is it any thing strange, my friends, that the bees have got the honey-comb pretty nearly right by this time? I will give you a little story, and one which has been very interesting to me, from page 150, Vol. II., *American Bee Journal*.

If a single cell be isolated, it will be seen that the sides rise from the outer edges of the three lozenges above mentioned, so that there are, of course, six sides, the transverse section of which gives a perfect hexagon. Many years ago, Maraldi, being struck with the fact that the lozenge-shaped plates always had the same angles, took the trouble to measure them, and found that in each lozenge the large angles measured 109° 28′, and the smaller 70° 32′, the two together making 180°, the equivalent of two right angles. He also noted the fact that the apex of the three-sided cup was formed by the union of three of the greater angles. The three united lozenges are seen in the figure below.

Sometime afterward, Reaumur, thinking that this remarkable uniformity of angle might have some connection with the wonderful economy of space which is observed in the bee-comb, hit upon a very ingenious plan. Without mentioning his reasons for the question, he asked Koenig, the mathematician, to make the following calculation: Given a hexagonal vessel terminated by three lozenge-shaped plates, what are the angles which would give the greatest amount of space with the amount of material?

Koenig made his calculations, and found that the angles were 109° 26′ and 70° 34′, almost precisely agreeing with the measurements of Maraldi. The reader is requested to remember these angles.

Reaumur, on receiving the answer, concluded that the bee had very nearly solved the difficult mathematical problem, the difference between the measurement and the calculation being so small as to be practically negative in the actual construction of so small an object as the bee-cell.

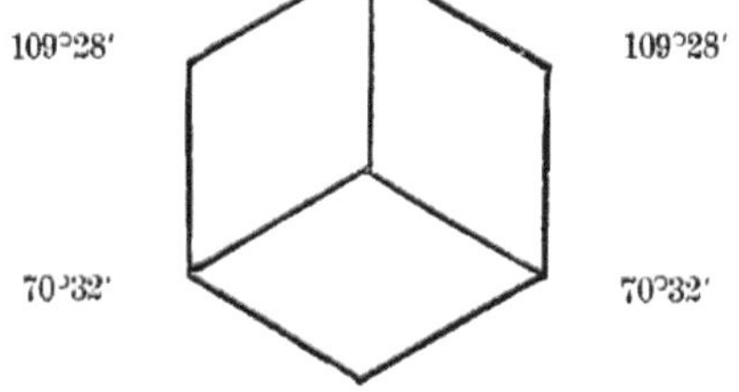

Mathematicians were naturally delighted with the results of the investigation, for it showed how beautifully practical science could be aided by theoretical knowledge; and the construction of the bee-cell became a famous problem in the economy of nature. In comparison with the honey which the cell is intended to contain, the wax is a rare and costly substance, secreted in very small quantities, and requiring much time and a large expenditure of honey for its production. It is, therefore, essential that the quantity of wax employed in making the comb should be as little, and that of the honey which could be stored in it as great, as possible.

For a long time these statements remained uncontroverted. Anyone with the proper instruments could measure the angles for himself, and the calculations of a mathematician like Koenig would hardly be questioned. However, Maclaurin, the well-known Scotch mathematician, was not satisfied. The two results very nearly tallied with each other, but not quite, and he felt that, in a mathematical question, precision was a necessity. So he tried the whole question himself, and found Maraldi's measurement correct namely, 109° 28′, and 70° 32′.

He then set to work at the problem which was worked out by Koenig, and found that the true theoretical angles were 109° 28′ and 70° 32′, precisely corresponding with the actual measurement of the bee-cell.

Another question now arose. How did this discrepancy occur? On investigation, it was found that no blame attached to Koenig, but that the error lay in the book of Logarithms which he used. Thus a mistake in a mathematical work was accidentally discovered by measuring the angles of a bee-cell a mistake sufficiently great to have caused the loss of a ship whose captain happened to use a copy of the same Logarithmic tables for calculating his longitudes.

DIFFERENT KINDS OF CELLS IN THE HONEY-COMB.

The bees build two distinct, regular sizes drone and worker cells. The worker-comb measures very nearly five cells to the inch, on an average. Some specimens average a little larger, and some a little smaller; but when the comb is at all irregular, it is quite apt to be a little larger. The best specimens of true worker-comb generally contain 5 cells within the space of an inch, and therefore this measure has been adopted for the comb foundation. If there are five cells to the inch, a square inch would give, on an average, about 25* cells, and 25 on the opposite side would make 50 young bees that would be hatched from every square inch of solid brood. As foundation is so much more regular than the natural comb, we get a great many more bees in a given surface of comb, and here, at least, we can fairly claim to have improved on nature.

The drone-comb measures just about 4 cells to the inch, but the bees seem less particular about the size of it than with the worker. They very often seem to make the cells of such size as to best fill out a given space; and we, accordingly, find them of all sizes, from worker size all the way up to considerably larger than 1/4 of an inch in width. Drones are raised in these extra-large cells without trouble, and honey is also stored in them; but where they are very large, the bees are compelled to turn them up, or the honey would flow out. As the honey is kept in place by capillary attraction, if the cells exceed a certain size, the adhesion of the liquid to the wax walls is insufficient, of itself, to hold the honey in place. Where drones are to be reared in these very large cells, the bees contract the mouth, by a thick rim. As an experiment, I had some plates made for producing small sheets of foundation, having only 3-1/2 cells to the inch. The bees worked on a few of these, with these same thick rims, but they evidently did not like the idea very well, for they tried to make worker-cells of some of it, and it proved so much of a complication for their little heads that they finally abandoned the whole piece of comb, apparently in disgust. Bees sometimes rear worker brood in drone-comb, where compelled to from want of room, and they always do it in the way I have mentioned, by contracting the mouth of the cells, and leaving the young bee a rather large berth in which to grow and develop. Drones are sometimes reared in worker-cells also, but they are so much cramped in growth that they seldom look like a fully developed insect.

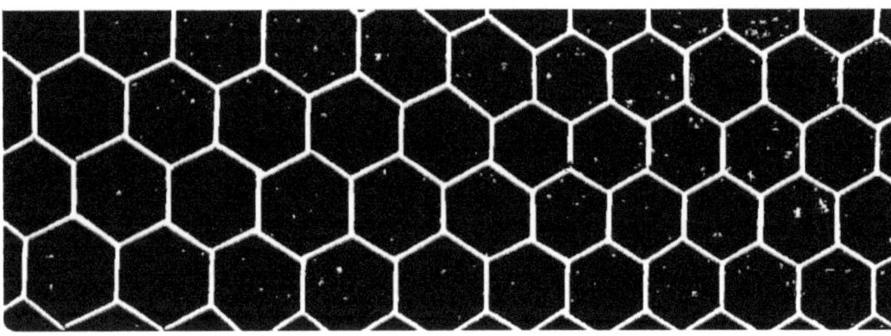

Drone-comb *Worker-comb*

Several times it has been suggested that we enlarge the race of honey-bees by giving them larger cells; and some circumstances seem to indicate that something may be done in this direction, although I have little hope of any permanent enlargement in size, unless we combine with it the idea of selecting the largest bees to propagate from, as given a few pages back. By making the cells smaller than ordinarily, we can get small bees with very little trouble; and I have seen a whole nucleus of bees so small as to be really laughable, just because the comb they were hatched from was set at an angle so that one side was concave and the other convex. The small bees came from the concave side. Their light, active movements, as they sported in front of the hive, made them a pretty and amusing sight for those fond of curiosities. Worker-bees reared in drone cells are, if I am correct, sometimes extra large in size; but as to whether we can make them permanently larger by such a course, I am inclined to doubt. The difficulty, at present, seems to be the tendency to rearing a great quantity of useless drones. By having a hive furnished entirely with worker-comb, we can so nearly prevent the production of drones that it is safe enough to call it a complete remedy.

HOW THE BEES BUILD THE COMB.

In this day and age of bees and honey, it would seem that one should be able to tell how the bees build comb, with almost as much ease as they would tell how cows and horses eat grass; but for all that, we lack records of careful and close experiments, such as Darwin made many years ago. In our house-apiary, there are dozens of hives where the bees are building right up close to the glass, at this very minute; and all one has to do, in order to see how itis done, is to take a chair and sit down before them. But the little fellows have such a queer, sleight-of-hand way of doing the work, that I hardly know how they do accomplish it.

In a little work published by Prof. Agassiz, about the year 1867, the renowned naturalist speaks as follows about the way in which bees build honey-comb:

"The bees stand as close as they can together in their hive for economy of space, and each one deposits his wax around him, his own form and size being the mold for the cells, the regularity of which, when completed, excites so much wonder and admiration. The mathematical secret of the bee is to be found in his structure, not in his instinct."

Notwithstanding the promptness with which the folly of such a statement was at once shown up in the bee-journals, it seems it never came to the eyes of Prof, A., or, at least, he never deemed it worthy of notice; for, in 1873, he gave, substantially, the same thing in a lecture at Cambridge, Mass., and it was praised and published in the Tribune and other papers, and sent broadcast all over our land. I believe all the bee-journals at once protested against giving the people such "twaddle" (if I may be excused for using the term), as science; but for all that, I think the learned professor never recalled his blunder, or even so much as admitted that he had never seen the inside of a bee-hive at all, but only guessed at it, or repeated what he had been told by someone.

About two years afterward, the great scientist, Tyndall, by some means got an inkling of the way in which Agassiz had "put his foot in it," and, in the Popular Science Monthly, wisely admitted that the bees did not stand in the cells to build their comb, but fixed them in this way: Says he, "The bees place themselves at equal distances apart upon the wax, and sweep and excavate" etc. Now, if Tyndall is teaching us other things in the same way, i. e.., delivering lectures on some subject on which he knows nothing, how much can we depend on anything he says? Oh why could not he and Agassiz, before attempting to explain the matter to the people, take the time to get a hive of real live bees, as did Darwin, and not be obliged to take anything at second hand? If they two were afraid of stings, any expert honey-raiser could afford them the facilities for a safe observation, and thus prevent their going into such folly, or falsehood, to call things by their right names, for they pretend to have knowledge where they have none. Take the money and buy a hive of bees, all ye that thirst for knowledge, and take it direct from God's own works, instead of receiving it second hand.

For particulars in regard to the North Pole, or as to whether the planet Jupiter is habitable, we may be obliged to listen to those who should know better than we do; but in our own industry no such necessity exists, for a swarm of bees is within the reach of all.

When distinguished persons have visited my apiary, I have almost invariably heard them mention the great discovery of Agassiz, in regard to the way in which bees build their comb; and when I explain that it was a great mistake, they usually think that so great a man as Agassiz, and one who always went to the ants and bees with his own eyes, must have been right, and that I had made a mistake somewhere.

I have occupied all this space, my friends, just to give you an illustration of how little real work some of the great scientists and lecturers are in the habit of doing, and of the importance of proving things for yourself, with your own eyes and hands.

If we examine the bees closely during the season of comb-building and honey-gathering, we shall find many of them with the wax scales protruding between the rings that form the body, and these scales are either picked from their bodies, or from the bottom of the hive or honey-boxes in which they are building. If a bee is obliged to carry one of these wax scales but a short distance, he takes it in his mandibles, and looks as business like with it thus as a carpenter with a board on his shoulder. If he has to carry it from the bottom of the honey-box, he takes it in a way that I cannot explain any better than to say he slips it under his chin. When thus equipped, you would never know he was encumbered with anything, unless it chanced to slip out, when he will very dexterously tuck it back with one of his fore feet. The little plate of wax is so warm from being kept under his chin, as to be quite soft when he gets back; and as he takes it out, and gives it a pinch against the comb where the building is going on, one would think he might stop a while, and put it into place; but not he; for off he scampers and twists around so many different ways, you might think he was not one of the working kind at all.

Another follows after him sooner or later, and gives the wax a pinch, or a little scraping and burnishing with his polished mandibles, then another, and so on, and the sum total of all these maneuvers is, that the comb seems almost to grow out of nothing; yet no bee ever makes a cell himself, and no comb-building is ever done by any bee while standing in a cell; neither do the bees ever stand in rows and "excavate," or anything of the kind.

The finished comb is the result of the united efforts of the moving, restless mass; and the great mystery is, that anything so wonderful can ever result at all from such a mixed-up, skipping-about way of working, as they seem to have. When the cells are built out only part way, they are filled with honey or eggs, and the length is increased when they feel disposed, or "get around to it," perhaps. It may be that they find it easier working with the shallow walls about the cells, for they can take care of the brood much easier, and put in the honey easier too, in all probability; and, as a thick rim is left around the upper edge of the cell, they have the material at hand to lengthen it at any time. This thick rim is also very necessary to give the bees a secure foothold, for the sides of the cells are so thin they would be very apt to break down with even the light weight of a bee. When honey is coming in rapidly, and the bees are crowded for room to store it, their eagerness is so plainly apparent, as they push the work along, that they fairly seem to quiver with excitement; but for all that, they skip about from one cell to another in the same way, no one bee working in the same spot to exceed a minute or two, at the very outside. Very frequently, after one has bent a piece of wax a certain way, the next tips it in the opposite direction, and so on until completion; but after all have given it a twist and a pull, it is found in pretty nearly the right spot. As nearly as I can discover, they moisten the thin ribbons of wax with some sort of fluid or saliva. As the bee always preserves the thick rib or rim of the comb he is working, the looker-on would suppose he was making the walls of a considerable thickness; but if we drive him away, and break this rim, we will find that his mandibles have come so nearly together that the wax between them, beyond the rim, is almost as thin as tissue paper. In building natural comb, of course the bottoms of the cells are thinned in the same way, as the work goes along, before any side walls are made at all; but the manner of thinning the bottoms of the cells in the foundation is quite another thing.

*The exact mathematical calculation make these numbers 29, 29 and 58, respectively, but ordinarily the numbers I have given in the context are more nearly correct.

A Correction from Thos. Wm. Cowan

A Correction from Thos. Wm. Cowan, Editor of the British Bee Journal, from Bee Culture 1898

Dear Mr. Root:- On page 144 you refer to the "number of cells or worker comb to the linear inch." Will you kindly look at my "The Honey-bee; its Natural History, Anatomy, and Physiology"? On page 180 you will see that I say, "The average size of a worker-cell between the parallel sides is 1/5 of an inch, or 0.2 (a printer's error makes it 0.02; but it is two-tenths of an inch). Then I go on, "We say 'average,' because considerable variation exists in different parts of the same comb, as both Reaumur and Huber found." I then go on to summarize the large number of measurements I took; and if you will read the details you will see what a variation there is. You say, "It has been said over and over again in bee-books and bee-journals, that there are five cells of worker comb to the inch, so that we have come to believe it;" also that Cook is the only authority you have run across who says worker-cells are a little more than 1/5 inch; but in my book you will find that, out of 36 measurements that were taken, I found the greatest aggregate diameters of any one series of ten cells to amount to 2.11 inches, which you see makes them considerably larger than 1/5 inch.

On the other hand, the least came to 1.86, which makes them smaller.

You will also see that, to reduce the possibility of error, I also measured a large number of series of 60 cells, which, if the cells are exactly 1/5 inch, would occupy a space of 12 inches. However, in almost every case the 12 inches was exceeded, although not always. Please also note that, on page 181, I say that cells worked by Carniolan bees are larger. Nearly the whole of the chapter is devoted to the measurements of combs and cells; and as I know these were most carefully taken, with most accurate instruments, I am certain of my facts. You refer to Cheshire; but has it occurred to you to test his figures? He tells us the length of the worker-cell is 15/32, whereas it is only 13/64, showing his cell to be nearly double the right length. His cell, drawn on paper, would look like this:

How would a bee like it? A similar error is made with drone-cells, which he says are 9/16 but which are only 9/32 inch long. He criticises Langstroth, who shows a cell with an acute angle, and says, "100° is the limit the bee can reach," and that no angles of less than 100 ° are found. I have been able to confirm Langstroth's statement by showing similar combs, and demonstrating that bees frequently work at a less angle, even, than 90°. I also show that, in the matter of angles, these differ considerably when carefully measured with a goniometer. I have for a long time considered that we should use the expression "average size" as being the more correct, as I have not believed in a worker-cell being exactly 1/5 inch. I see Mr. Weed uses the term "average worker-cell," which is about correct.

Thos. WM. COWAN.

Loomis, Cal., Feb. 24.

[I am glad to get this, even though I do have to confess that I did not give your book the careful scrutiny that I should have done. I remember looking into it, and finding the sentence that "the average size of a worker-cell between the parallel sides is 1/5 of an inch." Why I stopped and did not go further to take in all you said, I cannot say. I shall have to acknowledge-indeed, I do so most cheerfully-that you have gone into this question far more thoroughly than anyone else I know of. With regard to Cheshire, he who was so ready to point out the mistakes of others made a good many himself. If any writer lived in a glass house, he did. I am sorry to know that some of the glass seems to have been badly shattered. After all, he gave us much of value, even if he did make some glaring mistakes.

If I were not talking to Mr. Cowan's face, I believe I should say that, while his work is smaller, no one has pointed out an error in it, save the typographical one that he refers to above-ED.]

A Study of Natural Honey-Comb - C.C. Miller

Bee Culture – 1910

BY DR. C. C. MILLER.

Comb foundation is in such general use nowadays that it would be nothing strange to find bee-keepers who have never seen a frame of entirely natural comb. I have been making a study of some specimens – a dozen in number – that were built entirely at the sweet will of the bees, not even the least starter being in the case. They range in size from a piece of a few square inches to nearly a frameful.

POSITION OF CELLS.

Looking at brood foundation that I have, I find the cells placed with the angle at top and bottom.

In super foundation the angle is at each side, one of the cell-walls lying horizontally at the top and another at the bottom. I don't know why the two kinds differ.

The bees seem to copy after the first plan. Not very strictly, however. In only one case can the row of cells be said to be really in a horizontal row. In another specimen the row descends half an inch in about a foot. In the other cases the variation from the strict horizontal is still greater.

The cells run in a fairly straight row except in one frame where the line is somewhat wavy, apparently because there were four initial points of beginning, and the four parts were afterward joined together.

SIZE OF CELLS.

It is a common thing to say, "Worker-cells measure 5 to the inch, and there are, consequently, 25 cells on one side to the square inch." Neither of these statements is always true if we speak with any degree of accuracy. There are not always exactly 5 cells to the inch; and if there were, there would be, not 25, but 28-13/15 cells to the square inch. See Cheshire, Vol. I., page 176 – that is, if the cells were exact hexagons. The trouble is that they vary from this quite a little. On one piece of comb, measured horizontally, the average diameter of a cell was .201087 of an inch; in one of the diagonal directions it was .19853, and in the other .20357, the total average diameter being .201062 of an inch.

Upon reading those figures someone may think that I must have had some very nice instruments with which to take measurements. I had nothing but a common pocket-rule, and after I tell you how I did it you will see that a schoolboy could easily do the same.

Suppose I want to measure the diameter of a cell. Laying the rule upon it, and measuring merely that one cell, I could only say it was somewhere between 3/16 and 1/4 of an inch – not very exact. But if I measure 10 cells, and then divide by 10, I can come about ten times as near the exact measurement. The larger the number of cells I take in my measurement, the nearer I can come to exactness. Well, here's the way I do. I lay the rule upon the comb, with one end of the rule exactly corresponding with one of the cell-walls. Then I look along the rule till I see some notch which corresponds with some cell-wall. Then I count the number of cells in the given distance, divide the number of inches by the number of cells, and that gives the diameter of a cell. For instance, I find a notch of the rule at a cell-wall 9-1/4 inches from the end of the rule. I count the cells, and find there are 46. I divide 9-1/4 by 46, and I have .201087 of an inch as the diameter of one cell. Easy, isn't it?

But after I have the diameter of a cell it's just a little bit of bother to figure the area of the hexagon, especially as its three diameters are not all alike. A beautifully simple way of measuring the surface of a comb is given by A. Berchon, L'Apiculteur, p. 228.

Take the parallelogram ABCD. The line AC passes through the middle of 5 cells. Next to this vertical row of cells is another row of 4 cells, with a half-cell at top and a half-cell at bottom, making 5 cells in the row. So there are 5 cells in each vertical row in the parallelogram. The line AB passes alternately through the middle of a cell, coincides with a cell-wall, then through the middle of another cell, and so on. Each end of the line stops in the middle of a cell-wall; and if you put together the two half-cells at each end, the line measures 14 cells. There being thus 5 cells in each vertical row, and 14 in each horizontal row, there must be 5 X 14 = 70 cells in the parallelogram.

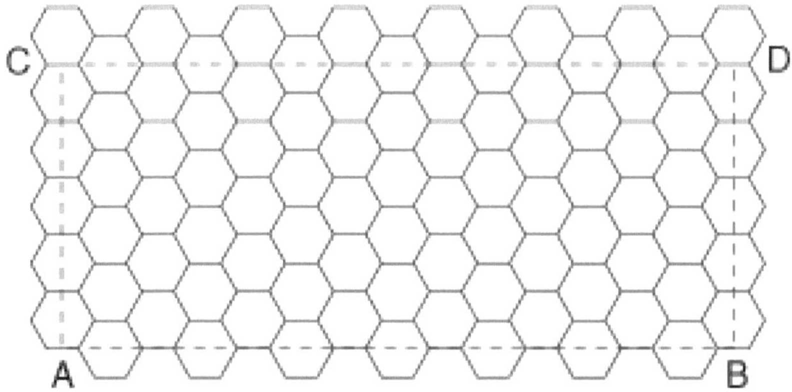

Instead of measuring from the center of one cell-wall to the center of another cell-wall I find it more exact to let the line AB begin at an angle of a cell and end at the corresponding angle in another cell.

It may be worthwhile to notice that the figure, copied from L'Apiculteur, has the cells running the wrong way, a side of a cell being at top and bottom of each cell, whereas it should be an angle.

In one piece of comb, measured horizontally, there were 42 cells in 8-1/2 inches, and measured vertically there were 38 cells in 6-11/16 inches. Multiply 42 by 38, and 8-1/2 by 6-11/16, then divide the former product by the latter, and you have 28.076 cells to the square inch in that piece of comb. In another comb there were 26.54 cells to the square inch — quite a difference in the two combs. T. W. Cowan (The Honey-bee, 181), took 36 measurements and found the diameter of a cell to range from .186 of an inch to .211. That's a much greater variation than in the two combs I have mentioned; but then, he made more measurements.

In a sheet of brood foundation I find 26.62 cells to the square inch. That's about the same as my sample with the larger cells; but it has smaller cells than some that Mr. Cowan found in natural comb. That shows it would be feasible to have foundation with larger cells, thus working toward a larger bee, if a larger bee would get more honey.

Of that I have some doubt.

Marengo, Ill.

The Structure of Comb – Part 1 - Bee World 1921-Betts

The Structure of Comb–Part 1–The Bee World–July, 1921–Pages 37-38

By MISS ANNIE D. BETTS, B.Sc.

From references by the classical writers it is clear that the comb of the honey-bee has been admired from very early times. This is not surprising; for it would be difficult to find any human engineering achievement, even in modern aeronautical practice, that surpasses the honey-comb as a solution of the problem of combining light weight and great strength. Even if the ancients did not fully realise this, yet the beautiful regularity of the hexagonal network of cell-mouths could not fail to impress them; and it is not astonishing that the first known research on the structure of comb deals with the hexagonal form of the cells. Its author was Zenodorus, of Sicily, in the second century B.C., shortly after the time of Archimedes. Zenodorus proved that, of the three regular figures that will completely fill a plane surface (namely, the equilateral triangle, the square, and the regular hexagon), the hexagon has the greatest content for a given circumference. Pappus (ca. A.D. 500) copied from Zenodorus, and remarked that the bees wisely choose that one of the three forms for the cell-mouth which they suspect will contain most honey for the same expenditure of wax in its construction. This suggestion, that the bees economise wax, grew later into a wonderful myth, far removed from the realities of the matter.

The ideal form of the bee's cell – seldom completely realised in actuality – is that of a regular six-sided prism, the base of which is formed of three rhombs of lozenges meeting in a point at the bottom of the cell (see Fig. 13).

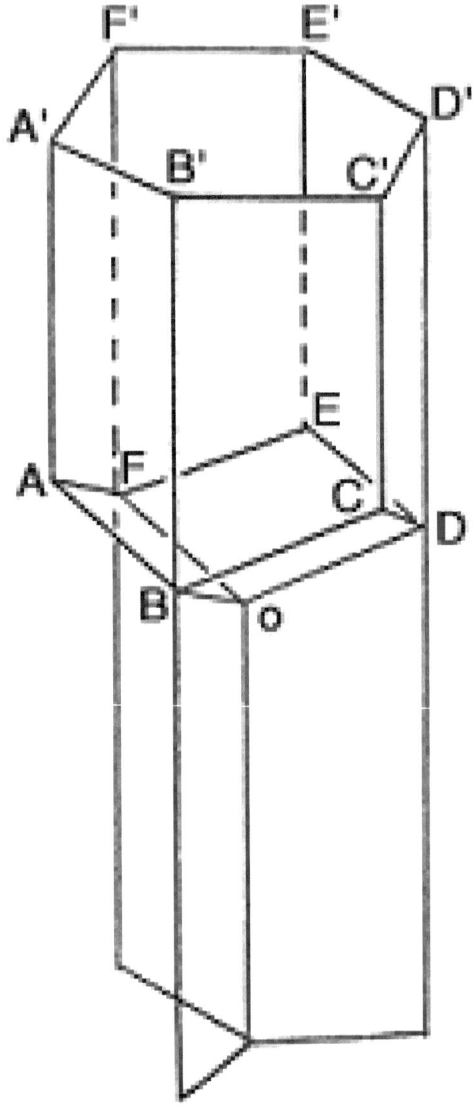

Fig. 13

A′B′C′D′E′F′ is the cell-mouth; A′A, B′B, etc., are the edges of the cell; ABOF, CDOB, EFOD are the three rhombs; O being the bottom of the cell. Let us now consider the other side of the comb. From O there starts a cell-edge similar to those at A′A, C′C, or E′E; so that the three rhombs each form part of the base of a different cell on the other side of the comb; A, C, and E being the bottom points of these three cells, and corresponding to O in the first cell. The edges B′B, D′D, and F′F are continuous right through the comb from one side to the other; a point that is probably of importance in connection with the well-known and hitherto unexplained "pitch" of the cells.

The plane passing through the points BDF is easily seen to lie in the exact middle of the comb (supposing the cells to be of equal depth on both sides); and if the pyramidal bases were replaced by flat ones in this plane, the cells would be unaltered in *volume*. Their *surface* would, however, be increased, as can be shown by a not very difficult calculation (here omitted to save space). The famous "problem of the bee's cell," around which the myth referred to grew up, is simply this: Find the shape of the rhombs, in a cell of the form shown in the diagram, such that the total area shall be a minimum, the other dimensions of the cell being unaltered. The answer gives us rhombs with angles respectively equal to 109 deg. 28′ 16.4″ and 70 deg. 31′ 43.6″. This result assumes that the walls and base-rhombs are all of equal thickness (which is not the case), that no wax is used to strengthen the edges of the cells (whereas about one-third of the total wax used goes for this purpose), and that all lines are straight and the cell quite regular (which is not so in actual comb). It will be seen, therefore, that this problem, however interesting to the mathematician, has but a slender connection with the bee's cell as it really is. This did not trouble the various investigators who were jointly responsible for the myth, because they were either mathematicians with no knowledge of bees, or else naturalists with an equally profound ignorance of mathematics. Consequently, none of them was in a position adequately to criticise the others' work:

The history of the subject is briefly this. After Pappus, no one seems to have studied the bee's cell till Kepler, the astronomer, in 1611, published a very good description of it. He was apparently the first to notice the rhombs of the base, and was evidently quite familiar with bees at work.

The confusion begins in 1712, when Maraldi, an Italian astronomer, studied the cell, measured roughly the angles A'AB, A'AF, and BAF, and found them approximately equal to one another. He then calculated that, if these angles really were equal, they must each be of about 109 deg. 28'. Maraldi is an "awful warning" to us all to express ourselves quite clearly, so as to avoid all danger of being misunderstood. By using somewhat involved phraseology, he succeeded in conveying to the French naturalist Reaumur (some years later) the idea that he had found this value of 109 deg. 28' by measurement! A feat which, as several writers have since remarked, was impossible with the instruments then in existence, even if the cells were regular, which they are not. Reaumur suspected that the bees economised wax, so asked a mathematical friend, Koenig, to work out the "problem of the bee's cell" above referred to. Koenig did so, and gave the larger angle of the rhombs as 109 deg. 26'. Later investigations showed that 109 deg. 28' was the correct answer (to the nearest minute) and that Koenig had made a slip in his arithmetic. Subsequently numerous foolish persons (prominent among whom was Lord Brougham) rushed in with triumphant observations to the effect that "the bee was right and the mathematician wrong," and made much theological capital out of the fact(?) Actually, of course, *it was Maraldi, not the bee, that was right*; but nearly everyone followed Reaumur's mistaken reading, and assumed that Maraldi had not calculated the angle, but had obtained it by measurement.

Many other investigators have studied the bee's cell; some of them were careful to examine specimens of comb, and to note the irregularities of the cells, but many considered the problem only as one in pure mathematics, and their results need not be further discussed here. One of the best and latest papers on the subject is that by H. Vogt (Breslau. 1911). He gives an excellent account of the history of the problem (though the literature he consulted comprises only about one-third of the total in existence), and also of the results of his own measurements. These were made mostly on three or four combs sent him for the purpose, and on plaster-of-Paris casts of cells of these combs. One feels that a greater variety of combs must be measured (lengthy and wearisome as the work would be) before quite reliable results can be obtained; but for the present Vogt's conclusions may be taken to be the best available. He shows that the edges of the cell, as well as the rim, are strengthened by wax deposited in the angles formed by each pair of walls. About one-third of the total amount of wax is thus employed. The form of the base is more pointed than it would be, were economy of wax the guiding factor in its construction. Allowing for the thickening of the edges, and for the differences in thickness between the base and the side-walls of the cell, Vogt states that, for greatest economy of wax, the larger angle of the rhombs should he about 116 deg.; actually this angle is (on the average) about 107 deg. Drone comb is in general more irregular than worker comb. The hexagonal network of the cell-mouths is nearer to perfect regularity than any other parts of the cell; this network, curiously enough, is more regular in drone than in worker comb. The rhombs are 1.59 times as thick as the side-walls in worker comb: for drone comb the figure is 1.58.

The Structure of Comb – Part 2 - Bee World 1921-Betts

The Structure of Comb–Part 2-The Bee World–August, 1921–Pages 73-74

By MISS ANNIE D. BETTS, B.Sc.

What, then, determines the shape of the bee's cell? It is certainly not economy of wax; for, if the shape of the cell were altered to that theoretically most economical, the saving of wax would only be 1/148th part of the total wax used (1/120th for drone cells). This is insignificant in comparison with the amounts wasted in other ways by the bees. So many theories have been put forward that space forbids a full account of them. One must, however, be mentioned, as it is probably a part of the true explanation.

If we suppose that the bees endeavour to build cylindrical cells with hemispherical bottoms, but start doing this so near to each other that the circles overlap, it is easily seen that cells of the form shown in the diagram will result. Such cells will exactly resemble those investigated by Reaumur and Koenig; a geometrical coincidence which is responsible for some of the confusion that has arisen in connection with this problem.

Some have thought, and no doubt to some extent rightly, that this is how comb is built. As we know from Huber's account, comb-building is a co-operative business. One bee does a little work, then passes on, and leaves another bee to continue the cell she has been working on. There can thus be no talk of actual interference of circles struck from equidistant centres; nevertheless, each bee is probably well aware when the cell-wall she is working on is thin enough. Consequently, the cells are formed as if geometrical interference were the controlling agency; and all similar cell-walls are finally of the same thickness, because individual bees differ but little in their sense of touch.

The walls, even of the cell-bases, are very thin, and the stresses in the comb must be taken chiefly by the thickened edges of the cells – the lines drawn in the diagram (BEE WORLD, July, 1921, p. 314 – Part 1). Comb may therefore be regarded as a framework of rods; and it can be shown that these rods are so disposed as to distribute the stresses due to the weight of the comb equally to all the internal rods at a given level. The same applies to the external hexagonal network; but the stresses in this are probably different from those in the interior of the comb. In comb turned through a right angle this is not the case; hence the importance of putting in foundation right way up. The distortion due to putting it wrong way up may not be serious; but mathematics tells us that some distortion is bound to occur, and it is better not to take the risk of getting badly-built combs.

As all who have allowed bees to build comb from starters know, a new comb has a characteristic form, being often narrower at the line of attachment than it is lower down. Looked at edgeways, the shape is similar, the comb being drawn out furthest somewhat above the middle, and being wedge-shaped at the edge. Comb built on foundation is also drawn out first in the middle of the sheet. I have never seen any explanation of these facts; so the following may be of some interest.

Comb, as we have seen, consists of a linkage of elastic rods (the internal edges of the cells) enclosed in a skin or network of hexagonal mesh (the cell-mouths). As the internal rods are so disposed as to be stressed to the same extent at the same level – those higher up the comb being more highly stressed than those lower down, on account of the greater weight below them – the interior of the comb may be regarded as mechanically equivalent to the fluid inside a hanging drop. The network of cell-mouths forms a skin enclosing the whole, just as the "skin" of a drop, subject to the forces of surface-tension, encloses the fluid inside it. We should therefore expect to find that naturally-built comb would assume the form of a fluid drop; as it actually does. Its flattened shape (instead of being a "solid of revolution," like the drop) is due to its bilateral symmetry – to the presence of the midrib. The tendency of comb to be narrower at the line of attachment than lower down is thus explained.

The above is of course only a suggestion, and will have to be worked out mathematically before it can be taken as proved. In the course of this research it may become possible to explain several other puzzling points about comb. Such are: the differences in the cappings of drone and worker cells and of honey cells; the unexpectedly pointed bases, as described by Vogt; and the "pitch," or upward inclination, of the cells. These last are probably connected; for if we suppose the more pointed base to be produced by the points ACE moving upwards, and the point O moving downwards (see diagram on page 314 of the last number – Part 1), the cell-mouth hexagons will not remain plane unless the edges A'A, C'C, E'E are either shortened or bent (if B'B, D'D, F'F do not alter in length). As it is impossible that A'A, etc., can he compressed without altering B'B, etc. (since they are all of the same extensibility), the cell-edges must bend. This of course assumes that the cells are at first of the form given by Reaumur's problem; this is a reasonable supposition, not because it gives the minimum surface but because it is the form which will result if the bees try to build cylindrical cells with hemispherical bases, and the parts in contact are flattened by mutual interference.

It is not, naturally, suggested that bees build comb in this way from any instinctive knowledge of the mechanical properties of the structure; they are probably merely influenced by the tendency of the comb to bulge and bend in undesirable places when they add too much to it at any point, compared with the progress of the rest of the comb. The problem is, in fact, one of statics, as Vogt says (though he speaks only of the cell and makes no allusion to the general form of the comb). That the bees are quick to notice any tendency to bulge or break down is certain from their behavior in repairing damaged combs; may it not well be that it is by this feeling that all is, or is not, well with the equilibrium of the comb that they construct it in the first instance to the pattern of the hanging drop? Their habit of thinning down the edge of old comb before adding to it would be explained by the tendency of the walls of the last row of old cells to distort, if they were of full length, when the weight of the builders was suspended from them at the increased temperature necessary for comb-building.

It has not been possible to give here anything like a full history of the researches on the bee's cell. The literature alone comprises about 150 references, mostly mathematical; and a fair-sized volume would be required to deal with the matter adequately. It is hoped that the above will, however, give an idea of the problem and its history and will prove of interest, especially in view of the present-day experiments with metal foundation. Should the suggestion above put forward as to the cause of "pitch" be correct, it should be found that comb built on metal foundation is devoid of pitch, if the foundation does not stretch or give at all. I shall supplement these notes in future after examining wax combs built on metal foundation.

The Structure of Comb–Part 3-Bee World 1921-Betts

The Structure of Comb–Part 3-The Bee World – September, 1921–Page 97

By MISS ANNIE D. BETTS, B.Sc.

Since writing my last notes on this subject, I have been able, through our Editor's kindness, to examine a partially built-out semicomb. The facts do not bear out the suggestion (*Bee World*. III., p. 319 – Part 2) that the "pitch" of the cells is connected with the deformation of the cell-bases from the form which geometrical symmetry indicates they should possess. Pitch – that is, the upward inclination of the upper cells in a partly drawn-out comb, and the outward (sideways) inclination of the cells nearer the two sides of the comb – is present in built-out semicombs also. An alternative explanation suggests itself to me; that is, that "pitch" is due to the attempt (not always successful) of the bees to keep the edges of the cell (AA′, BB′, etc., Fig. 13, p. 314, July issue – Part 1) continually at right angles to the surface of the comb A′B′C′D′E′F′. This would account for the direction of the pitch of cells in different parts of the comb, and also for the curvature of the edges AA′ etc., for the slope of the plane A′C′E′ to the horizon varies as the comb is drawn out (especially in comb built naturally, without foundation), and in such a manner that the cell-walls are bound to curve, as we find they do, if the bees always try to build them perpendicular to the face of the comb. When the comb is fully built out, the result is, of course. that many of these cell-edges are not perpendicular to the (now vertical) face of the comb; but at this stage even approximate perpendicularity is probably no longer necessary for the stability of the structure. The whole matter must, however, be worked out in detail before this suggestion can be accepted as proved.

I would wish to add a note on a question of priority. In the *Neues Schlesisches Imkerblatt* of 1920, in an essay by J. Huber, the view is advanced that the thick rims of the cells help in supporting the weight of the comb and its contents. As this is the first time this view has been put forward, to my knowledge, it may be as well to state that a similar conclusion was arrived at quite independently by the present writer in 1913, and that the articles in the last two issues of *The Bee World* were written before Mr. Huber's essay came under my notice.

The Building of Honey Comb - Bee World 1929-Wedmore

The Bee World – April, 1929 – Pages 52 – 55

E. B. WEDMORE.

In *Bee World* for December, 1928, the Rev. Yates Allen questions the well-worn instruction that foundation should be so placed in a frame or section that the sides of the hexagons are vertical. The reasons generally given in the guide books are twofold; firstly, that the bees prefer to build combs this way, and secondly, that a comb so built is stronger than one having angles at the sides and horizontal faces at the top and bottom of the cells. The first reason is built upon incomplete observation and the second on insufficient consideration.

I propose to show that the bees' preference is in no way related to mechanical strength or any effect of gravity, but mainly on the life of the support on which they commence to build the individual comb. Furthermore I shall show that comb has the same strength whether subjected to stress parallel to the sides of the cells or normal to them, and that it is no matter which way the foundation be mounted.

For brevity I shall describe comb built with vertical walls as "vertical way" and comb built with the tops and bottoms of cells horizontal as "horizontal way."

If anyone will read Huber, or examine for themselves, they will find that when the bees commence to build they accumulate wax until a wall like lump is built up on the support and then form the beginnings of cells by working on the wax from both sides. The hollows thus formed are worked until the walls are thin and stand out at right angles to the base. It would not be right to say that the bees could not make the walls stand out at some other angle or that they cannot build in any other way. The normal result, however, of extending similar hollows until they meet, the procedure so beautifully described and illustrated by Huber, is to produce walls at right angles to the base, and on examining a natural comb, built on any surface lying at any angle one finds that the bees commence by building out walls at right angles to the surface.

Now in most cases the bees commence building on a horizontal surface, as for example the underside of a branch or the underside of the top of a hole in a tree, or from a wax runner in the top of a frame. In these cases the first walls formed are vertical and this substantially secures the setting of the whole comb. It will be observed, however, that when a comb is started from the side wall of a box hive or from the face of a dummy, the first walls built are again normal to the surface and therefore horizontal, and this again determines the setting for the rest.

It will be observed that a rotation of only 30° converts a hexagon with vertical walls into one with horizontal walls, and occasionally a natural comb gets distorted this much so that one started vertical way is finished horizontal way or vice versa. Furthermore, one started from a surface sloping only 30° from the horizontal will be built horizontal way. When a swarm of bees starts to build in an old-fashioned skep, if the swarm does not depend from the centre the comb is frequently started along a face sloping about 30° and comes out horizontal way, more or less.

On such combs and on combs built from the wall the greater weight of bees on the longer free edge frequently causes a sag which rotates the hexagons, so that a comb started horizontal way is more than usually liable to distortion and to be completed vertical way. Thus there is more than one reason why one generally finds naturally built combs built vertical way.

Now as to the mechanics of the subject there is a sort of idea that the vertical walls in vertical way are better able to withstand the weight of comb below than are the zig-zag verticals found in horizontal way. The two constructions are shown in Figures 1 and 2.

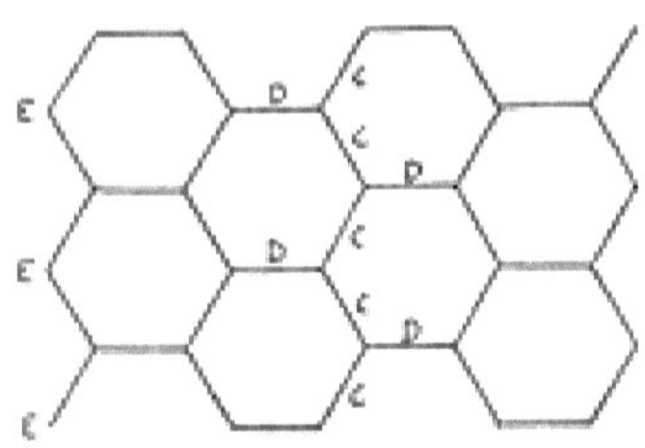

"Horizontal Way"
FIG. 2

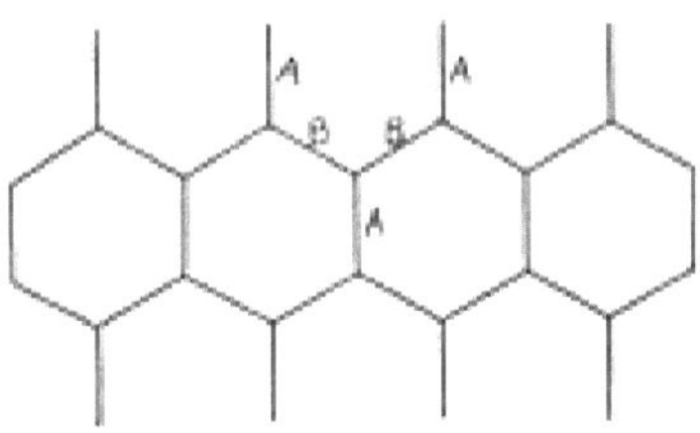

"Vertical Way"
FIG. 1

In Fig. 1 the weight of depending comb is supported by a row of vertical walls A, A, which share the load, and it is transmitted from each wall to the next by the walls B, B, of which there are two transmitting load from each A, in one row to the A walls in the row below. In Fig. 2 the weight is carried by the zig-zag verticals C, C, C, which in tending to straighten put considerable tension on the horizontal walls D, D. If the same vertical load in ounces be applied to support A, as to a support C, C, C, any mechanical engineer will advise that the strain on the individual walls C, C, will be greater than on A, A, in the ratio 2: radix 3. It will be found, however, on examining the number of walls per inch run of comb width that there are more verticals C, C, than A, A, again in the ratio 2: radix 3. In supporting a given weight of comb per inch run of width then, the stress per vertical will be reduced in Fig. 2 in just the same ratio as the stress per individual wall is greater. Thus the vertical walls have to resist the same individual forces in Fig. 2 as in Fig. 1.

The matter may be stated more simply another way. The forces are in every case transmitted from wall to wall at their meeting places, and at each meeting place or angle we have three walls meeting at equal angles and from symmetry the forces in the three walls must be equal. This will be found supported by any textbook on mechanics with a qualification which I shall deal with later. Meanwhile we conclude that the forces in the more or less horizontal walls B, B, in Fig. 1, and D, D, in Fig. 2 are equal to those in the verticals. This conclusion is supported by the fact that the bees make all these walls of the same thickness save that in deep combs, putting a great weight on the upper rows of cells, the walls will sometimes be found thickened at the top, and especially the points of attachment between the walls and the support. (The mid rib carries only about one quarter of the total weight).

Our conclusion leads, however, to another fascinating problem which I have never seen stated, although the solution of it accounts completely for certain characteristics of combs with which we are all familiar, so familiar in fact that we have taken them for granted.

The problem is what becomes of the horizontal stresses, for from the above description it is clear that the horizontal walls D, D, and those nearly horizontal B, B, are subjected to stresses as great as those of the vertical walls to which they are attached. What then happens to the stresses in the outermost horizontal walls when they do not reach to a rigid support to which they can be secured? We cannot get rid of the difficulty by finishing up with no outside horizontal walls as on the left hand side of Fig. 2, for the corners E, E, E, need pulling to the left to prevent the zig-zag vertical on that side from straightening under tension, just as much as if there were horizontal walls attached to F, E, E. The answer is to be found partly in the mid rib forming a continuous and common bottom to the cells in both sides, but not wholly, for clearly this mid rib cannot by itself offer support near the mouths of the cells, and the cells are not so rigid as to need more. Consider for example a comb of honey carrying on the cell walls say 10 ounces of honey or brood and bees per inch of width below a certain level (here, as above, width is measured along the face), then this stress is divided between 5 vertical walls at front and 5 ditto at back of comb in the inch width, giving therefore a stress of one ounce per wall per cell, back and front. This fixes the horizontal forces at this level, also at one ounce per wall, enough to crush the thin cell walls unless further support were given.

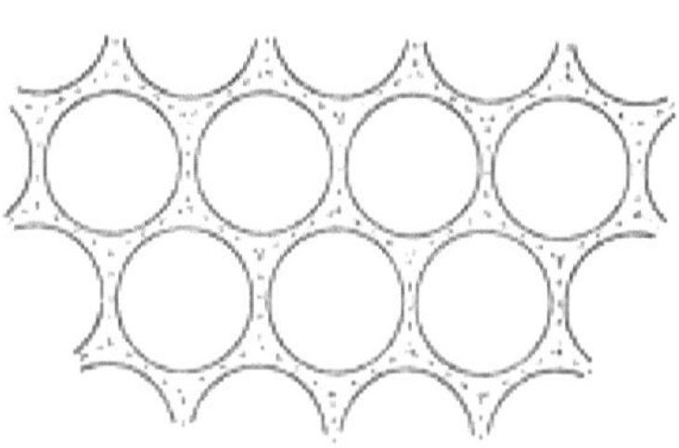

Face of Comb

FIG. 4

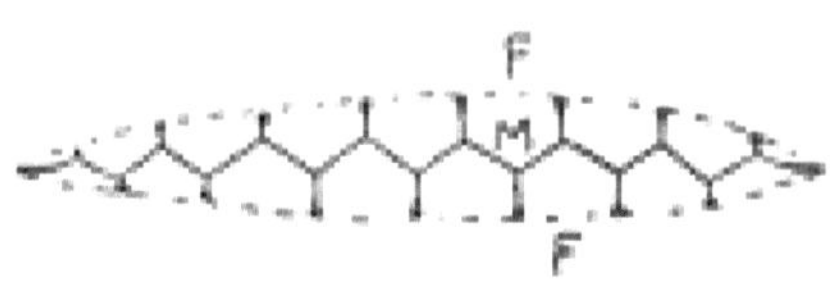

Section of Comb

FIG. 3

Now examine a comb, partly built, with free edges and see what a beautiful structure the bees provide to resist this force. As is well known the comb tapers to an edge on both sides and the edges of the cells on both sides are thickened up so that there is a relatively stiff open structure in both surfaces which both protects the mouths of the cells and assists in resisting the horizontal stresses. Fig. 3 shows a horizontal section of a small partially built natural comb in which M is the mid rib and F, F, the stiffened faces constructed as shown in Fig. 4 in which the thickening is slightly exaggerated. The tapering at the two sides serves to direct the forces back on to the mid rib. The form of the whole structure is beautifully calculated to serve its purpose.

As is well known, when the bees continue the structure shown, they proceed first to accumulate more wax on the mouths of the cells, and then they work in the thickened portion so as to draw out the walls further, proceeding uniformly over the whole exposed face so as to obtain the tapered edge and the smooth convex face. Any other face would not be so stiff.

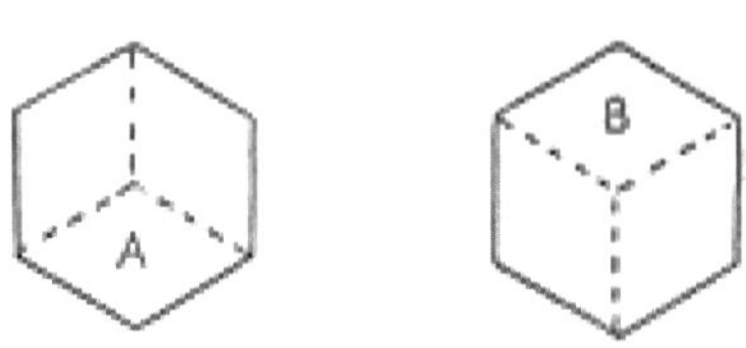

Cell Bottoms
FIG. 6

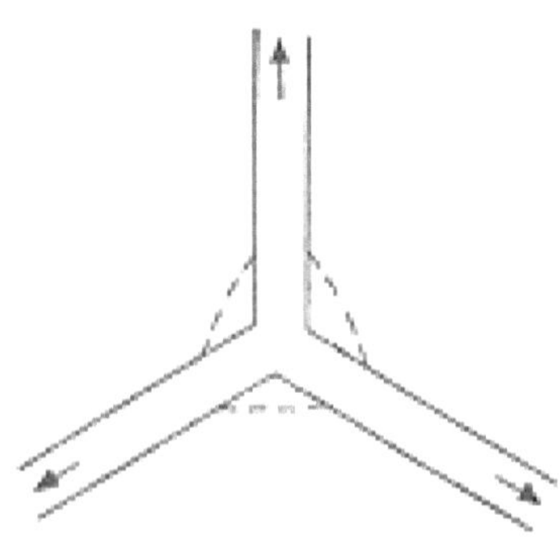
Corners of Cells
FIG 5

In saying above that the forces on these walls, meeting at a corner and making equal angles, were equal, I made a reservation. The statement made would be necessarily true if the walls were hinged together at the corner, but, in fact, there is some rigidity at the corner and especially at the base of the cell where the mid rib is. Now, further examination will show that the forces cannot be exactly equal in an actual comb because the stress due to the weight of the comb is a maximum on the top row of cells and nothing on the bottom edge. If there are 40 rows of cells the weight will diminish by about one-fortieth between each row and the next. This slight graduation of the forces is undoubtedly taken up by the slight stiffness of the structure as a whole. This is assisted also by the thickening of the corners of the cells, which thickening again serves another useful purpose which we may see by examining Fig. 5, which shows in full lines three walls meeting at angles of 120°. Any mechanic will tell you that if these three walls be subjected to tension in the direction of the arrows, even though the forces be balanced about the centre, there will be a tendency to crack at the sharp corners where there will be a concentration of force. This, however, can be got over by the engineer's device of rounding out the corners, a device used by the bees also, as indicated by the dotted lines.

We have seen how the stresses on the walls decrease as the bottom of the comb is approached. It is interesting to examine a finished comb. If in a walled chamber, it will be found built out to the walls and attached to them most of the way down. This serves three purposes. The horizontal stresses are well provided for. The cells at the edges are used to the full and some of the forces depending weight can be transmitted to the side walls. Below a certain level the edges are free and a bee space is left so that the bees have ready access from side to side and do not have to provide the stiff bracing to the side walls which would be partly wasteful. Yet they like to use the cells built around the bottom for breeding purposes for which they require to be full depth. Now if no provision were made these outside cells would be weak, exposing a maximum of thin and unsupported walls and angles. For this reason the bees make a start all round with the neat row of cells but make them very shallow and turn their mouths outward. The structure is worth examining. It provides a nice stiff mechanical edge and at the same time those numerous shallow pockets which, when swarming time comes, are so easily converted into queen cells.

I have noted on worker comb newly made in a skep and consisting only of white wax cells, walls at sides and bottom, equal in thickness and less than four thousandths of an inch thick. On now white drone comb I have noted side walls about six thousandths thick and base walls about nine thousandths thick.

Finally let me put this poser to the observant beekeeper. Fig. 6 shows alternative arrangements of the cell bottom, both obtained vertical way. Now the observer may consider that he, and perhaps the bees, would prefer arrangement A to arrangement B. Let him then try and arrange the foundation in a frame to give arrangement A throughout. He cannot do this.

A Biometrical Study of the Influence of Size of Brood Cell upon the Size and Variability of the Honeybee (Apis mellifera L.)-Grout

By Roy Grout. This Thesis was Submitted to the Graduate Faculty for the Degree of MASTER OF SCIENCE, Iowa State College, 1931.

ACKNOWLEDGMENTS

The author takes this opportunity to express his sincere appreciation and gratitude to all those who have in any way contributed to the task of which this thesis is the culmination. He especially wishes to acknowledge his indebtedness to Dr. O. W. Park for assistance in outlining the several problems involved; to Prof. G. W. Snedecor for assistance in statistical methods and interpretations; to Dr. C. J. Drake for providing facilities for the work; to Dr. G. C. Decker for his kindly interest and many helpful suggestions; to H. C. Dadant for his many helpful suggestions and for furnishing the foundation used in this experiment; and to my wife for assistance in the preparation and measurements of the parts of the worker bee.

INTRODUCTION

A Biometrical Study of the Influence of Size of Brood Cell Upon the Size and Variability of the Honey bee (Apis mellifera L.) by Roy A. Grout, 1931

The keeping of bees for the pleasure derived from them and for the benefits and profits arising from their products has been an important vocation and avocation for many centuries. Due to the interest of naturalists and observers, the study of the size, shape and type of the cells of a colony dates back several centuries before Christ. The first outstanding study of the effect of the size of cell upon the size of the honey bee did not occur until 1791 when Huber (32), at the suggestion of Bonnet, succeeded in raising worker bees in drone cells and drone bees in worker cells. In the former case, he did not notice any change in the size of the emerging worker bee, but in the latter experiment he observed that the emerging drones were smaller. This study initiated a series of observations of this phenomenon but only Zarudski, according to Michailov (43), noticed any increase in the size of the emerging bees. Martynov, Tuenin and Michailov recently have conducted microscopical investigations concerning this matter and all agree in their conclusions that the bees reared in drone cells are larger than their worker bee sisters.

About the beginning of the 20th century, several controversies arose which concerned the size of the honey bee and its dependability upon the size of the cell. The first of these concerned the effect of the age of the comb upon the size of its emerging bees. It is unquestionable that this had a great influence in bringing about the large cell controversy in France and Belgium. At about the same time, the problem of enlarging the bees became an important issue here in America and a long discussion concerning length of the honey bee probescis and its relation to pollination and honey production ensued. The following is a discussion of the above three controversies in the order given.

The controversy concerning the effects of the age of the brood comb upon the size of the emerging bees emanated from the supposition that the cast-off pupa skin, excrement and varnishing of the cell with the emergence of each generation tended to decrease the size of the cells. Many of the more prominent figures in the beekeeping profession took part in this controversy, the majority contending that the age of the comb had no effect upon the size of each succeeding generation due to the lengthening of the cell by the worker bees. As in the case of the worker bees reared in drone cells, microscopical investigations by Tuenin and Michailov showed that the cells of old combs were smaller and that there was an accompanying decrease in size of the emerging bees.

Baudoux (7) in Belgium was the first to conceive the idea of using a larger size of cell by increasing the size of the cell base on the artificial foundation given to the bees. Others who have worked along this line are Pincot, according to Gillet-Croix (26), and Lovchinovskaya (39). The work of the first two has not been of a very scientific nature but convincing to the extent that manufacturing houses are selling foundation with enlarged cells and claiming good results for the use of same.

The problem of raising larger bees and especially bees having longer tongues or a greater tongue-reach has been a topic of great interest in this country since the beginning of the present century. This problem reached its peak of interest when Root (59) observed one of the colonies in his apiaries foraging on the long corollatubes of the red clover (*Trifolium pratense*). He later sold on the market queens from the mother of this same colony with the guarantee that they would produce bees having tongues long enough to acquire nectar from red clover. However, these unusual characteristics were soon lost.

Honey bees are not native to America. Since their introduction into this country, an extensive hybridization has taken place. Recently, methods for controlling the mating of the queen bee by artificial means have been discovered but such methods, to date, cannot be used by the commercial beekeepers due to the intricacy of the technique and the low degree of impregnation.

In consequence of the difficulties encountered in controlled breeding and due to the increased use of foundation having enlarged cells in France and Belgium, the attention of a foundation manufacturer in this country has turned to the study of the effects of the enlarged cell upon the size and productivity of bees.

In this paper, a study has been made of the effect of cell size upon the size of the honey bee. No attempt has been made to study the production of colonies reared on large cell foundation due to lack of time, since an experiment of that kind should cover a period of two or more years. The writer realizes that while the crucial test for the commercial use of enlarged foundation is honey production, the present study should be a strong indication toward that end.

REVIEW OF LITERATURE

A Biometrical Study of the Influence of Size of Brood Cell Upon the Size and Variability of the Honey bee (Apis mellifera L.) by Roy A. Grout, 1931

Naturalists and observers have been interested in the size, shape and types of brood cells of a bee colony for many centuries. These studies date back at least to the fourth century B.C. to that ancient pioneer in the development of science, Aristotle. In a review of the literature of this subject we must also mention such outstanding naturalists as Swammerdam, Reaumur, Maraldi and the mathematician, Koenig, who observed the marvelously consistent shape of the cells of the honey comb and studied the mathematics of its structure.

According to Huber (31), Schirach, who discovered that the worker bee was possessed of the same sex as the queen bee, stated that nothing but certain physical conditions, such as a special food and a more spacious lodging, was needed for the worker bee larvae to become real queens. Desiring the support of an eminent philosopher in his contentions, he corresponded with Bonnet, who in turn corresponded with Huber, who confirmed Schirach's results.

It was the great naturalist, Huber (32), who, in 1791, at the suggestion of Bonnet, first attempted to investigate the effect of the size of cells upon the size of the honey bee. He removed from a hive all combs containing worker cells, leaving only drone combs. Moths invaded the hive, however, and ruined his first experiment. On his second attempt, he observed that the queen finally oviposted with reluctance in the drone comb, but that on the fourth day the bees disappeared from their cells. Upon placing in the hive a frame containing sealed drone brood in small cells, he observed that the bees set about to remove the drone brood from the cells so that the queen might have a suitable place to ovipost. Huber next attempted to rear worker bees in drone cells by grafting the worker bee larvae in the cells of a drone comb by removing the drone larvae and placing them with 1 day old worker larvae. This experiment was successful, but Huber, upon examining the pupa in an advanced stage, did not recognize any difference in the size of the bees. He then repeated the experiment using drone larvae instead of worker larvae. This time he observed that the drones reared in the drone cells were larger than those reared in worker cells.

Many observers since Huber have conducted similar experiments and have observed the phenomenon of worker brood reared in drone cells sealed with level cappings. Thus, Alley (1), in 1869, records placing a queen and three pounds of bees on drone combs with the result that worker bees were reared in level-capped drone cells. He does not state, however, whether he noticed any changes in the size of the emerging brood. Gundelach, according to Michailov (43), placed a swarm of bees in a glass-walled observation hive on drone comb and observed worker bees emerging from the large cells. Berlepsch (9), in 1876, conducted a similar experiment and confirmed the results of Gundelach. He also cited a similar result obtained by Bessel. Other investigations cited by Michailov (43) were made by Gunther, Klempin, Zarudski, Zesselski, Lehzen and Hanneman, and Buttel-Reepen. Pincot, according to Gillet-Croix (26), investigated the phenomenon at the biginning of the present century and Getaz (25) cites Drory of France and a similar case in Germany. None of the above observers, however, investigated the size of the bee by microscopical examination but depended entirely upon visual examination. Only Zarudski claimed that there was any increase in the size of worker bees due to their rearing in drone cells.

The first microscopical examination of the chitinous parts of worker bees reared in drone cells was conducted in 1901 by Martynov (41), who measured 100 proboscides of bees from the apiary of the Moscow Institute. He determined that the average length of proboscis of worker bees emerging from normal worker cells was 6.06 mm., while the length of proboscis of those emerging from drone cells was 7.01 mm., showing an increase of 0.95 mm.

A similar and more extensive investigation was undertaken by Michailov (43), in 1925, who chanced to find in one of his colonies a drone comb containing worker brood sealed with level cappings. Taking advantage of this opportunity, Michailov, who had previously stated that there was no great difference in the size of worker bees reared in drone cells, made a microscopical examination of 6 characteristics of 200 bees, 100 taken from the drone comb and an equal number from a normal brood comb, taken at approximately the same time. He determined by statistical methods that the bees reared in drone cells were significantly larger than their worker-cell sisters and drew the following conclusions in this respect:

1. That worker bees reared in drone cells weigh 11.36% more than worker bees reared in worker cells.

2. That the proboscis increases 4.83% in length due to the effect of the larger cells.

3. That the increase in the size of cell gives a corresponding increase in the size of the right fore wing, an increase in length of 2.69% and an increase in width of 2.06%.

4. That the increase of the size of the cell causes an increase in the sum of the widths of the third and fourth tergites of 4.37%.

5. There is no significant differences in the average number of hooks on the right hind wing between the two groups of bees.

6. That the bees reared in the drone cells are decidedly more variable than their worker-cell sisters except in the case of the number of hooks on the right hind wing where the variation is consistent in both groups and therefore unrelated to size of cell.

A further pursuit of the literature relating to the effect of the size of the brood cell upon the size and variability of the honey bee leads into several more or less distinct controversies and studies which bear upon the subject from various angles. The first of these controversies concerns itself with an extensive discussion of the length of proboscis and its relation to honey storing ability with special reference to ability to forage upon and pollinate red clover (*Trifolium pratense*). The second controversy has to do with the effect of the age of the brood comb upon the size of the honey bee. Both of the above controversies have influenced and have operated to bring about a controversy concerning the use of artificial foundation with an enlarged cell base. The last controversy is concerned largely with the study of the variability of social insects due to the variability of the brood cells. Since the above four controversies occurred at approximately the same time, the writer feels that the best method of approach to a review of their literature is to discuss each subject separately.

1. Length of proboscis and its relation to honey storing ability.

The length of proboscis, its relation to honey storing ability and, in particular, its relation to the acquisition of nectar from red clover rose to a peak in this country at the very beginning of the present century. Previous to that time a few observers had noticed honey bees working on red clover, and Rankin (56) had successfully attempted to breed a strain of bees having long tongues. These remarkable traits soon disappeared, however. Root (59) was the first great disciple of the long-tongued worker bee. He discovered in one of his apiaries a colony which was working on red clover. Upon measuring the tongue reach of these bees, he discovered that this colony had an unusually long tongue reach, 0.21 inches, whereas the average tongue reach was only 0.16 inches. He continued by raising queens from the mother of this colony and sold them for "red clover queens". This desirable quality shortly disappeared due to the inability to control the queen's mating. It is of interest to note that Kulagin (30) measured the length of proboscis of ten bees that were the progeny of four Root queens that had been sent to Russia by Titoff and found that the average length was 6.22 mm. as compared to an average of 6.21 mm. for the common black bees of Central Russia.

The contentions of Root regarding the tongue reach of these bees and their ability to acquire nectar from red clover raised a voluminous controversy which lasted until the middle of the year 1902 when it was dropped as suddenly as it commenced. Beekeeping savants such as Miller (46), Doolittle (22), Dadant (16), Gillette (27), Getaz (25), Cook (15), Swarthmore (64) and many others investigated and wrote concerning the subject. The greater part of the above mentioned believed that the longer proboscis was directly related to a greater storing ability in spite of the fact that very little scientific research was undertaken. The importance of their controversy lies in their contributions to the technique of measurement, the breeding and selection of bees and the influence of their investigations upon further scientific study along similar lines.

Previous to this time, Wankler (70) of Germany, had attempted to breed for length of proboscis and had invented and used by 1882 an instrument for determining the length of the bee's tongue. According to Gotze (29), Wankler was the first to show that the bees of different races may differ in the length of their respective proboscides. Likewise, Charton (13), of France, had invented the Charton glossometer in 1892 and by 1897 had presented some figures which seemed to indicate that bees store in proportion to the length of their proboscides.

In Russia, scientific workers have recognized the importance of the relation of the honey bee to the pollination of the red clover and have, for many years, made a very extensive study of the length of the proboscis of the honey bee and the relation of honey bees to seed production. The Russian territory is well suited to such a study since hybridization and shipping of bees from one part of the territory to another has practically never occurred. Klingain, according to Michailov (45), in experiments conducted from 1908 to 1913, showed in his first experiment that bumblebees pollinated 49.4% of the flowers in comparison to 45% in the case of Caucasian bees. In his second experiment he found that bumblebees pollinated 46% while the honey bees pollinated 31.1% of the flowers. He further calculated that it would be necessary to have one colony of Caucasian bees per acre of red clover in order to insure proper pollination.

Chochlov (14) stated that the minimum average length of the red clover corolla tube was 8.34 mm., that the bees push a part of their heads a distance of 0.65 mm. into the corolla tube, and that the nectar rises in the corolla tube a distance of 1 mm. Thus a bee would have to have a proboscis reach of 6.69 mm. to acquire nectar from red clover. Chochlov also found that the Abkhasian bee and the Kars bee of the Caucasus had a length of proboscis equal to 6.69 mm. The essential part of the work done by Chochlov is his establishing for the first time a technique for preparing the proboscis for measurement. His method consisted of anesthetizing with chloroform, killing in boiling water, boiling in potassium hydroxide solution (KOH) and preserving in oil of cloves after the parts had been washed in water for several days. Ewert (23) conducted a similar investigation of the length of the proboscis in relation to the depth of the corolla tube of red clover and essentially confirmed the results obtained by Chochlov.

A more extensive and accurate work on red clover was conducted by Gubin (30), who, contrary to Chochlov and Ewert, showed that the length of the proboscis would have to be from 7.9 mm. to 8.9 mm. in order to reach the nectar in the corolla tube. Since the bees with the greatest length of proboscis were found by Skorikow (63) to be Caucasian bees having a proboscis length of 7.55 mm., Gubin concludes that real red clover bees do not exist in Russia. He also showed that, in using the technique for preparing the proboscis for measurement as set forth by Choclov, the proboscis and other chitinized parts of the skeleton shrink, particularly upon boiling in potassium hydroxide solution, and that the submentum shrinks most (6.31%), the ligula, 3.82% and the mentum only 1.50%. He further showed that by placing the material in a 30% alcohol solution and running it up to a 70% alcohol solution, the shrinkage seldom surpasses 1.5%. Gorbatscheff (28), in 1929, differed with Gubin and claimed the the Caucasian race is the only one which makes use of the nectar of the red clover, but, according to Gotze (29), the material presented by Gorbatscheff is not capable of complete proof.

Additional tests in Russia regarding the relation of the honey bee to the pollination of red clover were made by Manokhin and Koorochkin, according to Michailov (45). Manokhin reported an increase of tenfold in the seed production of red clover influenced by the flight of Caucasian bees in comparison with an isolated field. One plot isolated from insects produced only 180 grams of seed per hectare and another produced but 225 grams; while a similar area, located near Caucasian bees, produced 46.4 kg. and another located near black bees produced 32.5 kg. of seed. The test by Koorochkin in 1926 showed that common local bees of Northern Russia produced seed in 0.99% of the flowers, Caucasian bees produced seed in 10.42% and bumblebees produced seed in 27.93%. Free blooming clover showed seed production in 38.93% of the flowers while a plot isolated from insects showed only 0.27% pollinated.

In 1929, a Russian worker by the name of Savelyeff (62) made an extensive investigation of the effect of various methods of killing, treating and preserving the chitinized parts of the bee skeleton with special reference to the proboscis. In a comparison of two methods, namely boiling fixation plus preservation in 70% alcohol and direct preservation in alcohol without fixation, he arrives at the conclusion that the differences obtained in a composite measurement of the proboscis (submentum plus mentum plus ligula) are not great enough to be due to any reason other than error in measuring. While the mentum showed an increase over the boiling-fixation method plus preservation in 70% alcohol, the submentum showed a decrease, while the ligula showed no significant dufference in length. Thus, contrary to Gubin, Savelyeff showed that the combined measurement of the three parts showed no significant difference due to the treatment. He agreed with Gubin that the use of a coefficient or constant to correct the length of the proboscis is improbable due to the different behavior of the parts. He differs with Gubin, however, in his conclusion that the correlation among the parts of the proboscis is the same as one would find in an examination of untreated material.

Alpatov (2), in 1930, reported the result of an investigation of proboscides preserved and measured in alcohol in comparison with proboscides boiled in a 5% solution of potassium hydroxide. He found that the treatment with potassium hydroxide solution shortened the total length of the proboscis by 2.6%. At the same time he investigated the tongue length of the Mingrel bees (*Apis mellifera* var. *caucasica* Gorb.) and showed that this race had the longest proboscis of any of the races of honey bees.

In Germany, Gotze (29), in 1927, declared that the probability of error with the Russian technique was so great that it was impossible to obtain correct results. In consequence, he measured the second member of the labial palpi as an estimation of the length of the proboscis. He also made a scientific examination of the ability of the bees to acquire nectar from several varieties of clover. Following the precedent of Ewert, he did not measure the ovary since the nectar stands above it if nectar is present in sufficient quantity. He found that if the nectar rising in the corolla tube reached a depth of 7.25 to 7.5 mm., the bees of one of his colonies, number 47, could regularly acquire nectar, while the rest of his bees could not work red clover unless the nectar rose beyond this point or the bee had a proboscis longer than the mean of its colony. He concluded that, in general, even those stocks having the longest average length of proboscis do not meet the requirements of clover storing ability; that, with each increase in the length of the proboscis, the ability to acquire nectar from red clover increases markedly; and that certain existing varieties of red clover can theoretically be used by those bees having the longest known proboscides. He maintains that a modification of the length of proboscis due to climate factors has yet to be proven by experimental data, in spite of the fact that Russian workers have shown that the length of proboscis seems to decrease going from south to north.

In regard to the question of how much better use the longer proboscis will prove to be, we must mention Merrill (42), who, in 1922, investigated the relation of length of proboscis, carrying capacity and colony strength to honey storing ability. He determined that a correlation between length of proboscis alone and storing ability could not be found, but that the length of the proboscis plus carrying capacity and colony strength was highly correlated with yield. He concluded that: (1) there is a distinct correlation between length of proboscis, carrying capacity and the amount of honey stored; (2) there is a distinct relation between the number of bees found in the colony in the spring and the size of the above named physical characters; (3) that while it is very strongly indicated that it would be advantageous to a bee to excel in all three of these physical characters, yet, if she is deficient in one character, the disadvantage may be overcome if she possesses one of the other characters to a marked degree.

Hutson (33), in 1926, continued this study, working with small numbers of bees, and confirmed the above results showing that there was no marked agreement between length of proboscis and honey stored, but that there was a marked agreement between the number of bees in a colony and the yield.

Higher yields from long-tongued races have been reported rather frequently. In this connection it is well to note that Merrill and Hutson investigated only modifications of one and the same race. Zander (73), in 1919, reported unusually high yields from Caucasian bees in Germany. The constitution of the honey was at the same time totally different, and Zander believed that this race prepared the honey in a different manner since the source must have been the same. Similarly, Tiadmann (67), in 1925, tells how *Cirsium oleraceum* had been used by the Krain bees while the native Hannover bees could not acquire nectar from this source. According to Gotze (29), other high yields have been reported by conscientious observers such as Tuschoff and Braun in 1927, but whether these yields were based upon length of proboscis or other properties is not known.

Further contributions to the question concerning the acquisition of nectar from red clover have been presented at intervals over a period of years in this country by outstanding figures such as Folsom (24), Phillips (51), Dadant (17), Demuth (20), Dietz (21), Pammell (48), Robertson (58), Pellett (50), and Burrill (11). Robertson believed that the bees perforated the corolla in order to reach the nectar. Dadant confirmed this statement. Pellett believed that bees could reach the nectar in some of the flower tubes, especially in years of drought. Pammell observed honey bees on red clover but he did not believe, after measuring the corolla tube, that the tube is shortened enough by drought to enable the bee to reach the nectar in it. Burrill cited many ways in which the honey bee could acquire nectar from red clover by means of perforations of the corolla tube due to other insects, honeydew secreted by aphids on clover, and sap leaks.

2. **Age of comb controversy.**

The controversy concerning the age of comb and its possible effect upon the size of the emerging brood dates back at least to the middle of the 19th century. Quinby (54), in his book published in 1865, states that old combs are good for a period of 10 to 15 years due to the fact that while the bottoms of the brood cells are filled with cast-off pupa skins, the cocoon, excrement and varnishing resulting from each generation, the side walls are only slightly thickened and the bees lengthen them to compensate for the thickening at the midrib. It is Quinby's opinion that the cells are unquestionably built larger than necessary. Dadant (18), Miller (47), Root (60), and others took part in this controversy and all believed that the age of the comb did not materially affect the size of the emerging bees.

This controversy first originated in Europe and then spread to our country. Riedenbach (57), of Germany, showed by filling new and old combs with water that although the midrib is noticeably thickened by the emerging generations, the cell volume is not greatly changed due to the lengthening of the side walls. Ludwig (40) coincided with Riedenbach in his contentions and by actual measurement showed that there is no difference existing in the general roominess of brood cells of old and new combs. Brunnich (10) believed that the danger of the cells of old brood combs becoming smaller with age was not as great as many beekeepers believed. Rambaldi (55) reports that in 1927 he had kept Palestine and North African bees on combs built from Root foundation having 856 cell bases per square decimeter during hundreds of generations and had noticed no effect upon the size of the bees.

In Russia the study of the effect of old combs upon the size of the emerging bees was first attempted by Tuenin (68), who showed, by weighing the emerging bees from combs from which 2, 6, 28, and 38 generations had emerged previous to the experiment, that the weight decreased with the number of generations from 0.12612 gms. to 0.10695 gms. and that the cell diameter decreased from 5.262 mm. to 4.99 mm. He concluded that as the number of generations that emerged from the cells became more numerous, the resulting bees became smaller as indicated by the weight of the emerging bees.

Michailov (44) continued this study by measuring 5 physical characters of the skeleton, namely, length of proboscis, length of the right fore wing, width of the same wing, the summation of the widths of the 3rd and 4th tergites, and the number of hooks on the right hind wing. He showed that the size of the cells as reduced by the emerging generations (5.89% in the diameter due to 16 and 18 generations) is accompanied by a significant reduction in the size of bees. By reducing the cell diameter by approximately 3% there was no significant reduction in the body size of the emerging bees. the depth of the cell showed no influence upon the size of the bee. He concluded that, in order to enlarge the bee by using artificial foundation, we should pay particular attention to the diameter of the cell and not to the depth of the cell. According to Rupp (51), who based his conclusions on the work of Michailov, a comb is too old to use for brood rearing when it is three years old based upon the figure of 5 to 6 generations per year.

3. The enlarged cell controversy.

With the invention of artificial comb foundation by Mehring, in 1857, a control of the size of the cells built by honey bees was first accomplished since it was discovered that the bees would build cells with the same dimensions as the imprint of the cell base upon the artificial foundation. The importance of his invention was the elimination of excess drone comb resulting in colonies composed almost entirely of worker comb. It also initiated a study of the exact size of the cells built by bees.

According to Dadant (19), Collin measured the dimensions of cells and stated in 1865 that there were 854 cells per square decimeter. Langstroth repeated the experiment and calculated that there were 838 cells per square decimeter and Charles Dadant confirmed his results. According to Baudoux (8), the following are results concerning the size of the cells of natural comb:

The house of Fratelli Piana in Italy calculated from measurements that there were 860 cells per square decimeter; another house in Italy measured comb from three different colonies and found that there was a considerable variation in the size of cells, namely 813, 807 and 854 cells per square decimeter. Baudoux measured combs taken from two different colonies and found that while the cells of one colony measured 854 cells per square decimeter, the cells from the other measured but 807 cells per square decimeter.

Concerning the size of cells built by different races of bees, Pincot, according to Gillet-Croix (26), reports that the Italian race builds 764 cells per square decimeter, that the bees of Burgundy build 798, that the common black bee native to France builds 854 and that a "degenerated common bee" builds 924 cells per square decimeter. Halleux, in 1890, according to Szezawinski (65), calculated that the black native bees build 845 cells per square decimeter. Rambaldi (55) records the North African bee as building 940 cells per square decimeter.

Baudoux (7), of Belgium, was the first to advocate the use of artificial foundation with an enlarged cell base. In 1893, he reports that a Mr. Fromont measured natural combs and found that the greater part had 825 cells per square decimeter in comparison with certain sheets of artificial foundation which had as high as 907 cells per square decimeter. Baudoux, struck by the reduction in the size of bees from an old skep containing combs having 912 cells per square decimeter, conceived the idea of raising bees in enlarged cells. He accomplished this by means of stretching normal foundation to the size he desired and had by 1896 sufficiently proved his point in Belgium, that a manufacturing company began to place upon the market artificial foundation having an enlarged cell base. It was Baudoux's contention that the nurse bees, following a natural instinct, filled the bottom of the larger cell more copiously with larval food, that this resulted in a larger bee, He also intimated that the larger bee would generate more body heat which would result in a greater quantity of brood.

By means of stretching foundation, he experimented with various sizes of foundation having 750 cells per square decimeter, 740, 730, 710, 700 and down to 675 cells per square decimeter. By means of a glossometer he determined the tongue reach of his colonies and by means of a thoraxometer, the diameter of the thorax. He found that with an increase of 50 cells per square decimeter in the size of the foundation, there was a corresponding decrease of 0.5 mm. in the tongue reach. His thoraxometer gives a diameter of the thorax as 3.7 mm., 3.9 mm., 4.1 mm., and 4.3 mm. for the bees reared in cells built from foundation having respectively 850, 800, 750 and 700 cells per square decimeter. He arrived at the conclusion that foundation having 700 cells per square decimeter gave a bee which was superior in all its measurements to those reared in combs built from the smaller sizes.

Independent of the work done by Baudoux, Pincot, according to Gillet-Croix (26), arrived at the idea of rearing bees in enlarged cells from a slightly different angle. Noticing the difference in size of the bees from a swarm placed on foundation and the bees of the parent stock reared in natural comb, Pincot came to the conclusion that this phenomenon was due to the natural cells being larger than those drawn from the foundation and actual measurements confirmed his theory. He then started experimenting with foundation having 736 cells per square decimeter and reports that during a two-year period 30 colonies using this size of foundation gathered approximately one-third more honey than did 30 colonies on normal foundation. In 1910 his apiaries were destroyed by a flood and Pincot was forced to abandon his experiments.

Lovchinovskaya (39), of Russia, reports, after present investigations in 1930, that an investigation concerning the effect of enlarged cells upon the size and activity of the honey bee was undertaken in that country in 1925.

For this purpose artificial foundation was made with an enlarged cell base of 5.85 mm. against 5.45 mm. in normal cells. The results of one experiment proceeding for 2 years with 10 colonies showed that the honey bees placed on the enlarged cells lived a normal life, both the worker bees and the queen worked normally. When only one frame of the enlarged cells was placed with nine frames of the normal size in a colony, the queen's attitude toward the enlarged cells was changed and she did not oviposit in the comb in spite of the fact that the bees worked upon it as they did on the other nine combs. When the reverse experiment was undertaken, with nine enlarged combs and one normal comb, the queen laid in the enlarged cells at once. Lovchinovskaya concludes that the worker bees regard with indifference the enlarged cells while the queen prefers the smaller cells.

He continued extensive investigations and showed that: (1) Bees from the enlarged cells weigh more than bees reared in normal cells. (2) The average weight of bees for a single year varies according to the conditions operating during the year. (3) The enlarged bees as well as the normal bees are heaviest during May and decrease in weight during the next three months. (4) The weight of bees leaving the hive is greatest during the first half of the day. (5) The weight of emerging bees is between 5% and 6% greater for bees emerging from the large cells. (6) That the emerging bees of both groups weigh more than the old bees. (7) In the case of bees returning to the hive, the weight of enlarged bees exceeds the weight of normal bees by 10.7%, while in the case of bees leaving the hive, the enlarged bees exceed by only 4.8%. (8) The load carried in the honey stomach of the enlarged bee is 52.6% heavier than that carried by the normal bee. (9) That the normal bee carried a load equal to 14% of her own weight while an enlarged bee carries a load equal to 20.4% of her own weight. (10) From the results of one season, the bees reared in the large cells gather more honey than those reared in normal cells, but the production cannot be judged from the data of one season.

4. Studies on the variability of the honey bee with special reference to size of brood cell.

The first paper giving data upon the variability in bees appears to be that of Koshevnikov (36), who, in 1900, studied the number of hooks on the hind wing. In 1905, the same author presented data which could be arranged in the form of a correlation table. Landacre (38), in 1901, counted the number of hooks on the hind wing but, according to Phillips (52), his data was not presented in statistical form. In 1903, Casteel and Phillips (12) made a biometrical study of the wing venation of the drone and worker bees and arrived at the conclusion that the uncreased variability of the drone was due in part to the increased variability of the cells in which they were reared.

Bachmetjew (6), in 1903, began a study of the variability of the hooks on the hind wings of honey bees. His results and conclusions have been strongly criticised in the literature, and Pearl (49), in 1910, wondered whether Bachmetjew was really serious or whether he was attempting to perpetrate a great biometrical joke. Phillips (53) recalculated the data presented by Bachmetjew in 1909 and discovered that the results were entirely normal and agreed with the results of other investigators.

Kellogg and Bell (35), in 1904, showed that there was greater variability in single wing veins than in the length or breadth of the entire wing and that there was a greater variability in the number of hooks on the hind wing than in the wing venation. This variability was as great in workers as it was in drones. Kellogg (34) later made a further investigation and concluded that, except for the number of hooks on the hind wing, drones were more variable than workers. He also found that the variability of drones reared in worker cells was greater than drones reared in drone cells, and stated that this greater variation was not due to special extrinsic factors such as size of cells.

In Russia many studies have been made upon the variability of the honey bee and the factors which influence variation. These investigations have been carried out on a large scale and have been calculated by statistical methods; they constitute a large part of our knowledge concerning the variability of the honey bee. Among these contributions are the works of Michailov, Alpatov, Tuenin, Choclov and others as cited by Alpator (3) in 1929.

Mention should also be made of biometrical studies on social insects, such as the work of Wright, Lee and Pearson (72), in 1907, on *Vespa vulgaris* from a single nest and by Thomson, Bell and Pearson (66), in 1909, on a general wasp population, In the first named paper, the authors re-examined the data of Casteel and Phillips and agreed that those data show a greater variability in drones than in worker bees, a condition which is reversed in the case of *Vespa vulgaris*. Other biometrical studies of note concerning social insects include the studies of Warren (71) on termites and the seasonal variation occurring in their forms, Alpatov and Palenitschko (4) who worked on different species of ants, and Arnoldi (5) who presented data concerning the variability of the ant *Cardiocondyla stambulowi* Forel.

EXPERIMENTAL

A Biometrical Study of the Influence of Size of Brood Cell Upon the Size and Variability of the Honey bee (*Apis mellifera* L.) by Roy A. Grout, 1931

A. Purpose of Study

The purpose of this experiment is to scientifically study the increase in size and variability of the worker bee as influenced by its rearing in brood cells constructed by worker bees on artificial foundation having an enlarged cell base. Three different cell sizes were used in this experiment. The number of cells per square decimeter were, 857, 763 and 706. The foundation having 857 cells per square decimeter is the standard commercial size manufactured in the United States while the two latter sizes approximate that having 750 cells per square decimeter which has been manufactured since 1896 by Jos. Mees Sons of Herenthals, Belgium and that having 700 cells per square decimeter which the same firm has manufactured since 1927.

B. Methods and Materials.

The foundation used in the experiment was furnished by Dadant and Sons, of Hamilton, Illinois. The sizes furnished were 857 cells per square decimeter (standard size manufactured in the United States), 763 cells per square decimeter and 706 cells per square decimeter. The two latter sizes were selected by Mr. H. C. Dadant and are approximately the same sizes as those being placed on the market by foundation manufacturers in Belgium and parts of France. The cutting of special dies and the manufacturing of the foundation were personally superintended by Mr. Dadant in order that the resulting cell bases should be true hexagons. The foundation received with the first shipment contained 7 vertical wires embedded in each sheet of wax. In addition to the vertical wires 4 horizontal wires were placed in the frames and embedded in each sheet of wax by hand. Some trouble was experienced with this foundation due to its warping between the embedded wires in warm weather. In the second shipment the foundation contained 10 wires embedded in the vertical position, which, when placed in a frame wired with 4 horizontal wires, did not warp and resulted in perfect combs when drawn out by the bees.

Some combs were used which had been constructed from special foundation placed in certain colonies during the summer of 1929 by Dr. O. W. Park. To facilitate recognition and handling of the combs, the system used by Dr. Park in marking the frames was followed in this experiment. The frames containing the standard-size foundation, having 857 cells per square decimeter, were marked "A" and one notch was cut in the top-bar. Likewise, the frames containing the foundation having 763 cells per square decimeter were marked "B" and two notches were cut in the top-bar; while the frames containing foundation having 706 cells per square decimeter were marked "C" and three notches were cut in the top-bar.

Since it is a well established fact that under normal conditions bees will extend the side walls of the cell and construct a comb containing cells of the same diameter as the imprint of the cell base on the artificial foundation, no control of size of cell other than special foundation was exercised.

Frames containing all three sizes of foundation were placed in each of 23 colonies of the Iowa State College Apiary early in the summer of 1930. In general, two frames of each size were placed in each colony. Individual colony records were kept and the queens were marked by clipping the right wings of those reared in an even-numbered year and left wings of those reared in an odd-numbered year.

An effort was made to collect the bees upon emergence from all three sizes of cells in a single colony at approximately the same time and under the same conditions. For this purpose a chart was made whereby the daily emergence of the bees from each size of cell was kept for all of the 23 colonies. Each frame was caged in a Root Nucleus Introducing Cage a day or two before the time of emergence and a selected area of brood was covered with an additional small screen cage insuring that the emerging bees would have no access to any nectar or honey. During the honey flow, it was often difficult to find an area of brood that did not contain some uncapped cells of nectar and honey, and bees were not collected from such combs.

Each sample collected from a brood comb contained at least 50 bees. During the summer of 1930, over 6000 bees were collected. During June of 1931, over 600 bees were collected. From these collections, approximately 3500 were selected as being most suitable for the experiment. The bees of this group were in sets of 150 bees consisting of three samples of 50 bees each taken from each of the three sizes of cells from the same colony, from the same mother and at approximately the same time.

After collecting each sample, the bees were slightly anesthetized, either with ether or calcium cyanide, and then killed by dropping into boiling water. This method of killing, as shown by Alpatov (3), caused the proboscis to be fully extended. The sample was then preserved in a 70% alcohol solution for further treatment.

The general plan of procedure for measuring the size of the individual bees of a sample consisted of the following treatment: (1) Determining the weight of the individual bee. (2) Dissecting the right fore wing, the third tergite, the fourth tergite and the proboscis of each individual bee. (3) Mounting these parts for measurement. (4) Measuring the parts.

Experiments showed that an individual bee taken from a 70% alcohol solution, dried for a few minutes on a filter paper and placed on a chemical balance lost weight faster than it could be accurately weighed. It was thought best, therefore, to take the individual dry weight of each bee. Further experiments were run which showed that, by removing the bees from the 70% alcohol solution, drying on filter paper for several minutes to remove excess preservative and placing the sample in a De Khotinsky Constant Temperature Oven Appliance at a constant temperature of 70 ° C for 48 hours, the individual bees of the sample no longer lost any appreciable weight. The sample was then placed in a desiccator containing concentrated sulfuric acid. Further experiments showed that after 72 hours the bees had become thoroughly dried and no appreciable loss of weight occurred.

The individual bees were then taken from the desiccator and weighed by means of an Eimer and Amend chemical balance accurate to 0.1 mg. A container of fresh calcium chloride was kept within the chemical balance at all times to dehydrate the contained atmosphere. It was also found that the repeated opening of the desiccator during a long series of weighings caused the individual bees to increase in weight. This necessitated the weighing of a test sample at intervals during an extended series of weighings to determine the average gain in weight of the individual bees. All weights given in this experiments have, therefore, been corrected for this factor.

After weighing, each bee was placed in a numbered vial containing tap water at room temperature and throughout the following treatment was recognized as a definite individual. After being in the water for 24 hours, the bees were soft enough for dissection. With the aid of a Spencer Binocular Microscope, containing a 3.5x ocular and a 55 mm. objective, and an ordinary dissecting set, the right fore wing, the third tergite, the fourth tergite and the proboscis of each bee were dissected. The dissected parts were then mounted directly upon numbered glass slides with Bueston's medium* and cover glasses were applied.

*Bueston's Medium for Mounting

Water	50 c.c.
Glycerin	20 c.c.
Gum Arabic	40 gm.
Chloral Hydrate	50 gm.

Dissolve Gum Arabic in water. When dissolved, add Chloral Hydrate. When this is dissolved, add Glycerin. Filter.

All linear measurements were taken by a projection method. The numbered glass slide was placed in a Leitz Simple Micro-Projector in a vertical position and projected upon a movable screen attached to the opposite wall. Upon the face of the screen was a horizontal and vertical scale and the screen was so constructed that the entire face could be rotated around its center in a plane perpendicular to the line of projection. This feature greatly facilitated measuring the projected parts since the measuring scale could be turned to any desired angle at which the part to be measured might happen to lie. The projection measurement apparatus was arranged so that a glass Spencer stage micrometer, having a scale 2 mm. in length ruled to 0.01 mm., placed in the Micro-Projector gave a corresponding projection of 2 mm. magnified 127 times on the scale of the movable screen.

The apparatus was calibrated by this method before and at intervals during each long series of measurements. It was thus possible to read directly the exact measurement of the part in hundredths of a millimeter. However, for the sake of convenience and in order to eliminate any personal equation involved in the reading of actual measurements of the parts of the bee, a reading was taken at the beginning of the part and another at its end, the true measurement being the difference between the two readings. Plate 1 diagrammatically shows the measurements taken on the right fore wing, the third tergite and the fourth tergite. Following the system used by Michailov (43), the widths of the third and fourth tergites were combined and the summation of the two widths was used throughout the computation.

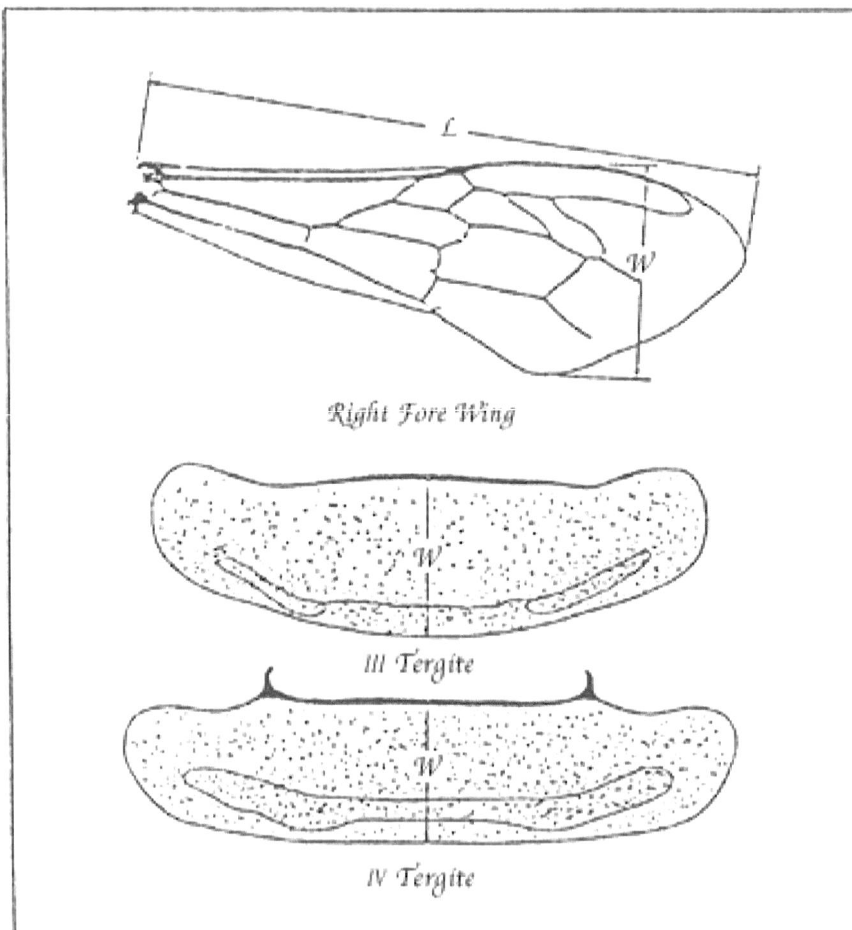

Plate 1. Diagram showing measurements of right fore wing and tergites 3 and 4.

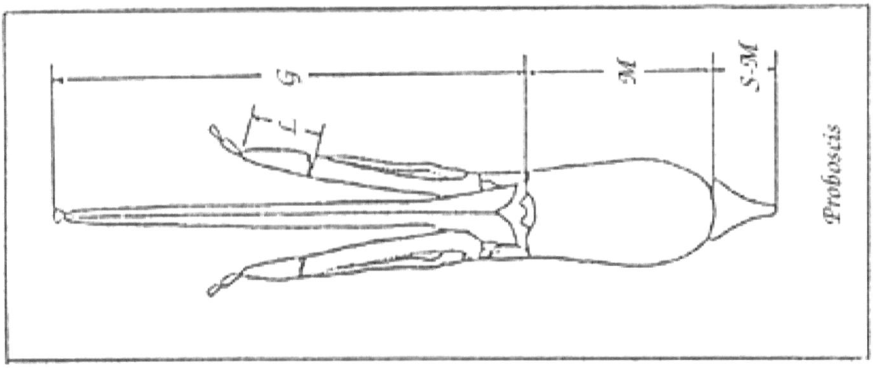

Plate 2

Plate 2 diagrammatically shows the measurements of the proboscis. In this manner the length of the submentum, the length of the mentum and the length of the glossa were obtained, the summation of the three lengths being the length of the proboscis. In only one group of bees was the length of the second member of the labial palpi taken.

Plate 2. Diagram showing measurements taken of the proboscis. G = from tip of labellum to anterior part of mentum. M = length of mentum. S-M = length of submentum. L = length of 2nd member of labial palpi. Length of proboscis = G + M + S-M.

The computation of the statistics was accomplished by recording the values of the measurements of each individual bee on a Hollerith Electric Tabulating and Accounting Machine. From the summations obtained in this manner, the arithmetic means, standard deviations, correlation coefficients, regression equations and other statistical constants were computed with the aid of a Monroe Calculating Machine. All formulas and methods used in the above computations are given by Wallace and Snedecor (69) in their bulletin entitled "Correlation and Machine Calculation" as revised by Snedecor in 1931.

C. Presentation of Data.

1. The size of the worker bee as influenced by size of brood cell.

A study of the three sizes of cells used in this experiment and their relation to each other reveal the following data. To facilitate an understanding of this and following data the size of the cell contained in a comb having 857 cells per square decimeter will be designated as size of cell "A"; the size of cell contained in a comb having 763 cells per square decimeter will be designated as size of cell "B"; in a similar manner the size of cell contained in a comb having 706 cells per square decimeter will be designated as size of cell "C". Between the sizes of cells "A" and "B" there is a reduction of 94 cells per square decimeter and between the "B" and "C" sizes there is a reduction of 57 cells per square decimeter, making a total reduction of 151 cells per square decimeter between the "A" and the "C" size. It was also thought advisable to investigate the increase of linear measurement of the cell and it was found that there was an increase of 5.98% in the diameter of the cells between the "A" and the "B" size, an increase of 3.96% between the "B" and the "C" sizes and an increase of 10.18% between the "A" and the "C" sizes of cells.

From the bees collected during the summer of 1930, data are presented in this thesis on the bees from three colonies. A sample containing at least fifty bees was collected from colony 25 from an "A" comb on August 21, 1930. One week later, on the 28th of August, two samples of bees were collected from a "B" and a "C" comb, respectively. The bees from colony 21 were collected within a period of two days, two samples being collected from an "A" and a "B" comb, respectively, on August 18, 1930, and a third sample from a "C" comb on August 20, 1930. The bees from colony 18 were collected over an extended period of time. One sample was taken from an "A" comb on August 30, 1930, another from a "C" comb on September 7, and the third from a "B" comb on September 23. The individual hive records of these three colonies show that the bees from each colony were the progeny of the same mothers.

From the samples of bees collected from colony 25, complete data were obtained on 44 bees of the sample from the "A" comb, 47 bees from the sample from the "B" comb and 45 bees from the sample from the "C" comb. Similarly, data are presented on 40 bees from the "A" comb from colony 21, 43 bees from the "B" comb and 45 bees from the "C" comb. In the case of colony 18, complete data were obtained on 41 bees from the "A" comb, 48 bees from the "B" comb and 50 bees from the "C" comb.

The influence of the increase in the size of the brood cells upon the size of various measurements taken on the parts of the individual worker bees of colony 25 is shown in Table 1. The measurements presented for comparison, given in column 1, are dry weight, length of right fore wing, width of right fore wing, sum of the widths of the third and the fourth tergites, length of proboscis, length of mentum, length of glossa and the sum of the lengths of the mentum and the glossa. In the second, fourth and sixth columns are given the arithmetic mean of each measurement and the standard deviation of the mean on the groups of bees from each of the three sizes of cells. In the third column are given the differences between the arithmetic means of the bees from size of cell "A" and the bees from size of cell "B" and the standard deviation of the mean difference. The differences between the means of the bees from size of cell "B" and the bees from size of cell "C" are presented with their standard deviations in column 5. Similarly, the differences between the means of bees from size of cell "A" and size of cell "C" are presented with their standard deviations in the seventh column.

The values of the mean differences that are statistically significant are starred. The test for significance was accomplished by dividing the mean difference by its standard deviation and comparing the resulting values with the corresponding "t" values given in Table 16 by Wallace and Snedecor (69).

Table 1. Influence of Size of Cell Upon Size of Bee. (Colony 25)

Measurement Taken	Size of Cell A M±	Mean Diff. of A and B M.D.±	Size of Cell B M±	Mean Diff of Band C M.D.±	Size of Cell C M±	Mean Diff. of A and C M.D.±
Dry Weight in mgs.	13.10000 ±0.0939	2.0302* ±0.1427	15.1302 ±0.1075	4.6878* ±0.2960	19.8178 ±0.2758	6.7178* ±0.2913
Length of right fore wing in mm.	9.6075 ±0.0229	0.0578* ±0.0277	9.6655 ±0.0155	0.0856* ±0.0234	9.76509 ±0.2758	0.1434* ±0.0288
Width of right fore wing in mm.	3.2836 ±0.0109	0.0345* ±0.0141	3.181 ±0.0090	0.0381* ±0.0120	3.3562 ±0.0079	0.726* ±0.0135
Sum width of 3 & 4 tergites in mm	4.8545 ±0.0148	0.1087* ±0.0223	4.9632 ±0.0167	0.0721* ±0.0235	5.0353 ±0.0165	0.1808* ±0.0222
Length of proboscis in mm.	6.5916 ±0.0190	0.614* ±0.0224	6.6530 ±0.0118	0.750* ±0.0188	6.7280 ±0.0146	0.1364* ±0.0260
Length of mentum in mm.	1.7477 ±0.0062	0.0261* ±0.0082	1.7733 ±0.0053	0.0020 ±0.0079	1.7758 ±0.0059	0.0281* ±0.0086
Length of glossa in mm.	4.2832 ±0.0147	0.0517* ±0.0172	4.5349 ±0.0090	0.0551* ±0.0152	4.3880 ±0.0125	0.1048* ±0.0192
Length of glossa & mentum in mm.	6.0316 ±0.0164	0.0771* ±0.0196	6.1807 ±0.0107	0.0551* ±0.0175	6.1638 ±0.0138	0.1322* ±0.0214

***Mean difference is statistically significant**

All values of the mean differences between the 8 characters of bees from size of cell "A" and bees from size of cell "B" are statistically significant and are therefore greater than would be the case if the bees were selected at random from the same population. In a comparison of the values of the bees from size of cell "B" and size of cell "C" it is shown that, with the exception of length of mentum, all means of the bees from size of cell "A" and size of cell "C" differ significantly in all eight cases.

In Graph 2 is presented a frequency diagram of the character dry weight for each of the three samples of colony 25. An examination of the three curves, each representing the frequency distribution of the bees from one size of cell, shows that not only do the arithmetic means and peaks of the curves differ widely but also the distributions, since the curves scarcely overlap. There is also a difference in the types of the curves. While the curves representing the frequency distribution of the dry weight of bees from size of cell "A" and size of cell "B" are quite similar, the curve representing the dry weight of the bees from size of cell "C" shows a much more extensive distribution.

The percent increase of the linear measurements on the various parts of the bees are shown diagrammatically in Graph 1 and the percent increase of all measurements is given below in tabular form.

Measurement Taken	% Increase from "A" to "B"	% Increase from "B" to "C"	% Increase from "A" to "C"
Dry weight	15.50%	30.98%	51.27%
Length of right fore wing	0.60	0.89	1.49
Width of right fore wing	1.05	1.15	2.21
Sum of widths of 3rd and 4th tergites	2.24	1.45	3.72
Length of proboscis	0.93	1.13	2.07
Length of mentum	1.49	0.11	1.61
Length of glossa	1.21	1.22	2.45
Sum of lengths of mentum and glossa	1.28	0.90	2.19
Avg. of 7 linear measurements	1.37	1.00	2.38

Graph 1

Diagram showing percent increase of linear measurements.

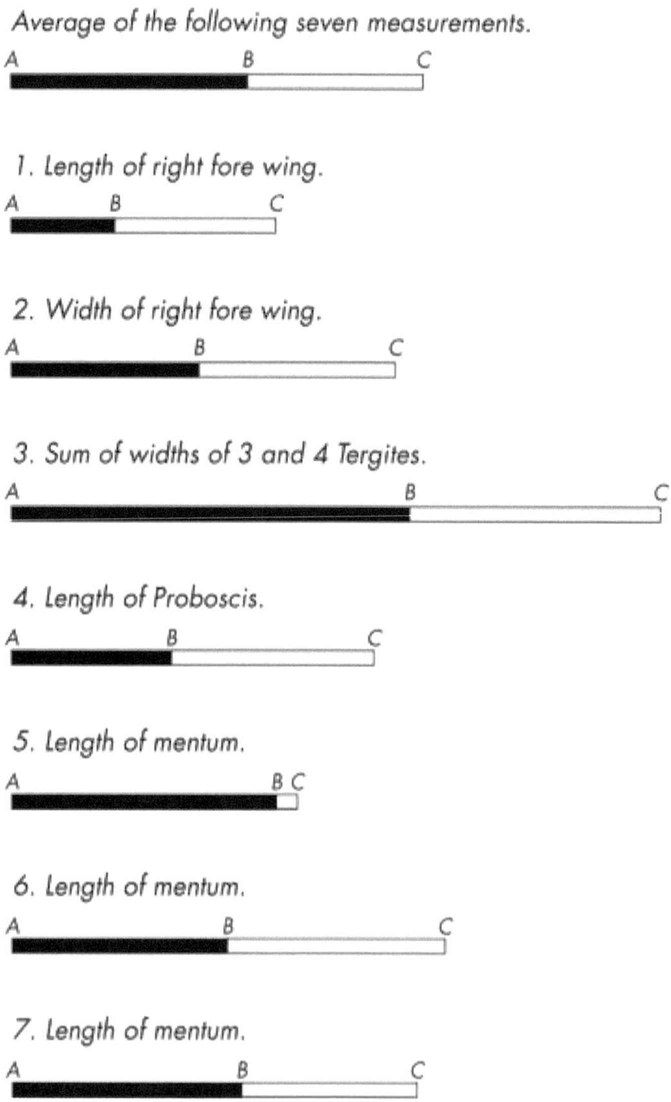

Graph 2

Dry Weight Frequency Diagram.

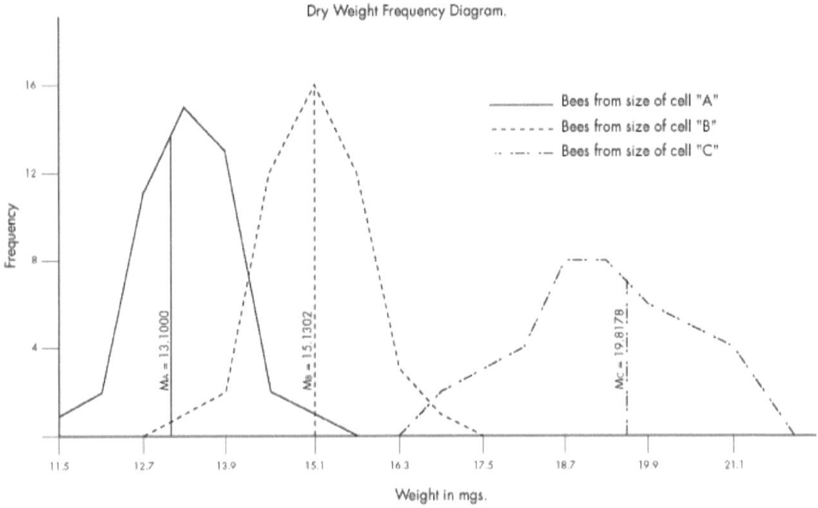

The frequency distribution of the character length of right fore wing is presented graphically in Graph 3. It is of interest to note that there is a trend in the peaks of the curves and the frequency distributions showing that not only do the means of the groups differ significantly but that the group as a whole tends to increase in size as the size of the cell, with which it is associated, increases.

The frequency distributions of the characters, width of right fore wing, sum of the widths of the third and the fourth tergites and the length of proboscis are presented graphically in Graphs 4, 5 and 6, respectively. The curves of Graphs 4 and 5 show that there is a trend both in the peaks of the curves and in the curves themselves toward a larger value of the measurement of the character with an increase in the size of the cell from which the sample is taken. The curves represented in Graph 6 do not show the familiar trend between the peaks of the two curves representing the length of proboscis of bees from size of cell "A" and size of cell "B". There is, however, a distinct difference between the peaks of the curves representing this character for bees from size of cell "B" and size of cell

"C". The trend of the distribution curves in all three cases is from a smaller to a larger length of proboscis as the size of cell is increased.

Graph 3

Length of Right Fore Wing - Frequency Diagram.

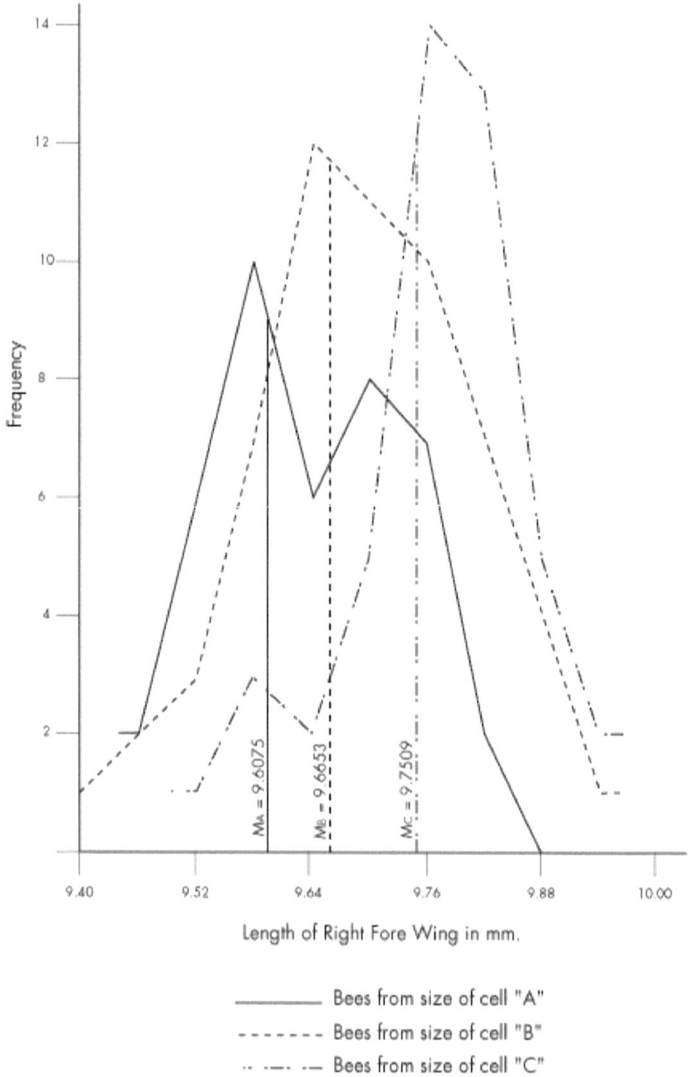

Graph 4

Width of Right Fore Wing - Frequency Diagram.

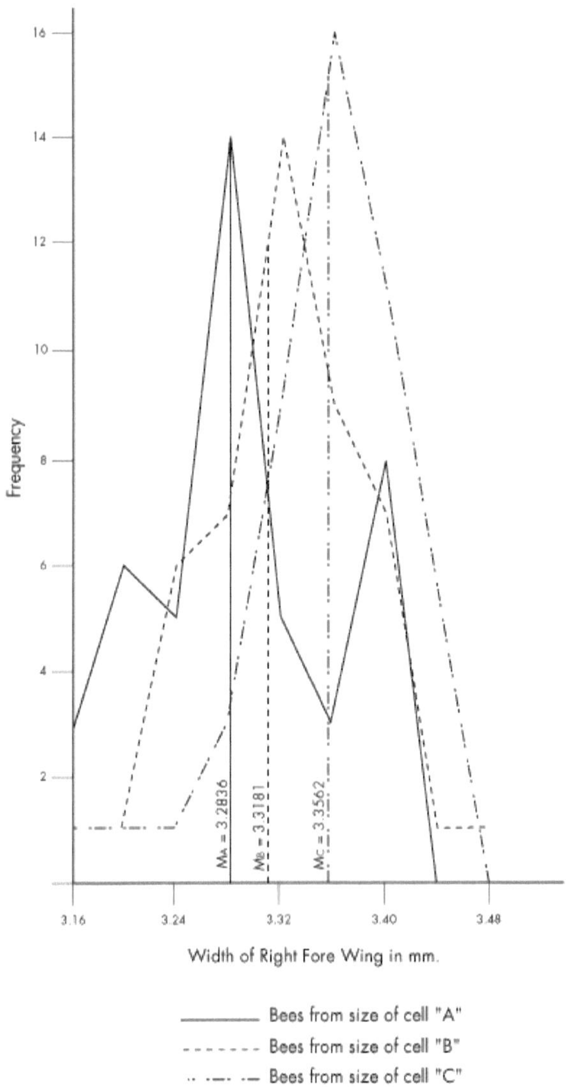

Graph 5

Sum of Widths of 3 and 4 Tergites - Frequency Diagram.

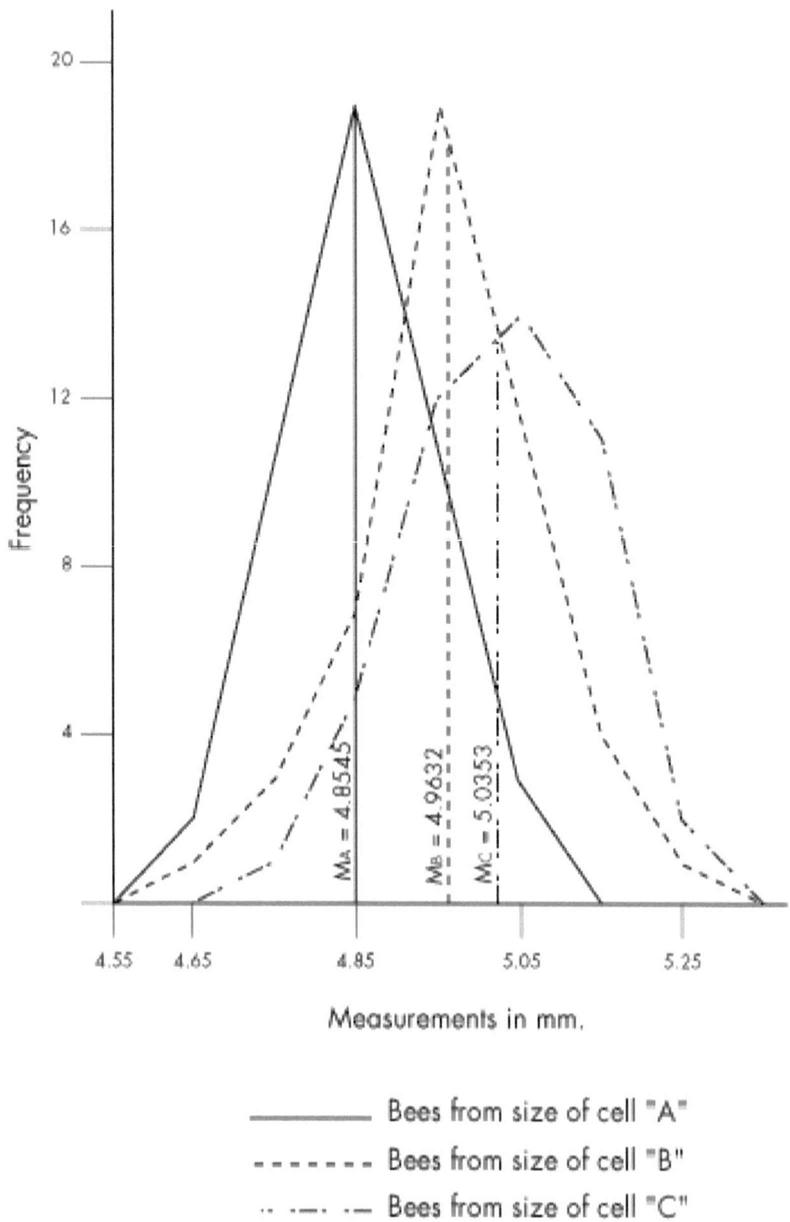

Graph 6

Length of Proboscis - Frequency Diagram.

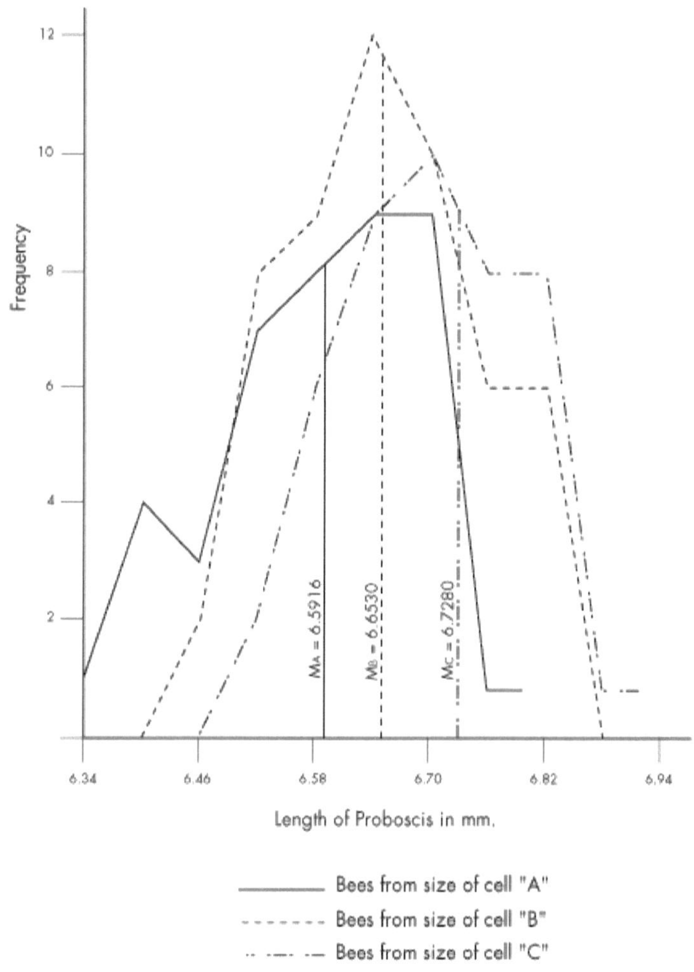

It is also of interest to investigate the percent increase of the arithmetic means of the various measurements of the bees as the size of the cell is increased. Dry weight, which is a measurement of mass and consequently volume, increases markedly from 15.50% to 51.28%, as the size of cell increases. The linear measurements show an average increase of 1.37% as the area of the cell base is increased 12.32%, an average increase of 1.00% as the area of the cell base is increased 8.07% and an average increase of 2.38% as the area of the cell is increased 21.39%. By comparing the average percent increase of the linear measurements to the increase of the diameter of the cell it is discovered that as the diameter of the cell is increased 5.98% there is an increase of 1.37% in the average linear measurement, that, with an increase of 3.96% in the diameter of the cell there is an average increase of 1.00% in the linear measurements and that an increase of 10.18% in the diameter of the cell is accompanied by a corresponding increase of 2.38% in the average measurements of the worker bees. It is of interest to note here that the ratios of the percent increases of the diameters of the cells are approximately the same as the ratios of the percent increases of the average dimensions of the worker bees.

Table 2. Influence of Size of Cell Upon Size of Bee. (Colony 18)

Measurement Taken	Size of Cell A M±	Mean Diff. of A and B M.D.±	Size of Cell B M±	Mean Diff of Band C M.D.±	Size of Cell C M±	Mean Diff. of A and C M.D.±
Dry Weight in mgs.	16.1020 ±0.2054	0.7895* ±0.2422	15.3125 ±0.1284	16115* ±0.1542	16.9240 ±0.0854	8.8220* ±0.2224
Length of right fore wing in mm.	9.6390 ±0.0241	0.0619 ±0.0360	9.5771 ±0.0268	0.1551* ±0.0522	9.7302 ±0.0175	0.0912* ±0.02996
Width of right fore wing in mm.	3.3210 ±0.0115	0.0094 ±0.0146	3.3304 ±0.0093	0.0140 ±0.0123	3.544 ±0.0080	0.0254 ±0.0138
Sum width of 3 & 4 tergites in mm	4.9212 ±0.0187	0.0065 ±0.0260	4.9275 ±0.0181	0.0605* ±0.0276	4.9880 ±0.0208	0.0668* ±0.0280
Length of proboscis in mm.	6.5778 ±0.0166	0.0130 ±0.0314	6.5909 ±0.0266	0.0794* ±0.0305	6.6702 ±0.0150	0.0924* ±0.0224

***Mean difference is statistically significant**

The influence of the increase in the size of the brood cells upon the size of the worker bees from colony 18 is shown in Table 2. The data presented in this table consist of the arithmetic mean and the standard deviation of the arithmetic mean of the bees from the three sizes of cells and the mean difference and the standard deviation of the mean difference between each of the three groups. The measurements presented in the table are dry weight, length of right fore wing, width of right fore wing, sum of the width of the third and fourth tergites and length of proboscis. The mean differences which are statistically significant are starred.

An examination of Table 2 shows that between the means of the measurements taken on the bees from size of cell "A" and size of cell "B" there is only one case where the difference is significant. This is in the case of dry weight and in a negative direction. Table 2 further shows that there is a decrease in the length of the right fore wing between size of cell "A" and size of cell "B", but the mean difference in this case is not significant. The other three measurements, width of right fore wing, sum of the widths of the third and the fourth tergites and length of proboscis show an increase in their respective means but the mean difference is not significant.

An examination of the arithmetic means of bees from size of cell "B" and size of cell "C" shows that the mean differences between the three groups are significant with the exception of the width of the right fore wing. In the case of the right fore wing there is an increase between the "B" and "C" groups but the increase is not statistically significant. An examination of the arithmetic means of the bees from size of cell "A" and size of cell "C" tells a similar story. The mean differences of all measurements are significant except for the measurement of the width of the right fore wing, whose means, while showing an increase from size of cell "A" to size of cell "C", do not show a significant mean difference.

The influence of the size of the brood cell upon the size of various measurements taken on the bees from colony 21 is shown by a comparison of the arithmetic means of the measurements of the bees from size of cell "A", size of cell "B" and size of cell "C", respectively, in Table 3. Data presented in this table includes the arithmetic mean and it's standard deviation for each measurement of the bees from each size of cell and the mean difference and its standard deviation for each measurement between the means of the bees of each group. The measurements taken on the parts of the bees of colony 21 and presented in Table 3 are dry weight, length of the right fore wing, width of the right fore wing, sum of the widths of the third and fourth tergites and length of proboscis. The mean differences which are statistically significant are starred.

Table 3. Influence of Size of Cell Upon Size of Bee. (Colony 21)

Measurement Taken	Size of Cell A M±	Mean Diff. of A and B M.D.±	Size of Cell B M±	Mean Diff of Band C M.D.±	Size of Cell C M±	Mean Diff. of A and C M.D.±
Dry Weight in mgs.	15.7350 ±0.1654	0.2443 ±0.2694	15.4907 ±0.2127	0.4449 ±0.2456	15.9356 ±0.1228	0.2006 ±0.2060
Length of proboscis	6.6710 ±0.0185	0.0741* ±0.0235	6.7451 ±0.0148	0.0269 ±0.0195	6.7720 ±0.0127	0.1010* ±0.0227
length of right fore wing	9.6930 ±0.0227	0.1442* ±0.0290	9.8572 ±0.0181	0.0657 ±0.0333	9.9029 ±0.0280	0.2099* ±0.0360
Width of right fore wing	3.3093 ±0.0133	0.0474* ±0.0155	3.3567 ±0.0079	-0.0003 ±0.0146	3.3564 ±0.0185	0.0471* ±0.0181
Sum of width of 3 and 4 tergites	4.9938 ±0.0169	0.1611* ±0.0282	5.1549 ±0.0226	0.0253 ±0.0282	5.1802 ±0.0169	0.1864* ±0.0239

***Mean difference is statistically significant**

An examination of Table 3 shows that the mean differences for the measurements between the bees from size of cell "A" and size of cell "B" are all significant except for dry weight. There is a decrease in the dry weight of the bees of these samples as the size of the cell is increased, but the decrease is not large enough to be significant. There are no significant mean differences between the bees from size of cell "B" and size of cell "C" although all measurements, except the width of the right fore wing, show an increase as the size of cell is increased. An examination of the mean differences of the measurements on bees from size of cell "A" and size of cell "C" show that in all measurements, except dry weight, the mean differences are significant and that in all measurements there is an increase accompanying the increase in the size of brood cell.

2. The variability of the worker bee as influenced by size of brood cell.

In Table 4 are presented the correlation coefficients of measurements taken on the parts of the worker bees from the three sizes of cells from colony 25. The measurements taken upon the individual bees are given in column 1. Sizes of cell is designated in the second column. In the third column are presented the correlation coefficients of length of right fore wing with dry weight for each size of cell. The correlation coefficients of width of right fore wing with dry weight and width of right fore wing with length of right fore wing, for each of the three sizes of cells, are given in column 4. The correlation coefficients of the sum of the widths of the third and the fourth tergites with dry weight, with length of right fore wing and with width of right fore wing for each size of cell are given in column 5. Similarly, the correlation coefficients of length of proboscis, length of glossa, length of mentum, and the sum of the lengths of the mentum and the glossa with corresponding measurements in column 1 are given in columns 6, 7, 8 and 9, respectively, for the bees from each of the three sizes of cells. Those values which are starred with one star are highly significant correlations, while those values which are starred with two stars are significant correlations but not highly so. The values which are not starred have failed to meet the requirements of significance. Significance of the correlation coefficients was determined by comparing the values obtained with significant values of "r" given in Table 16 by Wallace and Snedecor (69).

Table 4. Correlation Coefficients of Bees from Colony 25

Measurement Taken	Size of Cell	Length of right fore wing	Width of right fore wing	Sum width of 3 & 4 tergites	Length of proboscis	Length of glossa	Length of mentum	Sum of Len. Of mentum & glossa
Dry Weight	A	+0.4738*	+0.4255*	+03873*	+0.4460*	+03410**	+03135**	+0.4227*
	B	+0.3217*	+0.3780*	+0.2242	+0.0919	-0.0483	+0.0378	+0.0191
	C	+0.3135*	+0.2674	+0.3354**	+0.1939	-0.0591	+0.2634	+0.0569
Length of right fore wing	A		+0.7095*	+0.3011**	+07274*	+0.6902*	+0.3538**	+0.7404*
	B		+0.4427*	+0.2108	+02818*	+02934**	+0.2258	+0.3582**
	C		+0.6312*	+0.3635**	+06297*	+05216*	+0.3324**	+0.6086*
Width of right fore wing	A			+02444	+04716*	+0.3713*	+04060*	+0.1991
	B			+0.1164	+0.1776	+0.0861	+03314**	-0.0764
	C			+0.1831	+0.2918	+03113*	+0.1121	+0.2607
Sum width of 3 & 4 tergites	A				+02120	+02559	-0.0886	+0.9547*
	B				-0.1406	-0.0476	-0.0733	+0.9546*
	C				+0.2241	+0.1900	+0.2124	+0.9430*
Length of proboscis	A					+03977*	+0.9091*	+0.3011**
	B					+04967*	+08487*	+0.2108
	C					+0.5570*	+0.7881*	+0.3635**
Length of glossa	A						+0.0682	+04311*
	B						+0.0288	+0.5219*
	C						+0.1835	+0.3121**
Length of mentum	A							+0.6565*
	B							+0.8838*
	C							+0.8355*

***Correlation coefficient is highly significant.**
****Correlation coefficient is significant but not highly so.**

Concerning the data presented in Table 4, the following assertions can be made: (1) The length of the right fore wing is significantly correlated with dry weight for the bees from all three sizes of cells, but only in the case of bees from size of cell "A" is the correlation highly significant. (2) Dry weight is highly significantly correlated with width of right fore wing in the case of bees from size of cell "A" and size of cell "B", while in the case of bees from size of cell "C" the correlation coefficient approaches significance. (3) The correlation coefficient of the sum of the widths of the third and the fourth tergites with dry weight is highly significant in the case of the bees from size of cell "A", is significant but not highly so in the case of the bees from size "B". (4) The correlation of dry weight with length of proboscis, length of glossa, length of mentum, and the sum of the lengths of the mentum and the glossa is significant only in the case of bees from size of cell "A". (5) The correlation of length of right fore wing with width of right fore wing is highly significant. (6) The correlation of length of right fore wing with the sum of the widths of the third and the fourth tergites is significant, but not highly so in the case of the bees from the size of cell "A" and size of cell "C". The correlation coefficient in the case of bees from size of cell "B" is not significant. (7) A study of the correlation coefficients of length of right fore wing with length of proboscis and its integral parts shows a tendency for the correlation coefficients to be highly significant in the case of bees from all three sizes of cells. (8) There is no significant correlation between width of right fore wing and the sum of the widths of the third and the fourth tergites. (9) Concerning the correlation of the width of right fore wing with length of proboscis and its integral parts, there is a tendency for the correlation coefficient to be highly significant in the case of bees from size of cell "A". In the case of the bees from size of cell "B" the tendency is for the correlation coefficient not to be significant and in the case of the bees from size of cell "C" the tendency is for the correlation coefficient to be significant but not highly so. (10) The correlation of the sum of the widths of the third and the fourth tergites with length of proboscis and its integral parts is not significant. (11) As would be expected, the correlation of the length of proboscis with length of glossa, length of mentum and the sum of the lengths of the mentum and the glossa is highly significant. (12) Concerning the correlation between length of glossa and length is not significant. (13) As would be expected, the correlation of the sum of the lengths of the mentum and the glossa with its integral parts, namely, length of mentum and length of glossa, is highly significant.

Data are presented in Table 5 concerning the correlation coefficients of measurements on bees from colony 21. The arrangement of this table is the same as that of Table 4. The measurements which have been correlated are dry weight, length of right fore wing, width of right fore wing, sum of the widths of the third and the fourth tergites, and length of proboscis. Similar to Table

4, those values which are highly significant correlations are starred with one star, those values which are significant but not highly so are starred with two stars, while those values which are unstarred are not significant.

Table 5. Correlation Coefficients of Bees from Colony 21

Measurement Taken	Size of Cell	Length of right fore wing	Width of right fore wing	Sum width of 3 & 4 tergites	Length of proboscis
Dry Weight	A B C	+0.3346** +0.1602 +0.4966*	+0.2355 +0.2519 +0.2603	+0.4928* +0.1342 +0.6291	+0.1443 +0.4181* +0.3085*
Length of right fore wing	A B C		+0.7095* +0.4427* +0.6312*	+0.2732 +0.4551* +0.5483*	-0.0064 +0.1927 +0.4274*
Width of right fore wing	A B C			+0.2032 +0.1493 +0.3365**	+03257** +0.0879 +0.4068*
Sum width of 3 & 4 tergites	A B C				+0.1308 -0.0186 +0.4624*

***Correlation coefficient is highly significant.**
****Correlation coefficient is significant but not highly so.**

Concerning the data presented in Table 5, the following general assertions can be made: (1) In all three cases the correlations of dry weight with length of right fore wing, dry weight with the sum of the widths of the third and the fourth tergites and dry weight with length of proboscis are significant. (2) The correlations of length of right fore wing with width of right fore wing and length of right fore wing with the sum of the widths of the third and the fourth tergites are significant. (3) The correlation of width of right fore wing with length of proboscis is significant. (4) The following correlations are not significant: dry weight with width of right fore wing, length of right fore wing with length of proboscis, width for right fore wing with the sum of the widths of the third and fourth tergites and the sum of the width of the third and the fourth tergites with length of proboscis.

Table 6. Correlation Coefficients of Bees from Colony 18

Measurement Taken	Size of Cell	Length of right fore wing	Width of right fore wing	Sum width of 3 & 4 tergites	Length of proboscis
Dry Weight	A	+0.0817	-0.2170	+02613	+0.0074
	B	+0.5017*	+0.2876	+0.5345*	+0.2790
	C	+0.3704*	+0.0238	+0.3418**	-0.0058
Length of right fore wing	A		+03734**	-0.0895	+0.0744
	B		+0.3839*	+0.1112	+0.3168**
	C		+0.3703*	-0.1247	+0.2295
Width of right fore wing	A			-0.1013	+0.4816*
	B			+0.1779	+02810
	C			-0.2639	+0.6135*
Sum width of 3 & 4 tergites	A				+02492
	B				+0.1788
	C				+0.2699

***Correlation coefficient is highly significant.**
****Correlation coefficient is significant but not highly so.**

In Table 6 are presented the correlation coefficients of measurements on bees from colony 18. The arrangement of this table is similar to table 4. The measurements upon which correlations have been calculated are dry weight, length of right fore wing, width of right fore wing, sum of the widths of the third and the fourth tergites and length of proboscis. The system for indicating the significance of correlation coefficients is the same as that used in Tables 4 and 5.

Concerning the data presented in Table 6, the following general assertions can be made: (1) The correlation coefficients of dry weight with length of right fore wing, dry weight with the sum of the widths of the third and the fourth tergites, length of right fore wing with width of right fore wing and width of right fore wing with length of proboscis are significant. (2) The correlation coefficients of dry weight with width of right fore wing, dry weight with length of proboscis, length of right fore wing with the sum of the widths of the third and the fourth tergites, length of right fore wing with length of proboscis, width of right fore wing with the sum of the widths of the third and the fourth tergites and the sum of the widths of the third and the fourth tergites with length of proboscis are not significant.

From the data presented in Tables 1, 2, 3, 4, 5 and 6, it is evident that the samples of bees from colony 25 are more homogeneous than the samples of bees from colony 21 and colony 18. Consequently, a further study of the variability of the worker bee as influenced by size of brood cell will be concerned with the bees from colony 25.

Throughout the following presentation of data, the measurements on the parts of the bee will be designated as follows: (A) dry weight, (B) length of right foe wing, (C) width of right fore wing, (D) sum of the widths of the third and the fourth tergites and (X) the measurement upon which the regression is made.

Throughout the following presentation of data, the measurements on the parts of the bee will be designated as follows: (A) dry weight, (B) length of right foe wing, (C) width of right fore wing, (D) sum of the widths of the third and the fourth tergites and (X) the measurement upon which the regression is made.

In Table 7 are presented further data concerning the measurements of bees from colony 25 in which there is a regression of (A) dry weight, (B) length of right fore wing, (C) width of right fore wing and (D) sum of the widths of the third and the fourth tergites on (X) length of proboscis. The data consist of the standard deviations of the above-mentioned measurements, the standard regression coefficients, the multiple correlation coefficient, the standard error of estimate, the significance of regression and the regression equations for the bees from each size of cell. An analysis of variance of the length of proboscis between and within all three groups, namely, the groups of bees from size of cell "A", the group of bees from size of cell "B" and the group of bees from size of cell "C", is also presented.

Table 7. Statistical Constants of Bees from Colony 25

Correlation of (A) dry weight, (B) length of right fore wing, (C) width of right fore wing, and (D) sum of the widths of the third and fourth tergites on (X) length of proboscis.

Statistical constant	Sign	Size of Cell		
		A	B	C
Standard deviations of samples		±0.6228 ±0.1513 ±0.0721 ±0.0985 ±0.1263	±0.7372 ±0.1062 ±0.0616 ±0.1144 ±0.0807	±1.8502 ±0.1173 ±0.0530 ±0.1109 ±0.0982
Standard regression coefficients		+0.1587 +0.7481* -0.1154 -0.0464	+0.026 +02910** -0.0657 -0.2144	+0.0145 +0.7448* -0.1788 -0.0187
Multiple correlation coefficients	R	+0.7422*	+0.3540	+0.6445
Standard error of estimate		±0.0889	±0.0790	±0.0787
Significance of regression	Z	1.2406*	0.2064	0.9804*

Regression Equations

A. X = 0.3218 A + 0.6244 B – 0.2021 C – 0.0595 D + 1.1235

B. X = 0.0236 A + 0.2211 B + 0.0860 C – 0.1512 D + 4.9453

C. X = 0.0077 A + 0.5232 B – 0.3315 C – 0.0165 D + 1.8315

Analysis of variance of length of proboscis of A, B and C groups.

Source of Variatio n	Degrees of Freedom	Sum of Squares	Mean Square of Variance	Standard Deviation	Natural Log of S.D.
Between groups	2	4,251.81	2125.90**	46.1075	3.830977
Within groups	133	14,096.02	105.99	10.2951	2.331667
Total	135	18,347.83	135.91	11.6580	1.499310

Z=1.4993*

***Correlation coefficient is highly significant.**
****Correlation coefficient is significant but not highly so.**

An examination of the standard deviations of the measurements of the bees from the three sizes of cells shows that the variation is greatest in the case of length of right fore wing and length of proboscis with the bees from size of cell "A" and least in the case of the bees from size of cell "B". The standard deviation of the width of the right fore wing is greatest in the case of the bees from size of cell "A" and the least in the case of bees from size of cell "C". The standard deviation of dry weight increases as the size of the cell is increased. The standard deviation of the sum of the widths of the third and the fourth tergites is greatest in the case of the bees from the size of cell "B" and least in the case of bees from the size of cell "A".

Table 8. Statistical Constants of Bees from Colony 25

Correlation of (A) dry weight, (B) length of right fore wing, (C) width of right fore wing, and (D) sum of the widths of the third and fourth tergites on (X) length of mentum.

Statistical constant	Sign	Size of Cell		
		A	B	C
Standard deviations of samples		±0.6228 ±0.1513 ±0.0721 ±0.0985 ±0.0408	±0.7372 ±0.1062 ±0.0616 ±0.1144 ±0.0366	±1.8502 ±0.1173 ±0.0530 ±0.1109 ±0.0395
Standard regression coefficients		+0.2468 +0.1218 -0.2855 -0.2906	+0.0216 +0.1414 -0.3213 -0.1175	+0.1779 +0.3712 -0.1791 -0.0506
Multiple correlation coefficients	R	+0.5120**	+0.3783	+0.3735
Standard error of estimate		±0.0368	±0.0355	±0.0385
Significance of regression	Z	0.6211**	0.2802	0.2416

Regression Equations

A. X = +0.1618 A + 0.0329 B + 0.1616 C – 0.1205 D + 1.2745

B. X = -0.0511 A + 0.0487 B + 0.1907 C – 0.0376 D + 0.9342

C. X = +0.0380 A + 0.1250 B – 0.1336 C + 0.0180 D + 0.8394

Analysis of variance of length of mentum of A, B and C groups.

Source of Variation	Degrees of Freedom	Sum of Squares	Mean Square of Variance	Standard Deviation	Natural Log of S.D.
Between groups	2	245.69	127.35	11.2849	2.423466
Within groups	133	2019.86	15.187	3.8971	1.360231
Total	135	2274.55	16.85	4.1047	1.063235

Z=1.0632*

***Correlation coefficient is highly significant.**
****Correlation coefficient is significant but not highly so.**

The significance of the standard regression coefficients was tested by dividing the standard regression coefficient by its standard deviation and comparing the resulting value with the significant values for "t" as given in Table 16 by Wallace and Snedecor (69). The significance of the multiple correlation coefficients was determined by comparing the values obtained with significant values of

"R" given by Wallace and Snedecor (69) in Table 16. The significance of the regression for the bees from each size of cell and the analysis of variance of length of proboscis of all three groups was tested by calculating one-half the difference of the natural logarithms of the mean squares and comparing the values obtained with the significant values of "Z" as given in Table 6 by Fisher*.

An examination of the standard regression coefficients shows that only the standard regression coefficient of length of proboscis and length of right fore wing is significant for the bees from each of the three sizes of cells. The multiple correlation coefficient for the bees from size of cell "A" and size of cell "C" are highly significant, while the corresponding value for the bees from size of cell "B" is not significant. The "Z" test of the significance of the regressions further substantiates the significance of the multiple correlation coefficients by showing that the regressions of the measurements on the bees from size of cell "A" and size of cell "C" has been significantly accounted for.

A study of the standard errors of estimate shows that the standard deviation of the length of proboscis of bees from size of cell "A" has been reduced 29.61% due to the extension of statistical control over factors relating to length of proboscis. In the case of the bees from size of cell "B" the standard deviation of length of proboscis has only been reduced 2.11%, while in the case of bees from size of cell "C" the reduction is 19.86%. In the latter case, the standard deviation of length of proboscis has been notably reduced by the inclusion of dry weight, length of right fore wing, width of right fore wing and the sum of the widths of the third and the fourth tergites in the regression. A study of the regression equations for the bees from the three sizes of cells shows, in general, that length of the right fore wing is the dominating factor in these estimation equations of length of proboscis.

Table 9. Statistical Constants of Bees from Colony 25

Correlation of (A) dry weight, (B) length of right fore wing, (C) width of right fore wing, and (D) sum of the widths of the third and fourth tergites on (X) length of glossa.

Statistical constant	Sign	Size of Cell		
		A	B	C
Standard deviations of samples		±0.6228 ±0.1513 ±0.0721 ±0.0985 ±0.0978	±0.7372 ±0.1062 ±0.0616 ±0.1144 ±0.0616	±1.8502 ±0.1173 ±0.0530 ±0.1109 ±0.0828
Standard regression coefficients		+0.0297 +0.8553* -0.2471 -0.0532	-0.1404 +0.3625** -0.0105 -0.0022	-0.2640 +0.5759* -0.0060 -0.0681
Multiple correlation coefficients	R	+0.7035*	+0.3416	+0.5752*
Standard error of estimate		±0.0730	±0.0606	±0.0710
Significance of regression	Z	1.1284*	0.1637	0.7990*

Regression Equations

A. X = +0.0466 A + 0.5400 B - 0.3352 C + 0.0529 D – 0.1118

B. X = -0.1773 A + 0.2102 B - 0.0106 C – 0.0492 D + 2.7602

C. X = -0.01182 A + 0.4064 B - 0.0094 C + 0.0508 D + 0.3720

Analysis of variance of length of glossa of A, B and C groups.

Source of Variation	Degrees of Freedom	Sum of Squares	Mean Square of Variance	Standard Deviation	Natural Log of S.D.
Between groups	2	2700.16	1350.08	36.7434	3.603958
Within groups	133	887.22	66.75	8.1701	2.100480
Total	135	1157.38	85.758	9.2606	1.503478

Z=1.0632*

***Correlation coefficient is highly significant.**
****Correlation coefficient is significant but not highly so.**

An analysis of the variance between and within the groups of bees from all three sizes of cells shows that the variation between the groups is significantly greater that that within groups. This further substantiates the proof presented under section 1 of the presentation of data of a significant difference between the means of the length of proboscis of the three groups.

In Table 8 are presented statistical constants of measurements on bees from colony 25 concerning a regression of (A) dry weight, (B) length of right fore wing,

(C) width of right fore wing and (D) sum of the widths of the third and the fourth tergites on (X) length of mentum.

In Table 9 are presented data concerning the statistical constants of measurements on bees from colony 25. In this table the regression is dry weight, length of right fore wing, width of right fore wing and sum of the widths of the third and the fourth tergites on length of glossa.

Statistical constants of measurements on bees from colony 25 concerning a regression of dry weight, length of right fore wing, width of right fore wing and sum of the widths of the third and the fourth tergites on the sum of the lengths of the mentum and the glossa are presented in Table 10.

Statistical constants of measurements on bees from colony 25 concerning a regression of dry weight, length of right fore wing, width of right fore wing and sum of the widths of the third and the fourth tergites on the sum of the lengths of the mentum and the glossa are presented in Table 10.

A comparison of Tables 7, 8, 9 and 10 shows that the data presented in Tables 9 and 10 proffer the same conclusions as were drawn from the data of Table 7. From the data presented in Table 8, there is an agreement with the data of the other three tables in the variation of the length of mentum as indicated by the standard deviation of length of mentum of the bees of all three sizes of cells, the multiple correlation coefficient and the significance of the regression of the bees from size of cell "A", and the analysis of variance of length of mentum between and within the groups from all three sizes of cells. In no cases are the standard regression coefficients significant. In contrast to the data presented in the other three tables, the multiple correlation coefficients and significance of regression for the bees from the size of cell "C" are not significant. An examination of the regression equations shows that length of right fore wing has ceased to be a dominating factor in the estimation of length of proboscis.

Table 10. Statistical Constants of Bees from Colony 25

Correlation of (A) dry weight, (B) length of right fore wing, (C) width of right fore wing, and (D) sum of the widths of the third and fourth tergites on (X) sum of lengths of glossa and mentum.

Statistical constant	Sign	Size of Cell		
		A	B	C
Standard deviations of samples		±0.6228 ±0.1513 ±0.0721 ±0.0985 ±0.1088	±0.7372 ±0.1062 ±0.0616 ±0.1144 ±0.0735	±1.8502 ±0.1173 ±0.0530 ±0.1109 ±0.0926
Standard regression coefficients		+0.1207 +0.7652* -0.0920 -0.0556	-0.1181 +0.3659** -0.1364 -0.1430	-0.1620 +0.6737* -0.0710 -0.0824
Multiple correlation coefficients	R	+0.7035*	+0.3416	+0.6315*
Standard error of estimate		±0.0730	±0.0606	±0.0753
Significance of regression	Z	1.1284*	0.1637	0.9460*

Regression Equations

A. X = +0.2109 A + 0.5503 B - 0.1388 C - 0.0615 D + 1.2228

B. X = - 0.1778 A + 0.2533 B + 0.0106 C - 0.0919 D + 3.7551

C. X = -0.0801 A + 0.5314 B - 0.1241 C + 0.0687 D + 1.21145

Analysis of variance of length of glossa and mentum of A, B and C groups.

Source of Variation	Degrees of Freedom	Sum of Squares	Mean Square of Variance	Standard Deviation	Natural Log of S.D.
Between groups	2	3940.95	1970.48	44.3901	3.73016
Within groups	133	11347.13	85.32	92369	2.223206
Total	135	15288.08	113.25	10.6419	1569810

Z=1.0632*

***Correlation coefficient is highly significant.**
****Correlation coefficient is significant but not highly so.**

*Fisher, R. A., "Statistical Methods for Research Workers", second edition revised and enlarged. Oliver and Boyd, Edinburgh. 1928.

DISCUSSION

A Biometrical Study of the Influence of Size of Brood Cell Upon the Size and Variability of the Honey bee (*Apis mellifera* L.) by Roy A. Grout, 1931

The data presented in this paper show conclusively that it is possible to obtain a larger bee through the use of brood combs constructed from artificial foundation having enlarged cell bases. The author, therefore, corroborates the contentions of Baudoux (8) and Pincot, according to Gillet-Croix (26), that the bees reared in enlarged cells are larger than bees reared in normal cells. Even though the data at hand show that an increase in the size of the cells in which the bees are reared is accompanied by an increase in the dry weight, length of right fore wing, width of right fore wing, sum of the length of the third and the fourth tergites and the length of proboscis, the author hesitates to support them in their contentions that the larger bee is a better bee. Since the crucial test is honey production, further experimental data are necessary.

Baudoux's contentions (8) that the bee reared in the enlarged cell could fly further, would have a greater carrying capacity and could acquire nectar from flowers having a deeper corolla are not substantiated by the data of this experiment. The author is of the opinion that an extensive investigation of these characteristics should be made. Since in this experiment, all three sizes of cells were placed in the same colony in order that the bees emerging from the cells would be the progeny of the same mother, no indication of the increase in the ability of the bee to gather nectar was observed.

The data presented in this paper agree with data presented by Michailov (43 and 44), stating that an increase in the size of the cells in which worker bees are reared is accompanied by a corresponding increase in the weight, length of right fore wing, width of right fore wing, sum of the widths of the third and the fourth tergites and length of proboscis.

Since most writers agree that the principal point in the selection of bees and the acquisition of nectar is length of proboscis, it is of interest to note that in the case of colony 25 the length of proboscis was increased 2.07%

through the use of enlarged cells. In the case of colony 21 the length of proboscis was increased 1.51% and in the case of colony 18 the length of proboscis was increased 1.40%. The average maximum increase in the length of the proboscis for the bees of all three colonies is 0.11 mm. Whether this increase in the length of proboscis is significantly related to an increase in the honey production of a colony has yet to be proved.

The general tendencies of the statistical constants of measurements on bees from colonies 25, 18 and 21 reveal the following facts: (1) The measurement of dry weight varies greatly in certain cases and, in general, correlates poorly. Consequently, the author feels that better experimental control should be exercised in the measurement of this character than was used in the present work. (2) The measurement, length of right fore wing, is significantly correlated with all of the other characters. Of the four characters employed in a regression on the length of proboscis and its integral parts the length of the right fore wing is the only character which gives a significant standard regression coefficient and is dominant in those factors used in the estimation of length of proboscis in the regression equations. (3) The width of the right fore wing, while correlating significantly with the length of the right fore wing, shows no tendency to correlate significantly with the sum of the widths of the third and the fourth tergites and dry weight and its tendency to correlate significantly with length of proboscis is questionable. (4) The measurement of the sum of the widths of the third and the fourth tergites tends to correlate significantly with dry weight and length of right fore wing, but there is apparently no significant correlation with width of right fore wing and with length of proboscis. Consequently, the author feels that the sum of the lengths of the third and the fourth tergites is unrelated to length of proboscis and should be omitted in a study of those factors which are related or contribute to length of proboscis. (5) The variation of the character dry weight is greatest in the case of bees from size of cell "C" and least in the case of bees from size of cell "A". The variation of the width of the right fore wing is greatest in the case of bees from size of cell "A" and least in the case of bees from size of cell "C". The variation of the sum of the widths of the third and the fourth tergites is greatest in the case of bees from the size of cell "B" and least in the case of bees from size of cell "A". The variation of the measurements of length of right fore wing and length of proboscis is greatest in the case of bees from size of ell "A" and least in the case of bees from size of cell "B". Since length of proboscis and length of right fore wing have been shown to be significantly correlated, it is of interest to note that the variation in the three sizes of cells has the same trend.

Since Merrill (42) measured only the glossa in his study on the honey-storing ability of the bee and since the Russian method uses the entire length of the proboscis as measured in this paper, a study was made of length of glossa, length of mentum and sum of the lengths of the glossa and the mentum in relation to length of proboscis. While all three measurements correlate significantly with length of proboscis, it was found that length of mentum was not as good an estimation of length of proboscis as was length of glossa and the sum of the lengths of the glossa and mentum. This study substantiates the methods of both the above mentioned sources.

The inconsistencies arising in the data presented in this paper can be attributed only to a lack of proper technique in the treatment of the material and to the fact, that, while the bees were selected from a specific size of cell and were the progeny of the same mother, they were, nevertheless, selected at random from a large population.

Consequently, even though it connot be determined to what extent the above factors operated, the author feels that many of the inconsistencies of the data can be attributed to these causes.

Due to these inconsistencies arising in nearly all of the statistical constants presented in this paper, the author is of the opinion that the use of samples containing a larger number of bees selected from each size of cell would strongly tend to give consistently significant results. This opinion is substantiated by Phillips (53), who discounted the work of Merrill (42) on the basis that the correlations presented in his data were based on small numbers of bees taken at intervals during the season. Merrill (49) calculated that it was necessary to examine only forty bees of a colony in order to determine which colony would produce the most honey. The data at hand show that this is not always the case.

Contributing to the peculiarities of the correlations, particularly in the case of the correlation coefficients of length of proboscis with the lengths of its integral parts, is the fact that in certain cases there are spurious correlations. For example, even though length of glossa is significantly correlated with length of proboscis and length of mentum is significantly correlated with length of proboscis, length of mentum is not significantly correlated with length of glossa.

In this experiment, as has been previously stated, all three sizes of cells were placed in the same colony in order that the emerging bees would be the progeny of the same mother. While the worker bees extended the side walls of the cells of the combs in a normal manner and immediately proceeded to make use of the cells, at least for storage purposes, difficulties were experienced in getting the queens to oviposit worker eggs in the enlarged cells. This was especially true in the case of size of cell

"C". The combs containing size of cell "C" were often found to contain patches of drone brood and little or no worker brood. These observations showed that, while the worker bees apparently recognized no difference in the three sizes of cells, the queen bees showed a tendency to prefer the smaller cells for ovipositing. This observation agrees with experiments conducted by Lovchinovskaya (39), who showed that, when nine combs containing normal cells and one comb containing enlarged cells were placed in a colony, the worker bees apparently recognized no difference in the size of the cells but that the queen bee apparently recognized this difference and did not oviposit in the enlarged cells. In the reverse experiment the queen bee oviposited in the enlarged cells.

From a single brood count made by the author during the summer of 1931 on colonies entirely supplied with combs containing size of cells "A", colonies entirely supplied with combs containing size of cells "B" and colonies entirely supplied with combs containing size of cells "C", it was shown that the reaction of the colonies to each size of cell was apparently the same. While this observation corroborates the latter experiment of Lovchinovskaya (39) and the experiences of Baudoux (8), the author feels that further experiments, concerning the brood activities of colonies supplied with combs containing enlarged cells, should be made throughout a period of two or more seasons.

While this paper was being prepared, a correlation of dry weight, length of right fore wing, width of right fore wing, sum of the widths of the third and the fourth tergites, length of proboscis and length of the second member of the right labial palpus was calculated on bees from the three sizes of cells from colony 14. Of special interest in this correlation is the fact that only in the case of the bees from size of cell "A" was the correlation coefficient of length of proboscis with the length of the second member of the right labial palpus significant. The correlation coefficients of these factors in the other two groups were insignificant.

Gotze (29), in a recent paper, did not follow the Russian technique for measuring the individual parts of the proboscis and proclaimed that the probability of error in this technique of measurement was so great that it was quite easy, with small differences of individual colonies, to obtain incorrect results. He states that with all the colonies that he "accurately investigated" it was found that the length of the second member of the labial palpus varies with the total length of the proboscis. In his studies he obtained a judgment of the second member of the labial palpus and prescribed a definite formula by which the length of proboscis was estimated from the value of the measurement of the second member of the labial palpus.

A further comparison of the standard deviations of length of proboscis and length of the second member of the right labial palpus shows that the variability in the three sizes of cell differs between the two measurements. The variability of the length of proboscis is greatest in the case of bees from size of cell "C" and least in the case of bees from size of cell "B", while the variability of the length of the second member of the right labial palpus is greatest in the case of bees from size of cell "B" and least in the case of bees from size of cell "C". From this data it is shown that the variability of these two parts as influenced by size of cell is not at all consistent. From the combined study of the correlation coefficients and the standard deviations, it is shown that the second member of the right labial palpus is not related to length of proboscis and does not vary with it. Consequently, the author cannot agree with the conclusions of Gotze (29) upon this subject.

The author takes this opportunity to strongly recommend the technique of measuring the individual parts of the bee as described under "Methods and Materials" in this paper. With a little practice, from three to four measurements could be taken per minute, a speed which would be very difficult to obtain through the use of an ocular micrometer in a binocular microscope. A comparison of the measurements taken by means of the projection system with those taken by means of an ocular micrometer in a binocular scope showed that the projection system is just as accurate, if not more so.

SUMMARY

A Biometrical Study of the Influence of Size of Brood Cell Upon the Size and Variability of the Honey bee (Apis mellifera L.) by Roy A. Grout, 1931

1. The size of the worker bee as represented by the size of the various parts is significantly increased through the use of brood combs containing enlarged cells.

2. The average percent of increase of the linear measurements of the worker bee is directly proportional to the percent of increase of the diameter of the brood cell.

3. The number of bees used in a sample in this experiment is not large enough to give wholly consistent results, but these results are in general significant and indicative.

4. The measurement of the dry weight is not sufficiently controlled by the present experimental technique.

5. The correlations of length of right fore wing with the other measurements of the worker bee are highly significant.

6. Among the measurements taken in this work the length of the right fore wing gives the best statistical estimation of length of proboscis.

7. The correlations of width of right fore wing with the other measurements are significant but not highly so.

8. The correlations of the sum of the widths of the third and the fourth tergites with the other measurements are in general not significant.

9. The greatest variation of length of proboscis, length of right fore wing and width of right fore wing occurs in the bees from combs having 857 cells per square decimeter.

10. The least variation of length of proboscis and length of right fore wing occurs in the bees from combs having 763 cells per square decimeter, while the least variation of width of right fore wing occurs in the bees from combs having 706 cells per square decimeter.

11. The length of glossa and sum of the lengths of the glossa and the mentum are as highly correlated with the other measurements as is length of proboscis.

12. The length of the mentum is not as highly correlated with the other measurements as are the length of the glossa, the length of the proboscis and the sum of the lengths of the mentum and the glossa.

13. The second member of the right labial palpus is not significantly correlated with length of proboscis.

14. The least variation of the length of the second member of the right labial palpus occurs among those bees in which the variation of the length of proboscis is greatest and vice versa.

The Bee World – January, 1933 – Pages 37-41

Ursmar Baudoux

THE INFLUENCE OF CELL SIZE.

By Prof. U. BAUDOUX, Rucher-ecole experimental de Tervueren-lez- Bruxelles. Belgium.

In intensive bee-culture, the question of cell size is of great importance. It is inseparable from those of race-selection and of the improvement of the capacity for egg-laying and accumulating stores.

The end proposed is magnificent: To rear bees of extraordinary vigour, able to forage over a more extended flight-radius and *to visit a multitude of flowers the nectar of which is, at present, out of reach of their tongues*. The tongue of the bee is, indeed, the essential organ which we must develop, by selection first, and then by rearing the bees in large cells; for it is plain that *all the organs will participate in any enlargement of the body of the bee*.

About 1891, foundation with cells 920 to the dm^2 was introduced into our country. Beekeepers all adopted this size of cell. The experts of that time believed that it was advantageous to produce as many bees as possible on the least possible surface of comb. Thus there was a premature narrowing of the cells, and at the end of a few years the bees were miserable specimens.

It was then that, to combat so harmful a tendency, I published an article in *Progres Apicole* (June 1893) advocating the use of larger cells, as a result of experiments duly described. I had experimented up to the limit of 750 cells per dm^2 These sizes of cells were obtained by stretching foundation. Mr. Auguste Mees subsequently made them by stretching the sheets as they came off the cylinders, in 1893 to 1895.

In November 1895, at my suggestion and after I had reported on these facts to the *Assemblee Federale*, this body decided to give an order on behalf of its members to that manufacturer who would undertake to use a special machine (which would give perfectly hexagonal bases), for foundation with 750 cells to the dm^2 It was then that the first machine with 750 cells per dm^2 appeared; Mr. Mees ordering it in response to this desire.

The bees will not build more drone cells on the sheets of foundation thus obtained, than they do on foundation with small impressions. Moreover, if bees want to rear drones, they will not fail to do so, even if you give them foundation with 900 cells per. dm^2

[*Note*. A dm^2 is almost exactly 15 ½ in^2. The numbers per dm^2 must therefore be divided by 15.5 to give the number per in^2. Corresponding figures are: 930, 60; 852, 55; 775, 50; 697, 45; 620, 40; 465, 30. The count is for both sides of the comb.-Tr.]

The way was thus mapped out; and the results obtained were always satisfactory. But, unfortunately, one cannot please everyone. A long and underhand campaign was started by people without much experience against this idea of enlarging the cells. *When conditions were equal, however, my crops were always larger than those of my neighbours*. No one is so deaf as he that will not hear!

Encouraged by my experiences, I wished to do still better – to go to the bounds of possibility. But, to carry out my idea, I was in want (as in 1891) of the most important item – a machine which would give me larger cells still; and, to possess it, I had first to find out what point I ought to stop.

I therefore again proceeded by stretching foundation. Now, however, I was working with cells already enlarged; and any irregular cells would soon have been used for the rearing of drones. It was therefore necessary to work accurately. I invented a stretcher which gave me the number of cells I desired. It consists of a thick sheet of inclia rubber and a thin sheet of para, both held in a long clamp. The sheet of foundation, warmed, is placed between these; and, being rolled up, is drawn out longer. The operation is then repeated in the other direction. In this way, I was able to obtain sheets having only 741, 730 and even 675 cells per dm^2

Towards the end of 1927 appeared the new machine, giving 700 cells per dm^2 (also made by M. Joseph Mees, who thus perpetuated the traditions of his firm).

Just as the 750 machine sometimes gives 736, so the 700 machine sometimes gives 685. Those who are used to such work know how that slight differences are inevitable.

For the last five years, I have been replacing all defective combs in my experimental apiary at Tervueren with 700 foundation; and all new colonies are set up on 685. This year I placed some swarms on 675. This apiary is accessible to all. It is the instruction apiary, where the official practical and theoretical lectures are given. It is visited annually by hundreds of beekeepers, by parties from Associations coming specially for an excursion, by bee-experts, by sceptics; yet all these people go away convinced. Moreover, the use of 700 foundation is nowadays frequent in Belgium and even in the neighbouring countries: the more timorous use 750. Progress is slow, but nothing stops it.

To convince the more pessimistic, I reared bees on 950. I wished to show how the bee reared in a small cell degenerates, whether the smallness results from the age of the comb or from any other cause. Bees reared in 950 combs are so small that the most skeptical have to confess that the following principle is justified: – *The size*

of the bee is correlated with the capacity of the cell. Small cell, small bee; big cell, big bee. And the size remains the same during the whole of the bee's life.

The drone reared in a worker cell is small and puny. I shall show later that, when reared in a larger cell, he reaches a formidable size.

Those who contradict the principle cited above are people who want to say something in spite of their ignorance and their lack of observation.

In order to keep a rigorous check on the results obtained, I was obliged to invent: –

No. of Cells per sql. Dem.	Origin of the Bees	Space in mm. through which bees pass:		
		Small Bees	Main Body	Large Bees
1050	Belgian Congo Native or Italian	3.4	3.5	3.6
950	"	3.6	3.7	3.8
850	"	3.8	3.1	4.0
750	"	4.0	4.1	4.2
700	"	4.1	4.2	4.3
685	"	4.2	4.3	4.4

The *thoraxmeter*, either with slide or dial (Figs 1, 4); the automatic *glossometer*, with dial and large fluid surface (Fig. 2); and the *sacmeter*, in which the honey taken by a group of bees is indicated by a capillary tube, easy to read off, outside the apparatus (Fig. 3). I shall give, in a future article, the results obtained with these instruments, each of which has been the subject of a thorough and most interesting study. Here I will only state the few essential points needed to make all clear to the reader. I take the liberty of giving, with some measurements of cells, the size of some bees from the Belgian Congo. These confirm the other results. As can be seen, the bees reared in cells of one size are not all of the same dimensions. For example, the 850 lot. Some of these pass through 3.8 mm., most of them through 3.9

mm., the largest through 4.0 mm. But, if the size of gauge through which the "850s" can pass is placed at the entrance of the "700s," not one bee of this size could pass through.

Whence the differences of dimensions of bees from the same size of cell? Probably from their individual constitution, or from an unequal distribution of food. These differences occur, no matter what the size of cell.

The automatic glossometer has revealed many unsuspected things. The tongue of the bees gets longer by 0.5 mm. for each 50 cells less per dm^2 of comb. Thus, selected bees from 750 comb have tongues of 7.7 mm. – a fact confirmed several times. The same colony was placed on 700 combs and examined two months later. The tongues were then 8.2 mm.; this measurement also was confirmed several times. There is always an advantage with the large cells. The experiments related in *L'Apiculture Belge* were made elsewhere, with conclusive results. The measuring instrument has a float and is of an admirable sensitiveness. I can measure to 1/20th mm.; it shows the useful effect of tongue length.

The sacmeter will be an infallible instrument of measurement for use on a group of bees. I am awaiting the results of fresh tests before publishing the measurements. It has been found that the sacs of bees reared in 685 are much larger. The figures will be checked in some forthcoming experiments. The wings are 1 mm. longer than those of bees reared in 850 comb.

Whatever people may say, Italian bees build, in their native land, the same size of cell as our native bees. I have received natural combs from Italy, and ascertained that the stories of combs with 764 cells per dm^2 are – well, stories! There are people who will deny plain facts and who will at the same time accept other statements as gospel truth. I have measured 51.7 and 53.5 mm. per ten cells, just as with our native bees. I have made a longer investigation of this subject, which I will submit later on.

The influence of cell capacity is real, and no possible test has been neglected that might aid in making sure of this. All these matters have been set forth clearly, and beekeepers are now in possession of the data which have cost me much time, work, and above all patience to bring to a successful issue. They will profit by using the largest cells possible. *The combs are drawn out more quickly, they are utilisable for longer in the brood-nest and they are easier to extract.*

It might be thought that since a larger bee has been obtained, there remains nothing to be done. It would, however, be flying in the face of all progress to put "perfect" at the end of any task. One cannot stop upon the road.

In spite of the results attained, I am asked: Will it be possible to fix naturally a race of large bees? To that I reply this should be a task for us all for it is difficult to eliminate from one's neighbourhood all the elements harmful to such an enterprise. Nevertheless, the good seed spreads, and it is always possible to improve one's district by "radiation." All my work is directed toward that end.

In the first place, do the bees reared in 700 combs possess dimensions rendering them capable of building larger cells? To that I reply; Yes. They build worker cells of 736 and drone cells of 470, instead of 85O and 530. I have had occasion to confirm these facts many times already, but I was determined to make a public experiment in the presence of expert beekeepers. Two skeps (*cloches*) were populated with bees reared in combs with 700 cells per dm^2, and also a nucleus hive with empty frames. A little melted wax served as guides. The same beekeepers, at a meeting called specially later, were able to confirm the measurements given above. The centre to centre measurement was carefully tested; it was 36 mm. I shall return to this presently.

There is thus an enormous advantage, and that for a first generation from parents reared in 850 comb. But what will happen when the bees participate in the heredity of queens reared by "700" bees (therefore larger) and of drones reared in cells 470 per dm^2? I think that the bees resulting from these crosses (which will necessarily be made in the future) will have even more aptitude for maintaining larger cells and a larger race.

Consider the size of the drones reared in worker cells (5 mm.), ordinary drone cells (5.5 mm.), and cells of 470 comb (6 mm.). At the first glance the difference of size and of length of these drones (on pins at the apiary for demonstration purposes) is striking. As with the bees, their large size, once acquired, is there for life. These 6 mm. thorax drones are show objects. They will serve to extend selected strains and to maintain a naturally large race. It will be desirable to measure the choice drones and to eliminate undersized ones. There is some work there for investigators; any results sent to me will always be welcome.

The two colonies in skeps enabled me to determine that the worker combs (empty brood combs) were 22.8 mm. thick, and the drone combs 29 mm. The spacing of frames from centre-to-centre is 36 mm. Consequently, I conclude that a centre-to-centre measure of 39 mm. is necessary for convenient manipulation of the hive and to avert congestion of the spaces by the large bees.

This question of cell size is necessarily linked with that of hive capacity. Supported by the reasoning of Dadant, Voirnot, Layens, etc ., and in order to avoid entering upon a multitude of controversial questions, I shall state, as they do: the brood-nest ought to contain about 63,000 cells for laying in, in addition to the space needed for the rest of the bees' requirements. The honey chamber, properly so-called, ought to be more or less large, according to the local flora.

(1). I amplify the horizontal hive by giving it 20 frames, 38 x 38 cm. [nearly 15″ X 15″]; 11 frames for the brood, with a surface of 14.44 dm^2, giving 158.84 dm^2 of 700s, or 111,188 cells. This is the ideal hive for all, and especially for beginners – and they are not wanting (my goodness no!) even among the old hands!

All the frames abut on an entrance which can be as long as the hive; whence fresh air, more easy access to the storage space, rapid spreading of the nectar over a large surface and thus its rapid evaporation, easy inspection of any part, whether brood-nest or honey chamber. This hive is the one that seems to me to give least trouble.

(2). I amplify the vertical hive also thus: 12 frames, 30 X 45 cm. and 12 shallow frames 15 X 45 [11-1/2 and 5-3/4 X 17-5/8 nearly]. The 12 frames of the nest, each of 13.5 dm^2, give 162 dm^2 X 700, or 113.400 cells. For comparison, 12 Voirnot frames contain 111,708 cells and 12 Dadant frames, 115,668 cells, if 850 foundation is used.

The increase in size is justified, since otherwise the brood-nests are too soon crowded, whence danger of swarming. *Big cells, big bees, a big hive*.

To use 700 foundation, wire your frames vertically (7 wires for Dadant, 5 or 6 for Voirnot or Layens). Never any wires horizontal or crossed. It is better to put in one more vertical wire, to obviate buckling of the foundation. Having the frames ready, get them built: –

(1) By swarms. Feed if there is no flow, so that building may not be checked.

(2) By colonies that have swarmed, or by casts, whose young queens do not clamour for drone cells.

(3) By honey-storing colonies, outside the brood-nest, in the middle of the flow. The bees will not alter a single cell.

(4) Get the combs built, feeding heavily, at the end of July and in August. At this time the bees will not build drone cells.

Be careful not to introduce a frame with a sheet of foundation into the middle of the nest of a strong colony disposed to swarm; for such will alter some of the cells to make drone cells, even if you give them 900 cells per dm^2

I mention these matters specially, because it is those who make clumsy mistakes who serve as megaphones for the people who find it is not to their interest to encourage this innovation.

Do not let us, at the present day, be of the company of those beekeepers who ignore entirely the facts of contemporary practice, and who argue about things which they have never taken the trouble to try or to examine. It is enough to drive one mad. I hope that, fortified by all these little details, you will be able to prove to these people, by plain facts, what can be done by strong bees, reared in cells of 700 to the dm^2

For those who are not convinced, Tervueren is always there.

The Bee World - January, 1934 - Pages 2-5

THE INFLUENCE OF CELL SIZE. (With Illustrations and Table of Data by M. Baudoux, Tervueren, Belgium).

In April last we published a description by M. Baudoux of his work on the enlargement of the bee by the use of large-cell foundation (THE BEE WORLD, XIV, p. 37). Through his kindness, we are now able to supplement this general account with a Table, giving the average dimensions of bees reared in comb of the various sizes, and with illustrations of the bees (from No. 1, 650; to No. 9, 1050).

The data in the Table are the averages of a large number of observations - the work on which it is based was begun in 1891 - and deserve close study. The Table appeared in *L'Apiculture Belge* of November in a slightly different form, some of the items not being carried to the third decimal place as in the copy placed at our disposal. We have taken the liberty of omitting the column *cubic content of a sq. dm. of comb*, since it can be obtained instantly from the *comb thickness* by moving the point. The smoothness with which the data plot on squared paper against cell-number per sq. dm. or any other of the quantities that may be selected, is an assurance-if, in face of M. Baudoux's careful and ingenious methods of work, any were needed-that the results are to be relied upon.

As regards the separate items. *Span* (distance between the tips of the expanded wings) is equal to twice the length of the fore wing *plus* the distance between the wing roots (headed *wing* root *distance*). *Thorax* is the depth (or width-M. Baudoux informs us that they are equal) of the thorax, as measured by the thoraxmeter (see THE BEE WORLD, XIV, p. 38). *Tongue* gives a series of values which are intended to be illustrative of the gain to be expected from the use of large cells; the actual values will vary with the strain of bee used. [We may add: This will be so with other quantities-for example, wing length. All varieties of bee are not geometrically similar.] *Body length, wing width* and *thickness of the comb* need no comment. *Cell width* is measured as the distance between two parallel edges. The *cubic content of a cell* is of interest, in that comparison of it with the values for the drone comb approximately confirms Mullenhoff's results (announced some 50 years ago) that the drone cell has a volume double that of a worker cell. The ratio for M. Baudoux's cases lies between 1,823 and 1,938. *Weight at emergence* is somewhat higher than the normal weight of a field bee when "empty"; it is, at the author reminds us, the only weight that can be used for exact comparative purposes, since older bees' weights vary so much with the state of their alimentary canal.

The item *volume of sac* calls for more comment. M. Baudoux considers that the maximum load taken by a bee when robbing, or visiting a large supply of syrup provided for experimental purposes, is carried partly in the stomach; and that the bee can regurgitate the stomach contents at will. The honey sac alone, he states, is unable to contain the whole of the loads sometimes measured, for there would not be room for them in the abdomen. The figures he gives in this column therefore represent, not the maximum possible load of the bee of that size, but the volume of the honey sac only; and are based on the diameter of the anterior part of the abdomen, less the double thickness of the body-wall (taken as 0.2 mm. in all). We understand, however, that experiments with the sacmeter and actual measurement of full sacs have confirmed the values given, in a general way.

Dimensions of Workers and of their cells.

Cells per sq dm	Span	Wing Root Width	Wing Length		Thorax	Tongue	Body Len	Cell width	Comb Thick-ness	Cell Vol	Sac capa-city	Wing Width		Wt at Emergence
			Fore	Hind								Fore	Hind	
650	24.000	3.000	10.500	7.500	4.48	8.00	16.00	5.960	23.40	360	47.10	3.20	2.40	145
700	23.100	2.887	20.106	7.218	4.312	7.70	16.40	5.750	23.00	328	41.95	3.08	2.31	139
750	22.242	2.780	9.731	6.950	4.152	7.41	14.82	5.555	22.60	301	37.40	2.96	2.22	134
800	21.428	2.678	9.375	6.696	4.00	7.14	14.28	5.375	22.20	277	33.40	2.85	2.14	129
850	20.657	2.582	9.037	6.455	3.856	6.88	13.77	5.210	21.80	256	29.90	2.75	2.06	124
900	19.928	2.491	8.718	6.227	3.720	6.64	13.28	5.060	21.40	237	26.85	2.65	1.99	119
950	19.242	2.405	8.418	6.013	3.592	6.41	12.82	4.925	21.00	222	24.20	2.56	1.92	115
1000	18.500	2.325	8.137	5.812	3.472	6.20	12.40	4.805	20.60	206	21.90	2.46	1.86	111
1050	18.000	2.250	7.875	5.625	3.360	6.00	12.00	4.700	20.20	192	19.90	2.40	1.80	108

All dimensions are in millimeters and weights in milligrammes.

Dimensions of Droned and of their Cells

Cells per sq dm Workers/ Drones	Span	Wing Root Width	Wing Length		Thorax	Body Len	Cell width	Comb Thick-ness	Cell Vol	Wing Width	
			Fore	Hind						Fore	Hind
650/403	32.500	4.400	14.000	10.000	6.400	18.000	7.56	26.60	656	4.35	3.26
700/434	30.185	4.235	13.475	9.625	6.160	17.325	7.29	26.20	605	4.18	3.13
750/465	30.028	4.078	12.975	9.268	5.931	16.682	7.04	25.80	558	4.02	3.01
800/496	28.928	3.928	12.500	8.928	5.714	16.071	6.81	25.40	515	3.87	2.90
850/527	27.886	3.787	12.050	8.607	5.518	15.492	6.60	25.00	476	3.73	2.80
900/558	26.903	3.653	11.625	8.303	5.314	14.946	6.41	24.60	441	3.60	2.70
950/589	25.977	3.527	11.225	8.017	5.131	14.432	6.24	24.20	410	3.47	2.60
1000/620	25.110	3.410	10.850	7.750	4.960	13.950	6.09	23.80	383	3.36	2.50
1050/651	24.300	3.300	10.500	7.500	5.800	13.500	5.96	23.40	365	3.26	2.44

All Dimensions are in millimeters and weights in milligrammes.

In connection with measurements of the tongue length, M. Baudoux in 1926 raised the question (L'Apiculture Rutionelle, December, 1926) whether measurement of the tongue under the microscope really gives a correct idea of its length when in use, seeing that it is a very extensible organ. His glossometer is designed to give values more nearly representative of the practical possibilities of the tongue. He makes due provision (as he recognised to be necessary in the same journal, January, 1927) for avoiding the errors due to surface tension, which draws the fluid up the sides of any surface wetted by it-and is of assistance to bees when they are working on flowers with deep corolla-tubes.

M. Baudoux finds that the size of the drone cells built by bees of a given series always bears a constant ratio to the size of the worker cells. There areas are as 50:31, or 1.61 approximately. The drone cells in the Table therefore are arranged in increments of 31. M. Baudoux, in his notes on the Table in *L'Apiculture Belge*, calls attention to the fact that the interval between measurements of the bees of one series and those of the next is larger for the larger bees. This is natural; for 50 is a larger fraction of 650 than it is of 1050. If, however, the quantities are compared by plotting all the data against some length-for example, the cell width-it will be found that the increase of nearly all the quantities is even and regular throughout the series (see below).

If big bees are merely small bees whose every dimension has been multiplied in the same proportion, and if the materials of which they are constructed are in every way similar for the corresponding organs, then we ought to find that any two measures of length bear a constant ratio to one another, whether the bee be large or small. That this is the case is easily seen by plotting against some length, as suggested above. The span, wing root length, lengths and widths of wings, length of body, and the cell width, all plot as very nearly straight lines against any one of their number; and if one of them be divided by another (say, body length by wing length), the result is a quantity which is the same (or nearly so) for all the different sizes of bees. The *comb thickness* does not however conform to this rule, for reasons which must remain for future consideration.

From geometrical principles, one would expect that under these conditions the weight of the bee would vary as the cube of her length or other linear dimension. Very surprisingly, this is not so. *The weight of the bee is proportional to her length*, not to her (length). This is most unexpected; as M. Baudoux states, it must mean that the enlarged bee is not as solid as the bees of the smaller series. M. Baudoux, we understand, intends to test this point. His results will be awaited with much interest.

This matter of the specific gravity of the bee, and its decrease with increase of size, is not only of theoretical interest. If we suppose that the head, thorax and abdomen share alike in the lightening process, we shall-if we continue to enlarge the bee-arrive finally at an insect which cannot fly as fast or lift as great loads as smaller bees. Big wings demand big muscles to move them; as far as can be seen, the mass of the flight muscles must increase as the cube of the wing length. If the weight of the body increases in this proportion, the bee will continue to be an efficient flyer; but if-as is the case with M. Baudoux's bees-it does not, then one of two things must be happening. Either the bee will have less flying muscle than she needs to work her long wings; or the flying muscles will make up a greater proportion of her total weight. In either case, theory would indicate that very much enlarged bees should be less efficient nectar-carriers. That the limit (where this begins to occur) has not yet been reached is shewn by the excellent practical results which M. Baudoux obtains.

M. Baudoux finds that the sp. gravity of his bees is about 0.525-a little more than half that of water. This agrees well with the usually received value for insects

(0.5), and with that calculated from Armbruster's figure for the volume of a bee (195 cmm.). This would give sp. gr. 0.525 if the bee weighed 102 mg. Actually, most field bees weigh less than this when "empty," and their sp. gr. will often be nearer 0.4.

[It may be noted that the weight of insects reared in a state of nature does actually conform fairly closely to the cube law. Weighing and measurement of a large number of bees, wasps and related insects, also of Syrphid flies of various sizes, has shown that the weight varies approximately as the cube or the 3.5 power of the wing length. (The latter result is probably due to the insects being unduly heavy with eggs or a full colon). Even wasps or bumble bees of different sizes-the nearest approach in nature to the bees reared by M. Baudoux-follow the same law. It may be remarked that all these insects also have the thorax weight a more or less definite percentage of the body weight; for good flyers, with few exceptions, about 40 to 55%. This includes legs and leg muscles, of course. Poor flyers that use their big legs a great deal, such as the digger wasp *Ammophila*, have a high thorax percentage weight; but when allowance is made for this, it is some guide to flying power. The drone's thorax, for example, habitually accounts for more than half of his weight.

It might be thought that, by ensuring that the colony had ample stores, any stinting of the larvae could be prevented, and so their weight increased; but we think it likely that the beekeeper will not be able to intervene here. The nurses will most probably go on giving the grubs, not what they could eat if allowed, but what they are considered to need!

In this connection it is worth while comparing v. Rhein's work on the rearing of giant workers by feeding them lavishly with older-worker larval food in roomy artificial cells (THE BEE WORLD, XIV, p. 141, December, 1933). His largest specimen weighed 175 mg. at emergence. If its weight followed the same laws as that of M. Baudoux's series of bees, it would correspond to bees reared in cells 450-500 per sq. dm. It is probable, however, that this is not the case. A comparison of v. Rhein's figures of a giant and a normal worker shows that the abdomen of the giant is 1.6 times longer than that of the normal bee at the same pupal stage; and that its thorax is only about 1.14 as wide (at most). This probably means that a good part of the abnormal weight is due to the development of the ovaries and spermatheca which is a feature of such giants; so that the giant would (it of normal worker construction) be little if at all heavier than the bees reared in 650 comb by M. Baudoux. That is to say, the unexpectedly light weight of the bees we are considering may be unavoidable, if they are to remain workers and not become half-queens. The nurses may be obliged to stint them in order to prevent development of queen organs and probably also queen instincts. If they did not do this, the use of very large cells would probably result in the rearing of bees which produced an unusual proportion of laying workers or were otherwise abnormal. So far, there appears to be no risk of this; for M. Baudoux's big bees have proved themselves most efficient honey-producers, and show no sign of queenlike indolence.]

We understand that M. Baudoux welcomes criticism, as long as it is of a constructive nature, and offered by persons who have tried, or are prepared to try, the large cell foundation for themselves. We are convinced that this method of improving the bee deserves to be considered very seriously.

A.D.B.

The Bee World – July, 1935 – Pages 81 – 82 - THE BEE LABORATORY

Science is nothing but trained and organised common sense.-Huxley

RECENT WORK ON THE INFLUENCE OF CELL SIZE.

H. Gontarski has studied the influence of cell size on the phaenotypic variation of the honey bee (Zts. f. Morph. u. Oekol. d. Tiere; Abt. A. Zts. f. wiss *Biologie*, Bd. 29, H. 3, 1935). He finds that the greatest percentage increase in size takes place if cells of 5.74 mm. diameter are used; that is, foundation with 700 cells per sq. dm. He points out that the variation in size of individual bees in a colony is partly due to phaenotypic variation due to the cell size, and partly to differences called forth by the quality and quantity of food - differences, in fact, that are similar to, but less than, those which cause a worker larva to become a queen when suitably fed and nursed. His big bees had the normal worker number of ovarian tubules - maximum 20 - so did not show any hyper-trophy of the ovaries nor had they developed spermatheca. They were in other anatomical respects also normal workers. There is thus little fear that such bees would be more queenlike in their instincts, and perhaps not such keen foragers as smaller bees.

The author thinks that the influence of the cell is mainly exerted indirectly, by providing more room for food; but does not deny a possible direct influence as well, due to more room for growth.

In another paper (*Zts. f. vergl. Physiol. Bd.* 21, H.5. 1935) he deals with cell dimensions, and shows that they depend partly on the size of the bees and partly on the condition of the colony at the moment and its inclination or disinclination to build.

Bees appear to perceive the size of the cells in foundation with the aid of their legs. They probably "realise" the cell as a whole, and there are limits to the size of cell-base to which this realisation can extend; so that foundation with bases larger than drone comb is drawn out quite irregularly. Bees probably have an instinctive tendency to continue unfinished work as it was begun-this is important for ensuring the carrying out of the work of the hive in a steady and regular manner.

There is a certain size of cell-base at which the bees tend to switch over from worker to drone cells. This appears to be larger the more the strain of bees tends to non-swarming propensities.

The queen is evidently able to recognise the size of cells, for she delays laying when given comb with cells of abnormal sizes. Two stocks were given empty frames in April, in which they built drone comb. The cells were noted to be very irregular in size; the queens laid in the smallest cells first (and the largest last), suggesting that they only reach the "mood" for laying unfertilised eggs gradually, as the season advances.

The impulse to lay "drone" eggs seems to depend, not on the absolute size of the cells, but on a perception of the difference between the two sizes of cell, which acts as a kinaesthetic stimulus to the queen's nervous system and thus to her spermatheca.

Large cells have been a frequent subject of discussion in many bee journals of late. R. Graftiau (*L' Apiculteur*, February) mentions that one can get combs of large cell foundation beautifully drawn out if one gives them to weak stocks on 4 or 5 combs only, is well as in swarms, natural or artificial. S. Sevin (the same, January) compares the surplus and winter stores of stocks on various cell sizes, to the advantage of large cells.

The last-named author also discusses (February) the number of cells per sq. dm. and a formula for deducing them from the width of a cell. He makes the error – with which everybody should now be familiar, since it is annually exposed when made in connection with the timing of motor races and other fast sporting events- of carrying his calculations to more significant figures than the attainable accuracy warrants. (Incidentally, Armbruster published a note on this formula in 1932; *Arch. f. Bienenk*, XIII 7/8)). The formula as usually given is inconvenient to use, as it must be worked out on paper at some length. The following may therefore be of use to those buying foundation and desiring to know its size.

If W be the width of a cell between parallel walls, its area is $\frac{W^2\ 3.5}{2}$ and if W be measured in mm., there are $40{,}000/W^2\ 3.5$ cells on the two sides of a sq. dm. of comb. This is - much more nearly than we need go in practice - $23{,}094/W^2$.

If there are X cells in a dm. length, $W.X=100$; and $N=2.3094\ X^2$. Or we may write, $N=(1.51\ X)^2$.

N is therefore roughly equal to the square of 1-1/2 times the number of cells in a dm. length, or to the square of the number in 15 cm.

When we proceed to inches, the rule is equally easy, since N is the square of the number of cells in 1.52 X 3.94 inches, or nearly 6 inches. Thus, if there are 26 cells in 6 inches, N is 26 X 26, or 676 cells per sq. dm.

The rule is therefore: – *Count the cells in 15 cm., or in 6 ins., and square this number.*

To obtain the number per sq. inch, count the cells in 1-1/2 inches and square. This figure is approximately 15.5 times less than that for a sq. dm., since a sq. dm. = 15.5 sq. inches.

Frequent Variation in Cell Size - Bee World 1935

The Bee World – November, 1935 – Page 124

A correspondent of *Le Rucher Belge* (August - September) remarks on the frequent variation in size of cells in commercial foundation (especially now that larger sizes are sold). Beekeepers who are careless and do not make sure that they have got all their foundation of one size, are thus apt to get several sizes in one colony. The bees may in consequence avoid breeding in the sizes they do not 'fit,' and if this does not occur, will have workers of several sizes. Combs in such cases should be sorted into sizes at wintering time.

The Bee World – December, 1935 – Page 138--PRESS MIRROR.

Writing in *L'Apiculture Francaise* for October, Dubois de Szczawinski maintains that Baudoux's work has been misunderstood. It was not the large cell in itself for which he was working, but the selection of a better bee. The writer collaborated with Baudoux for years and speaks with authority on his aims and methods.

We have a great respect for the work of the late M. Baudoux, who was a most careful and painstaking experimentor; but we cannot agree that the judgment of the apicultural world on his work is mistaken. It is easy to understand that anyone reared, as he undoubtedly was in the tradition of Lamarck should believe that it is possible to improve the bee permanently by giving her the chance to grow larger in each succeeding generation. To a Darwinian or a Mendelian, however, this is very uncertain doctrine; and the bees themselves appear to confirm this criticism, since they tend to retrogress in the size of cells they build when let alone. It is possible that, as has so often happened in the history of science and technical progress, as well as in other departments of human activity, the end at which Baudoux was aiming is not the one for which his name will - most deservedly be remembered by future generations of beekeepers. Baudoux's intention may have been misunderstood by those who lay stress on the phaenotypic enlargement of bees by the use of large-celled foundation-but it is probable that his *real contribution to apiculture* has not been misunderstood.

Influence of Size of Brood Cell Upon the Size of the Worker Bee - ABJ 1936

*American Beekeeping Journal – April, 1936 - By Roy A. Grout,** Illinois*

The late Monsieur Ursmar Baudoux of Belgium was the first to conceive of the use of an artificial foundation having an enlarged cell base to increase the size of the emerging bee. In the year 1893 he was amazed on discovering bees from an old skep which were very much smaller than normal. He then conceived the idea of raising bees in larger cells. He accomplished this by means of stretching regular foundation to the size he desired and had by 1896 sufficiently proved his point in Belgium that a manufacturing company began to place upon the market artificial bee comb foundation having enlarged cell bases. It was Baudoux's belief that the nurse bees following a natural instinct filled the bottom of the enlarged cell more copiously with larval food, and that this caused an increase in the size of the worker bees He also intimated that the larger bee would generate more body heat which would result in a greater quantity of brood.

By means of stretching foundation, he experimented with various sizes of foundation having 750 cells to the square decimeter, 740, 730, 710, 700 and even 675 cells per square decimeter. (This is in contrast to the U. S. standard size which is 857 cells per square decimeter.) By means of an ingenious glossometer of his own make, he determined the tongue reach of his colonies and by means of an equally ingenious thoraxometer of his own make, the diameter of the thorax. He found that with an increase of 50 cells per square decimeter in the size of the foundation, there was a corresponding decrease of one half millimeter in the tongue reach. His thoraxometer gave thorax diameters of 3.7 mm., 3.9 mm., 4.1 mm. and 4.3 mm. for worker bees reared in brood cells built from foundation having respectively 850, 800, 750, and 700 cells per square decimeter. He arrived at the conclusion that artificial foundation having 700 cells per square decimeter gave a bee which was superior in all its measurements to those reared in combs constructed from the smaller sizes of artificial foundation.Independent of the work done by Baudoux, a Frenchman by name of Pincot arrived at the idea from a slightly different angle. Pincot noticing the difference in size of worker bees from a swarm placed on foundation and the worker bees of the parent stock reared in natural combs, came to the conclusion that this was due to the fact that the natural brood cells were larger than those drawn from the foundation and actual measurements confirmed his theory. He then started experimenting with foundation having 736 cells per square decimeter and reports that during a two year period thirty colonies using this size of foundation gathered approximately one - third more honey than did thirty colonies on normal foundation. Unfortunately, in 1910 his apiaries were destroyed by a flood and Pincot was forced to abandon his experiments.

While the experiments of these two cannot be considered of a very scientific nature, each claimed larger bees resulting in a greater yield of honey. Their activities, particularly those of Baudoux, were convincing to the extent that a firm in Belgium has offered enlarged cell foundation for sale since 1896. Other manufacturers have followed in this course. France was next and more recently Italy and England have manufacturing concerns offering for sale enlarged cell foundation and claiming better results through its use. Consequently, interest in this country has been focused upon this matter.

A Russian worker by name of Lovchinovskaya reporting on experiments started in 1925 using enlarged cell foundation showed that bees reared from enlarged cells weighed more, had a greater load capacity and that from the results of one season produced more honey.

During the period 1930 to 1932, the first scientific study was made to determine the effect of rearing in enlarged brood cells upon various parts of the worker bee. These experiments were carried on at Iowa State College under the direction of Dr. 0. W. Park.

Three different sizes of foundation were used having respectively 857, 763 and 706 cells per square decimeter. The foundation having 857 cells per square decimeter is the commercial size manufactured in the United States. The foundation having 763 cells per square decimeter closely approximates that having 750 cells per square decimeter which has been manufactured since 1896, by Jos. Mees Sons of Herenthals, Belgium, and the latter size closely approximates that having 700 cells per square decimeter which the same firm has manufactured since 1927. Care was taken to eliminate all warp and sag in the finished comb and no control of size of brood cell other than size of foundation was used.

To facilitate recognition and handling of the combs, and for convenience in presenting date, the frames containing the standard size foundation, that having 857 cells per square decimeter, Were marked "A" and the cell size was referred to as size of cell "A". Similarly, frames containing foundation having 763 cells per square decimeter were marked "B" and the cell size was referred to as size of cell "B". Likewise, the frames containing foundation having 706 cells per square decimeter were marked "C", and the size of cell was referred to as size of cell "C".

In general, two frames of each size of foundation were placed in the same colony. Individual colony records were kept and the queens were marked by clipping the right wing of those reared in an even numbered year and the left wing when reared in an odd numbered year.

It is of interest to mention that difficulties were experienced in getting the queens to oviposit worker eggs in the enlarged cells when all three sizes were placed in the same hive at the same time. This was particularly true in case of size of cell "C". While the worker bees apparently recognized no difference in accepting the three sizes of cells, the queen bees showed a preference for the smaller cells for ovipositing. This result was confirmed by similar experiments carried out by Lovchinovskaya.

An effort was made to collect the bees upon emergence from all three sizes of cells in a single colony at approximately the same time and under the same conditions. To determine the date of emergence, a chart was used whereby the daily emergence of bees from twenty-three colonies was recorded. Prior to emergence, each frame was caged in a Root introducing cage and a selected area of brood containing no nectar or honey was caged with an additional screen cage.

Each sample of bees contained at least fifty specimens. Following the method outlined by Alpatov of Russia the bees were slightly anaesthetized and then killed by dropping into boiling water. They were then preserved in a 70 percent alcohol solution to await further treatment. The measurements taken on each individual bee were dry weight. length of right fore wing, width of right fore wing, the sum of the widths of the third and fourth tergites and the length of proboscis.

In Plate 1 are shown the measurements taken on the right fore wing, the third tergite and the fourth tergite. (In explanation the two latter parts are the two largest plates on the top of the abdomen.) Plate 2 shows the measurements taken on the tongue or proboscis. In the series of graphs which are illustrated, each character, namely, dry weight, length of proboscis, length of right fore wing, width of right fore wing, left fore wing and the sum of the widths of the third and fourth tergites is plotted for each size of cell. In all cases there is a distinct trend towards a larger character of the worker bee as the size of the brood cell increases.

PLATE 1 **PLATE 2**

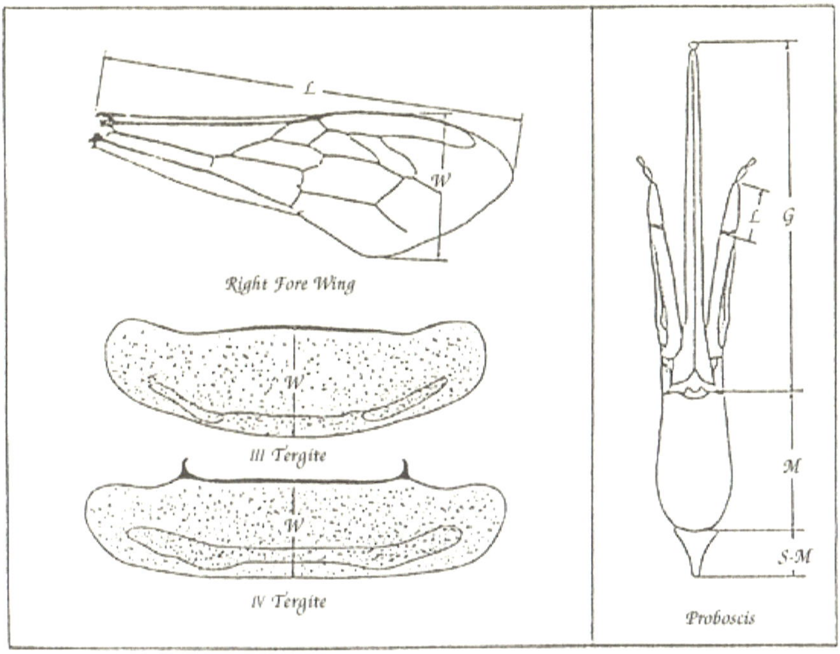

In table 1 are given the averages of the five measurements for the bees from each size of cell and the percent increase of these measurements. It is of interest to note that the length of proboscis increased 2.07 percent as the size of foundation was increased from 857 cells per square decimeter to 706 cells per square decimeter.

We, therefore, find that the size of the brood cell is definitely a factor in determining the size of the adult worker bee. It is also apparent that larger bees are obtained through the use of artificial bee comb foundation having enlarged cell bases. It is reasonable to state when we compare a 2.07 per cent increase in the length of the tongue, that size of brood cell alone is not sufficient to produce a much larger bee. It is much more reasonable to state that selection and breeding of bees plus the application of such factors as size of brood cell should accomplish marked results in producing larger bees.

Measurement Taken	% Increase from "A" to "B"	% Increase from "B" to "C"	% Increase from "A" to "C"
Dry weight	15.50%	30.98%	51.27%
Length of right fore wing	0.60	0.89	1.49
Width of right fore wing	1.05	1.15	2.21
Sum of widths of 3rd and 4th tergites	2.24	1.45	3.72
Length of proboscis	0.93	1.13	2.07
Length of mentum	1.49	0.11	1.61
Length of glossa	1.21	1.22	2.45
Sum of lengths of mentum and glossa	1.28	0.90	2.19
Average of 7 linear measurements	1.37	1.00	2.38

Table 1

From the results we have obtained we cannot agree with Baudoux either in the results he obtained or the consistency of his results. While he records increases of 11.9 per cent to 25 per cent in length of proboscis as the size of the brood cell increases from 850 cells per square decimeter to 700 cells per square decimeter we are only able to find an increase of 2.07 per cent as a maximum. However, we believe that our results compare favorably with results obtained by Michailov of Russia who, on measuring the tongues of worker bees reared in worker cells as compared with those of worker bees reared in drone cells, found an increase of 4.82 per cent. While this is a greater increase than ours, it must be considered that the size of cell was increased slightly more than twice as much as in this experiment which in all probability would account for the difference. Our results also compare favorably with those obtained by the same worker on worker bees reared in new combs as compared with worker bees reared in old combs. Here Michailov records an increase of 1.05 per cent in the length of the tongue.

Since we have made the statement that size of brood cell alone is not sufficient to produce a much larger worker bee, we must consider the fact that the crucial test for the commercial use of enlarged foundation is greater honey production. While this experiment should be a strong indication toward that end, the exact relation of this increase in the size of adult worker bees to a greater yield of honey has yet to be proved. During the past four years, we have been conducting an experiment in a commercial yard with from fifteen to twenty colonies containing brood combs constructed from each size of foundation, making an apiary of sixty colonies maximum. To date we have not been able to find any significant increase in the honey production due to the use of enlarged cell foundation. This experiment is still being continued in a location more favorable for honey flows and we expect to have some definite results in the near future.

*Rewritten from Journal Paper of the Iowa Agricultural Experiment Station. Ames. Iowa. Project No. 129.

**Formerly. Research Assistant in Apiculture, Iowa Agricuitural Experiment Station, Ames, Iowa

ABJ, April, 1936 – Page 180 By P. C. Chadwick, California.

The article by Dr. Lloyd R. Watson in your February issue seems to me to have left an opening for some discussion as to the advisability of enlarging the type of our present honey bee by select breeding.

His reference to the breeding of the primitive horse from inferior types to our present forms is a matter of selection consistent with nature and is sound from that standpoint. However that there are limitations to the distance nature will permit some forms of life already apparently complete for the purpose created, must be recognized. To my mind the honey bee is in that class.

Dr. Watson speaks of "the little wild honey bee," the inference being that there is a larger tame honey bee. Personally I do not regard the present honey bee as we find it today, as a wild bee.

Take a colony from the most remote part of an infrequented woods, bring it to your apiary, house it as other colonies are housed or let it remain in a section of its original tree home and you will find it to be no more wild than the colonies long in your possession.

We beekeepers are in the habit of going off on an impossible tangent once in a while so far as nature is concerned. For years our bee journals have been printing reams of articles on the question of a non-swarming strain of bees. It has always seemed to me there was a lot of time wasted advocating such an improbable accomplishment, because nature would hardly yield to an arrangement that in itself might destroy the species. If accomplished it would be tantamount to breeding the mating instinct out of domestic animals. If it were possible to keep a single colony from swarming for a period of 99 years, they would more than likely swarm on the hundredth year if conditions were met that made it desirable from nature's standpoint.

Dr. Watson says, "but the genetical potentialities of Apis melifica now, after 3,000 years of continuous history, are as a closed book." With this I heartily agree, but maintain the thought that another 3,000 years may pass and find little difference in Apis melifica at the end of that period, because the honey bee would seem to be in perfect balance with the requirements of its natural mission, as it is.

Even if man should seek to change it, nature would likely intervene in such a way as to preserve the necessary balance. For there is some reason to believe that in the plan of nature the honey bee was not only created to conform to the necessity of its mission as a pollenizing agent, but that the plants and their bloom may have been fashioned to conform to the convenience of the bee. At any rate there is a barrier that seems to have been deliberately placed by nature to prevent any wide deviation of the bee in size and action from what nature designed that it should be, this being accomplished by limiting the size of the bee to that of the cell in which it is developed, beyond which it cannot go. A wise move on the part of nature, designed to prevent this all important pollenizing agent from developing beyond a size necessary for the adequate service for which it is intended.

Dr. Watson further says: "She is perfect, indeed, from the viewpoint of nature." If such is the case (and I believe Dr. Watson has exactly expressed the matter), I doubt if by any process man will be able to tear down such a perfect accomplishment. If it were possible to increase the size of the honey bee to that of the bumblebee, would it be a benefit or a detriment to nature?

The Size of Brood-Comb Cells - 1938

The Bee World – September, 1938 – Pages 106-107 By ANNE BETTS.

Many beekeepers who have been interested in the work of Baudoux and others on the influence of the brood-comb cells on the size and efficiency of bees bred in them, must have wished that it was easier to estimate the number of cells per square inch or square dm. of comb - without, that is, an elaborate calculation or the use of a graph (particularly as few of the latter have been published, and none, we believe, in English. See THE BEE WORLD, XVI, p. 82, for data from which a graph can be drawn).

This difficulty has also, it seems, been felt in Switzerland, and Herr M. Justrich, of the race-breeding organisation, takes up the subject in the August *Schweizerische Bienenzeitung*. He points out the technical difficulties which beset the measurement of the width of a cell and suggests the very happy idea that, instead of measuring the width, say, of 10 cells, we should count how many cells go into 10 cm. (1 decimetre). This number, multiplied by itself and again by 2.309, gives with very fair accuracy the number of cells on the two sides of a square decimetre of comb, which is the number generally used to express the size of cells. He gives a table, ranging from 12.8 to 20 cells per decimetre and (correspondingly) from 378 to 923 cells per sq. dm.

Now it is undoubtedly easier to most people to multiply three figures together than to multiply two together and then divide the product into a third. But even the easier process will call for a large piece of paper and perhaps a good deal of head-scratching on the part of beekeepers whose school days are some way behind them, and who do not habitually do any calculations other than the keeping of simple accounts.

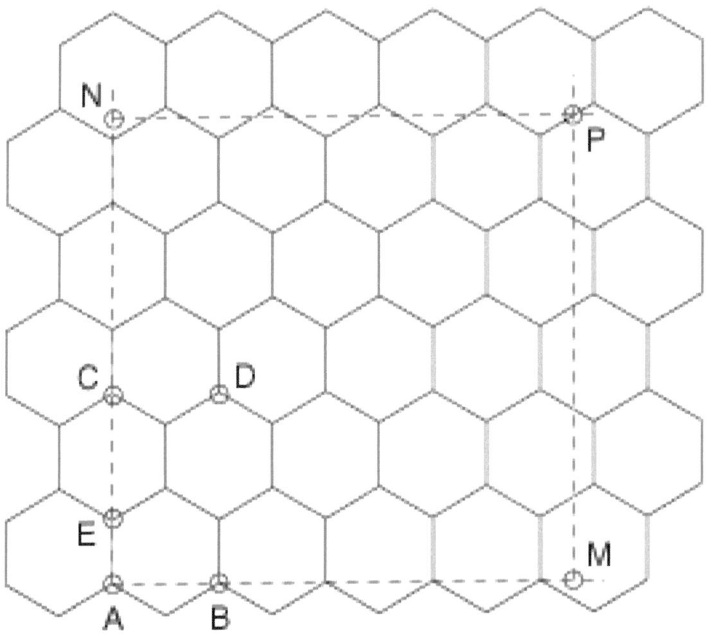

Figure 1

We therefore meditated a little on Herr Justrich's idea, and are glad to report that it has led to a method which does away with the use of large numbers in multiplication. These large numbers are the result of the fact that the area of a six-sided cell, expressed in terms of one of its sides or of its width, inevitably contains the square root of 3 as a factor. This quantity is incommensurable – that is, the decimal expressing it never comes to an end – but the first few places are: 1.7320508. Hence we must take at least 1.732 when calculating with it, to get a sufficiently accurate result. It is possible, however, to get rid of this troublesome four-figure number by using the method which follows.

Place the comb on the table in front of you so that it is the same way up as in the hive, with the cells arranged "right way up." Choose one of the upright sides of a cell and start from its bottom end (marked **A** in the diagram, and surrounded with a small circle to distinguish it). Measure a distance (10 cm. or one decimetre) across the comb, parallel to the top bar; and count how many cells there are in this length. (In the diagram, the cells are drawn much too large, especially if we imagine the diagram to be enlarged, so that **A M** really is 10 cm. and not about 6.5 cm., as it has been drawn; but the principle is of course the same whatever the size of the cells). In our diagram there are about 4.3 cells in **A M**.

Now consider the vertical line **A C**, consisting of the side at **A** and the whole height of the cell in the next row above. For shortness, we shall call this line a "height." Measure **A N** vertically upwards, and of the the same length as **A M** (in practice, therefore, 10 cm.), and count how many "heights" like **A C** go into **A N**. In the case we have drawn there are about 2.4 of them.

Now consider the space **A C D B**. The portion just below **C D** is the same as the portion (outside **A C D B**) below **A B**; and it is thus easy to see that **A C D B** is equal to the area of *two* cells.

Our measurements and counts have therefore shown us that there are, in 10 X 10 cm., or 1 sq. dm., 4.3 X 2.4 areas like **A C D B**. Each of these areas is two cells, and there are also two cells on the other side of it; so that the number of cells per sq. dm. at once comes out as 4 X 4.3 X 2.4, or about 41, in this particular "comb."

In an actual case of comb built on foundation, we found **A M** to contain 17.9 cells, while **A N** contained 10.7 "heights," nearly. The number of cells per sq. dm. was, therefore, 4 X 17.9 X 10.7, or 766 cells (to the nearest whole cell).

If the bees' cells were always perfect hexagons, **A C** would be exactly three times **A E** (the side of a cell), and **A B** would be **A E X** the square root of 3, or **A E X** 1,732 nearly. But in practice comb is seldom regular, particularly when built on foundation made in a roller-mill, so that it is better to count both **A M** and **A N**, and not to depend on the theory that the number of cells in **A M** must be 1.732 the number of "heights" in **A N** (apart altogether from the trouble of dividing by 1.732). For example, in the actual case above, there would have been only 10.3 "heights" in **A N**, had the comb been quite regular, and therefore about 737 cells per sq. dm, instead of 766.

Since it is not very easy to guess the fractions of **A C**, we have drawn a "height" on a large scale in Fig. 2, and marked on it the various fractions. It will be noticed that those who like to use the duodecimal system have an easy task, since the natural features of the comb make it very easy to see the 1/6th, 1/3rd, 1/2, etc., points.

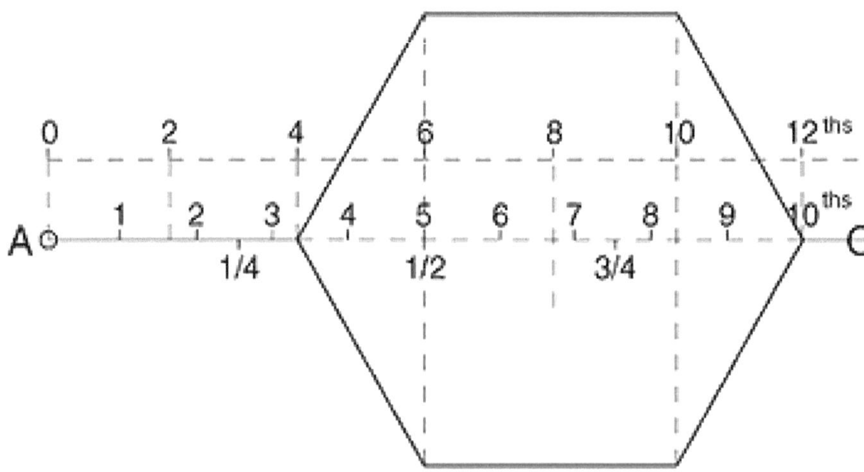

Figure 2

In most practical cases, one could take the nearest whole cell in each length. In the case mentioned, this gives 18 X 11 X 4, or 792 cells – an error of only 26 cells per sq. dm., or about 3.4%. The arithmetic can then often be done in one's head.

Those who wish to use square inches instead of the metric system can easily do so. Take 2″ for the lengths **A M**, **A N**. Then **A N P M** will be 4 square inches, and it is only necessary to multiply together *the figures obtained by counting*, in order to get the cells per square inch. In the case of actual comb quoted 5.5 X 9.1, or 50 cells per square inch. (Since there are 15.5 square inches in one square decimetre, this gives 775 cells per square decimetre – a little too big, due partly to the estimate being rougher with a smaller length, and partly (perhaps) to variation in the cells in the different parts of the comb we measured. If it is desired to get the same accuracy as with the decimetre estimates, 4″ (which is nearly 10.2 cm.) should be used; but we must then divide the result by 4.

The truth is, comb varies a good deal; so that, for practical work – for example, when wishing to breed from the colony which builds the largest cells – the rough method (taking the nearest whole cell or "height" in each case) is probably good enough.

The making of a table connecting cells per decimetre (along **A M**) and number of cells per square decimetre is very easy, if we can assume (as Herr Justrich has done) that the cells are regular hexagons. We have only to take the number of cells in 1 dm., square it, divide it by the square root of 3 and multiply by 4 – a series of operations which can be done very fast on a slide rule and quite accurately enough to get good results. If readers would like such a table, we shall be pleased to publish one later on.

The Bee World – June, 1944 – Page 46

MISCELLANY

Geometry of the Ideal Bee's Cell. The bee's cell was the object of much discussion by amateur mathematicians (and some who were more eminent) in the 18th and 19th centuries. A good deal of nonsense was written on the subject, for many of the writers did not realise how seldom bees build cells conforming to the ideal pattern, nor that the resulting cell-form depends much more upon the stresses in the semi-plastic wax inside the warm building-cluster than upon such considerations as an instinct for using the least amount of wax possible. See *B.W. 3, 37*, for the history of the problem.

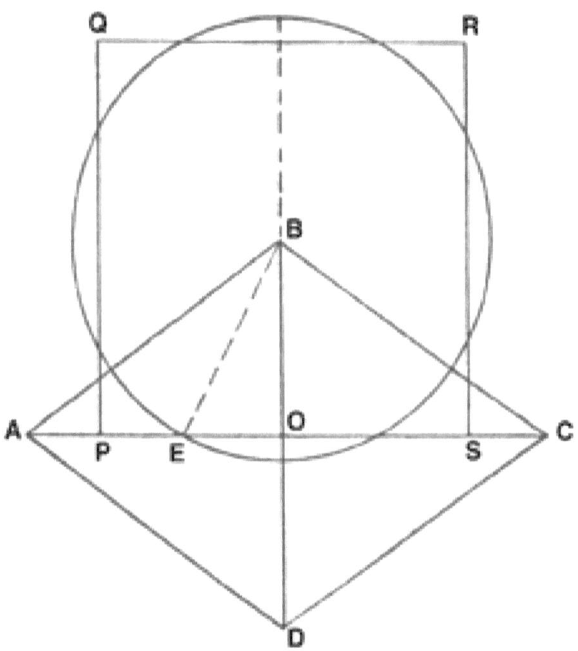

None of the investigators seem, however, to have hit upon the interesting approximate solution of the problem of "squaring the circle" provided by the so-seldom-realised ideal cell. ABCD is one of the rhombs or lozenges from the base of such a cell, AC and BD its diagonals, meeting at O. Bisect the angle ABO by line BE. With BE as radius and centre B, describe a circle, cutting AC at E. *The square on BD (or 2 x BO) is nearly equal to the area of this circle.* The proof is easy, recalling that BO, OA and AB are in the proportions of the square roots of 1, 2 and 3 to one another; while OB/BA, being the cosine of the angle ABO (or 2 x angle EBO) is, by a well-known rule, equal to 2(OB/EB)2-1.

Now (2.OB/BE)2=(area of square PQRS divided by square of radius BE)="pi," nearly. This, from the above, is = 2(OB/BA+1) = 2(1/3 of the square root of 3+1)=2×1.57735=3.1547.

This is not a very close approximation, to "pi," being about 4% out. But it is closer than the Egyptians' value, the square of 16/9, or 3.1605. The accurate value, to 4 places of decimals, is 3.1416; and the usual rough approximation 22/7, is about 0.4% out.

A.D.B.

To Obtain the Number of Cells per Sq. Dm. - Bee World 1948

The Bee World – June, 1948 – Page 47

To obtain the number of cells per sq. dm. in a comb of given cell dimensions, use the formula: Number of cells per sq. dm. = 2.3 X N2, where N is the number of cells in one decimetre, measured horizontally (across the vertical sides of cells "right way up.") This formula is due to Signor G. Muzzati, and is published in the January number of *L'Apicoltore d'Italia*, (XV, No. 1, 1948).

The Efficiency of the Use of Enlarged Cells - 1965

XX Jubilee Apimondia Congress – August, 1965 – Pages 675-677-THE EFFICIENCY OF THE USE OF ENLARGED CELLS HONEY-COMBS IN THE CONDITIONS OF ROMANIA C. ANTON-ESCU, ROMANIA

"The large cell" preoccupied and continues to preoccupy many beekeepers and scientists in numerous countries. Etienne de Meyer (Belgium), in the first report presented to the XI-th International Congress of Apiculture (Paris 1-17 August 1937) said:

The large cell is only a selection, better said a means of selection..." In his opinion the possible limit might be that of 575-600 cells per dm2, that is, cells with a diameter of about 6.3-6.2 mm. He also demonstrated that the bee "adapts itself" to the cell put at its disposal, and that only an artificial honey-comb with cells of a corresponding size would permit to maintain the bee's size and body dimensions.

The second report presented to the same Congress by Dr. A. Zappi Racordati (Italy), on the theme "The problems of the large cell in Italy", mentioned among others:

"To obtain larger bees it would not be sufficient to use honey-combs with cells larger in size than those normally built by the bees, but, it will be necessary to find a way of obtaining as large as possible bee queen and drones. Periodic replacement of the honey-combs is always a very important measure which permits to guarantee to the bee the best conditions of development: it would thus seem necessary to make this replacement also in case the larger cells are adopted"...

"We have in Italy too, enthusiastic adherents for enlarging the bees by means of the Baudoux method. Measurements made with the greatest accuracy by the National Institute of Apiculture, using a single method (that of De Meyer and Trannazer) on a considerable number of working-bees honey-combs, belonging to the pure Apis ligustica race, have clearly proved that size of cells – even in relatively limited zones – varies within wide enough limits, more exactly between 860-760 per dm2".

Under the influence of the above mentioned reports, I set myself the purpose to experiment on the efficiency of large cell honey-combs. This was possible beginning with the year 1941 when I was able to build honey-combs with cells of 5.65 mm and 5.85 mm diameter, that is, 726 and respectively 678 cells per dm2.

The experiment (in the year 1941) comprised 4 local race (Region of Ploiesti) bee families living in Dadant beehives with 12 standard frames and stores, and working in honey-combs with cells of 5.4 mm diameter (794 cells per dm2). The experience lasted 4 years. Two of the bee-families from the experimental group were completely shaken off into new hives equipped with artificial honey-combs and cells of 5.65 mm diameter. For the other two families, the artificial honey-comb frames with enlarged cells were intercalated in the brood-nests, during the intense Acacia (Robinia pseudoacacia L.) and linden-tree (Tilia) harvest. All the bee-families of the experimental group behaved almost identically: the majority of the cells were transformed in drone cells and a small part occupied by brood-nests of working-bees, but with thickened walls.

During the following years (1942-1944) the experiment was repeated with the same number of bee-families and identical unfavorable results. During the years 1954-1958 I adopted a new working method: the artificial honey-combs with enlarged cells were used for building up and honey storing during the intense harvest period for other colonies and after the honey extraction they were used for the colonies of the experimental group (during the following season) for brood breeding, using as for the first experiment, the shake-off method. The honey-yield of the colonies from the experimental and control groups, in the conditions of stationary bee-breeding, is shown in table 1.

Bee colony group	No. of bee colony	Year 1954 (kg)	Year 1955 (kg)	Year 1956 (kg)	Year 1957 (kg)	Year 1958 (kg)	Avg for 1954-1958 (kg)
Experimental group (cells 5.65mm)	6	19.0	15.6	34.9	6.8	20.0	19.4
Control group (cells 5.4mm)	6	17.1	13.7	29.3	5.8	17.4	16.6
Avg increase % (experimental group)	---	11.1	13.8	19.1	17.2	19.5	16.9

Table 1.

An analysis of the figures in table 1 shows that the yield-increase obtained from the experimental group bee colony during the 5 years of experiment varied between 11.1 and 15.5% or in average by 16.9% per year which can be considered a positive factor for the efforts which every beekeeper is undertaking in order to increase production and investment rentability in apiculture.

Experimentation with enlarged cells frames (5.65 mm) was successfully undertaken by beekeeper Vladimir

Cudelca of the Region of Bacau (city of Roman), in the middle of the province of Moldavia, with a group of 5 bee colony during the years 1957-1963. According to his statement, the increase of production he obtained was over 20% yearly, which might mean that also in other regions of our country the enlarged cell can and must contribute to the productivity of the apiaries in our country.

Further researches (Eng. E. Mirza and collab., 1962) revealed data which are supporting a large-scale use of enlarged cells honey-combs. These researches are showing that in the bee-population of our country the cell size of the natural honey-comb presents a great variability, depending on the zone. Thus in the Transylvanian highland the average horizontal diameter is 5.50 mm, the differences, as compared to that average, ranging between the limits of 5.24-5.88 mm. It is to be mentioned that in the mountainous zone the size varies with in the limits of 5.35-5.88 mm.

Conclusions

1. Experiments with a large cell honey-comb in the conditions of the Socialist Republic of Romania (5.65 mm) show that a large scale introduction of such honey-combs represents an important reserve for the increase of the bee-hive's productivity in all sectors. To this aim it is necessary that the honey-comb should be build up first – during intense harvesting – in other colonies or in the respective colonies for honey-storage, and only afterwards it should be used for brood breeding.

2. Recent researches concerning the cell size of the natural honey-comb built by the bee-population of our country show that there is a great size-variability and that in the mountainous zone these sizes greatly overstep the artificial honey-comb cell size which we have experimented (5.88 mm). Consequently, in a zone which

represents approximately a third of the territory of the S. R. Romania, the use of honey-combs having cells with a horizontal diameter of 5.65 and even 5.88 mm constitutes a problem of acute actuality.

3. In all countries where research-work is showing that, in certain zones the bee-colonies are building natural honey-combs, it is to be recommended that the bee-colonies should have at their disposal honey-combs with cells as nearly as possible to the size of cells which bees are building naturally.

REFERENCES

The XI-th International Congress of Apiculture – in "Romania Apicola", No. 11-12/1937.

Etienne De Meyer – The work of Baudoux, in "Romania Apicola", No. 2/1938.

Antonescu C. – The efficiency of the use of enlarged – cells honey-combs in "Apicultura", 1959.

Mirza E., eng. and collab. – Data concerning the size of cells in S.R.R. honey-combs, in "Apicultura", 1961.

Preference of Varroa jacobsoni Oudemans for Different Cell Types and Some Factors Affecting Reproduction

Apiacta, Feb. 1984 – pages 165-167-DE RUIJTER, A. (Netherlands)

Reproduction of *Varroa* only occurs inside capped worker and drone brood cells. Prior to reproduction, female mites leave the adult bees and enter brood cells just before capping. The mites enter the larval food and stay there until the bee larva frees them by eating the food. At this time, the cell has been capped. More mites are found in drone cells than in worker cells. There seems to be chemical attraction which is stronger in drone larvae than in worker larvae. In queenless colonies, more mites were found on drone pupae in drone cells than on drone pupae in worker cells. In Brazil, more mites were found in large worker cells than in smaller ones. These two observations indicate that features of the cell are important in respect of distribution of *Varroa* on the comb.

In our first experiment, we tried to establish this by offering different types of cells on one comb to Varroa-infected colonies.

Feeding on the larva inside the capped brood cell initiates oviposition. It has been demonstrated that application of juvenile hormone to honey bee larva, just after cell capping, stimulates the reproduction of *Varroa*. Oviposition starts 60 hours after cell capping. An egg is laid about every 30 hours. As a rule, the first egg is female, the second male, the third and following eggs are also female. Females develop from fertilized eggs and are diploid, males develop from unfertilized eggs and are haploid.In our second experiment we studied factors influencing the pace of oviposition and the success of reproduction.

Experiment 1.

Material and methods

We put drops of melted beeswax in a worker comb in a regular pattern into certain cells: in every fourth horizontal row of cells, every fourth cell was made 3 to 4 mm less deep than neighbouring cells. This comb was placed inside the broodnest of a *Varroa* infected colony. When most of the brood was capped, we examined the contents of treated and untreated cells. We scored the number of adult *Varroa* females that had entered the different cells. This was done separately for five age classes of the bee brood, according to Ifantidis.

Table 1. Number of Cells Examined and Mean Numbers of Adult Female Mites per Cell for Each Age Class of Bee Brood

Age class of the bee brood*	Untreated Cells		Treated cells		Prefer-ence** (B/A)
	No. of cells	Mean no. of adult female mites per cell (A)	No. of cells	Mean no. of adult female mites per cell (B)	
I	153	0.013	25	0.080	6.1
III	199	0.005	44	0.091	18.2
IV	133	0.038	10	0.200	5.3
II	52	0.019	9	0.222	11.6
III	49	0.041	11	0.564	8.9
III	106	0.283	18	0.556	2.0
IV	159	0.264	50	1.280	4.8
II	47	0.0532	6	2.667	5.0
III	26	0.731	3	0.667	0.9

*I=prepupa, II=pupa with white eyes, III=pupa with red eyes, IV=pupa early tanning, V=pupa with brown thorax.

** The mean number of mites in treated cells divided by mean number of mites in untreated cells is given as "preference".

Results and discussion

The preference is always bigger than one (mean=6.97), except for one observation where the number of treated cells was very low. This means that adult female mites enter protruding cells more frequently than normal worker cells, though all cells contain worker brood.

Conclusion

Apparently chemical attraction is not the only factor that determines the distribution of mites in available cells; cell type is also important. The mechanism by which the mites discriminate between different cell types is not clear. Probably. the behavior of the bees that carry the mites to the brood cells is different towards different cell types. The process of leaving the bee and entering the brood cell needs further study. Intervention in this process might lead to biological control.

Experiment 2.

Table 2. Mean Offspring of Females Confined in a Newly-capped Drone or Worker Cells and then Put in a Newly-capped Worker Cell

Period in brood cells:	No. of cells	Egg/ larva	Protonymph		Deutonymph		Adult
			mobile	immobile	mobile	immobile	
24 hrs drone + 9 days worker	n=19	0.0	0.5	0.6m	0.9m	1.8m	0.9f 0.8m
48 hrs drone +9 days worker	n=10	0.2	1.0	0.1m	1.3m	1.1m	1.7m
48 hrs drone +7 days worker	n=15	0.7	0.7	0.2m	1.7m	1.1m	
m=male f=female							

Material and methods

Open worker brood cells were marked by a sheet of transparant plastic, attached to the frame by two drawing pins. The comb was then put back into the colony. After 4 hours, newly-capped cells were numbered and opened by using a razor blade. A female *Varroa* mite was introduced into each cell, the cells were carefully closed and the comb placed back in the colony again. After a desired period the cells were opened and the content studied.

We tried to increase the stimulation of oviposition by transferring mites 24 or 48 hours after cell capping into newly-capped cells. In this way, the mites were able to feed on round larvae and spinning larvae twice and would get a double amount of any factor in the haemolymph of the bee.

Results and Discussion

After 24 hours extra in a worker cell or drone cell, a male is produced in most cases. Apart from an adult female, in many cases there are 3 female deutonymphs. Compared to 10 days, the intervals between the successive eggs must have been shorter.

After 48 hours extra in a worker or drone cell, no males are produced at all. In many cases there are three deutonymphs (means are 2.4 and 3.0). This again suggests that the interval between successive eggs is shorter than in the normal situation.

Conclusion

Oviposition, as well as sex determination are affected by stimuli during the first two days inside the capped brood cell.

Distribution of Varroa Jacobsoni in Brood Combs of Honey Bee Colonies, and Resultant Effects on Colny Development - Apiacta 1984

Apiacta #2, Feb., 1984, D. DE JONG; BRAZIL, Roger A. MORSE; USA

The honey bee parasite, *Varroa jacobsoni*: must enter the cells of drones or worker bees in order to reproduce. The selection of an appropriate cell is of utmost importance to the mite and, ultimately affects both the rate of mite population increase and the damage to individual bees and to the colony as a whole.

Some factors examined which influence this selection are the age, sex, and caste of the larva, presence and proximity of queen cells, and architecture.

Independently of these influences, it appears that the mites are clumped in their distribution among cells, implying that they do tend to enter cells already occupied by other mites. This may be a mechanism to help decrease the incidence of inbreeding among the mite progeny.

Nevertheless, the mites are not greatly clumped in worker cells. Few bees in the infestations we see in Brazil are visibly injured, making evaluation of damage difficult. However, there is a consistent loss in weight and a drastic decrease in lifespan, even of bees infested with a few mites. But the bees can emerge from their cells, releasing the mites and their progeny to continue their infestive cycle.

The Effect of The Size of Honey Bee Cells on The Rate of Infestation by Varroa Jacobsoni

Apiacta #2, Feb., 1984, D. MESSAGE, L. S. GONCALVES, BRAZIL

Varroa jacobsoni, an ectoparasite mite of *Apis*, causes serious damage to colonies of *A. mellifera* while in those of *A. cerana*, the original host, this is not the case. Koeniger et al. (1981) observed that in *A. cerana* the mites only reproduce in drone cells, while in *A. mellifera* there is reproduction in both worker and drone cells. It's possible that the resistance of *A. cerena* to this mite is due to this biological characteristic.

Based on this information and knowing that the cells of workers of Africanized bees are smaller than those of European bees, we examined the infestation rate and reproduction of *Varroa* in small pieces (+/-300 cells) of Africanized and Italian built combs.

The worker combs of Africanized bees had a mean of 4.50 to 4.77 mm and a volume of 0.187 ml, while for Italian bees the values were 5.07 to 5.11 and 0.255.

The mean weight of 18-19-day old Africanized worker pupae was 94.6 mg, and their infestation rate was 4.8%, while for Italians the mean weight was 111.7 mg, and 11.5% of the cells were infected.

Study of the Preference of the Mite Varroa Jacobsoni for Apis Mellifera Drones

XXX International Apicultural Congress, Apimondia, Nagoya, Japan, Oct. 10-16, 1985, ISSA, M. R. C.; DE JONG, D.:, GONCALVES, L. S. (Brazil)

The mite *Varroa jacobsoni* infests brood and adult honey bees (Grobov, 1977). The mite is known to enter brood cells when the larva is 5 to 6 days old (Issa and De Jong, 1981).

This research concerns the biology and behavior of the mite with emphasis on understanding the mechanisms involved in the marked preference observed for drone brood. It has been previously demonstrated that the mite prefers drone brood to worker brood for reproduction (Haragsim, 1973; Marin, 1977). This also occurs with Africanized honey bees. The percentage of infested cells was 7.55% for worker larvae and 40.57% for drone larvae; the infestation rates were 0.089 for worker larvae and 0.62 for drone larvae. Generally, about 40% of drone cells are infested, while for workers, the average is close to 10%. One, two, or three invading mites are commonly found in drone cells, while in workers, most cells have only one mite.

This preference of the mite for drone larvae has also been verified in the laboratory. When larvae of drones and workers in the same developmental phase were reared on Petri dishes whose bottom was covered with a beeswax layer with a shallow cavity in the middle where the mites were placed. The percentage of infested larvae was 15.5% for workers and 67.4% for drones; the rate of infestation (number of mites/number of larvae) was 0.24 for workers and 0.97 for drones.

Adult drones were also more infested than adult workers, with a rate of infestation of 0.291 for drones and 0.060 for workers. The adult workers were then divided into 2 groups by age: young workers (3 to 6 days old) and adult workers (17 to 21 days old). The rate of infestation was 0.103 for young workers and 0.049 for adult workers.

infestation was studied in hives with egg-laying workers and no difference in infestation was detected between bees with developed ovaries and bees without developed ovaries. Infestation in the worker-laying hives was lower than in hives with no laying workers. The mean infestation rate of 6 hives was 0.045 before egg-laying workers appeared.

Several substances were tested for their effect on mite preference both in hives and in the laboratory. Among these substances (juvenile hormone, hemolymph of drone larvae, drone larval extract in water, methanol and acetone, lyophilized drone larvae, and food from drone larvae) only food from drone larvae tested in the hive gave positive results.

When the behavior of the mite was studied in terms of the search for the host, we observed that the mite perceives the presence of the larva when it is quite close. Most mites perceive a larva at a distance of 3 mm, but some of them reacted to the presence of a larva at a distance of 4 to 5 mm. Perception of the larva is accompanied by agitation of the first pair of legs. This same agitation is also observed when the mite meets an obstacle in its path, when the mite may or may not climb onto it.

The mites that penetrate the brood cell reach the larval food and remain there until the larva consumes all the food (De Jong *et al.*, 1982). When the mite are removed from the larval food and the food remains are cleaned off, they start to move again over a time varying from 2 minutes to 18 hours.

One of the stimuli to which the mites reacted most clearly was an air draft. The mite left no doubt about its positive response by making a 180° turn to reach the source of stimulation (site from which the draft came). The air drafts tested were produced by human respiration, aquarium aerator, chilled air, and draft from a container of carbon dioxide.

REFERENCES

DE JONG, D.; MORSE, R. A.; EICKWOTH, G. S.(1982) Mite pests of honey bees. Ann. Rev. Entomol. 27: 229-252.

GROBOV, 0. F. (1977) Varroasis, in bees. *In* Varroasis, a honey bee disease. Bucharest. Apimondia, pp. 48-90.

HARAGSIM, 0. (1973) The mite *Varroa jacobsoni* as a threat to beekeeping in Europe. Imkerfreund 28: 312-317.

ISSA, M. R. C.; DE JONG, D.(1981) Estudo do peso e idade larval de abelhas *Apis mellifera* infestadas pelo acaro *Varroa Jacobsoni*. Suplemento Ciencia e Cultura 33(7): 674.

MARIN, M. (1977) Diagnostico y tratamiento de la Varroasis. *In* Varroasis, enfermedad de la abeja melifera. Apimondia Publishing House, Bucharest pp. 18-21.

The Influence of Cell Size on Infestation Rates by the Mite Varroa Jacobsoni - Apimondia 1985

XXX International Apicultural Congress, Apimondia, Nagoya, Japan, Oct. 10-16, 1985, 41. DE JONG, D.; MESSAGE, D.; ISSA, M. (Brasil)

An important phase in the life cycle of the honey bee parasite, *Varroa jacobsoni*, is the choosing of a bee brood cell for reproduction. That the mites prefer drone brood over that of workers is well established. Recent work by our group has demonstrated that the presence of the male bee larva is not the only important factor. This makes sense biologically since the longer development period of the drones allows for a greater reproduction by the mites. However the cues that the mites use to make their choice are unknown.

Worker brood cells which protrude from the comb surface due to irregularities in the comb midrib are preferred (De Jong and Morse 1984). When a piece of European bee comb is implanted within an Africanized bee comb, the larger diameter cells of the European comb attract more mites even though the larvae in the two cell types come from the same queen (Message and Goncalves 1984). Drone larvae transplanted into drone sized cells had more mites than those transplanted into worker sized cells, which demonstrates that cell size also mediates the attractiveness of drone larvae (Issa, De Jong, and Goncalves 1984).

Further studies are planned to elucidate the cues which Varroa uses for cell choice.

Effects of Comb Cell Diameter on Parasitic Mite Infestations in Honey Bee Colonies

Author(s): ***ERICKSON JR ERIC H, RICHARDSON GARY V, KEHL KEVIN L, ARP DENNIS L, CAMERON BRETT E***

Interpretive Summary:

The Varroa mite is an economically important external parasite of honey bees. These mites reproduce in brood cells where they feed on hemolymph of developing bee pupae. Bee longevity is reduced and eventually entire colonies die. Thousands of domestic honey bee colonies are being lost annually to Varroa. This study was undertaken to determine the effect of brood comb cell diameter on the population dynamics of Varroa, and the validity of using Varroa fecal accumulations (FA) as a field diagnostic for Varroa. Thirty-six full-strength colonies, each with one of four types of brood comb, were monitored quarterly over a two-year period from initial infestation with Varroa to colony decline/death. There were four surviving colonies at the end of the study period, all in the small cell treatment group. There were no significant differences between treatments in the incidence of Varroa, Varroa FA, colony survival or honey production due to the absence of significant differences between treatments due in large part to wide variability between colonies within treatments. However, trends for the incidence of Varroa, and Varroa FA, as well as for colony survival and honey production suggest that reduced cell diameter may have a limited impact on Varroa and HBTM population dynamics, and on colony performance. The results suggest that cell diameter could be a useful Varroa management tool when used as part of an integrated pest management system. These studies further demonstrated that FA can be used to diagnose Varroa infestations.

Contact:

USDA, ARS, HONEY BEE RESE

2000 EAST ALLEN ROAD

TUCSON

AZ 85719

FAX: (520)670-6493

Email: medley@tucson.ars.ag.gov

Approved Date: 1999-07-19

TEKTRAN

United States Department of Agriculture

Agricultural Research Service

Updated: 1999-08-13

Square Decimeter Measurement Conversion Chart

Cell Width (cm for 10 worker cells)	Rhombic (Old World) Square Decimeter Measurement	USDA Square Sq dm Measurement w/Cell Width Deviation where known	Baudoux Square Square Decimeter Measurement	Rietsche Square Square Decimeter Measurement
5.96	482	650-5.0	650	
5.9	574/575			
5.8	594			
5.75	604		700	
5.7	615	700-5.7		
5.6	637	725-5.6		750
5.55	649		750	
5.5	655	750-5.5		
5.444	675			
5.4	684	800-5.4		780-800
5.37	692		800	
5.35				
5.3	711	830-5.3		
5.21	737		850	
5.2	740	850-5.2		830-850
5.17				
5.15	751 (Vogt) 753			
5.1	763 (Castillion)			
5.06	780/781		900	
5.0	789 (Maraldi)	920-5.0		
4.92	821 (Georgandas) 827		950	
4.9	820 (Castillion) 832 (Reaumur /Klugel)	950-4.9		
4.83	838 (Lang/Dadant) 856/857 (L'Abbe Collin/Grout)			
4.8	860 (Fratelli Piana) 867/868		1000	1050
4.7	870 (Swammerdam) 905	1050-4.7	1050	
4.66	920/921			
4.58	953 954 (Maraldi)			
4.5	987			
4.4	1032			
4.3	1081			
				Lusby - July 1997

Appendix D: Chemical & Varroa Effects on Honey bees

The Effect of Synthetic Pyrethroid Insecticides on Honey Bees in Indiana: Laboratory Studies and a Survey of Beekeepers and Pesticide Applicators

A Thesis Submitted to the Faculty of Purdue University by William Eugene Chaney. In Partial Fulfillment of the Requirements for the Degree of Doctor of Philosophy. August 1988

ACKNOWLEDGMENTS

I wish to express my appreciation to my major professor, and friend, Dr. Rich Edwards, for his support in my graduate work, my professional career, and my personal life. I would also like to thank Dr. Eldon Ortman for his support as head of the Department of Entomology. A special thanks to Dr. Marlin Bergman for his encouragement and to Dr. Harry Potter for his help in the survey portion of this work. My gratitude also to Dr. Virgil Anderson who made a statistical difference in my graduate education and to Dr. Tom Jordan for stepping in at the last minute.

Most importantly, I would like to thank several outstanding individuals without whom this work could have never been done. They are, in no particular order: Carl Geiger, Terry McCain, Suzie Appel, Mark Johnson, Mike Gale, Steve Mroczkiewicz and Connie Holderfield.

Abstract

Major Professor: C. Richard Edwards.

Insecticides are an important component of the row crop production system in Indiana. Concern for the safe use of these products has lead to a system of regulating the application of pesticides that is designed to protect the public, the environment and the applicator. One non-target organism that is affected by some pesticide applications is the honey bee. Because of its social nature, the impact of pesticides on bees is sometimes expressed as detrimental effects on the colony to which the exposed bee delivers her contaminated nectar or honey.

This study looked at several aspects of the honey bee/pesticide problem, including one class of insecticides about which there is controversy concerning their impact on bees. This class is the synthetic pyrethroids. These studies found that the relative toxicity to adult bees of the four products examined was: permethrin> flucythrinate > fenvalerate > fluvalinate, in decreasing toxicity. The toxicity of these products was also shown to increase at 18° C and 12° C as compared to their toxicity at 25° C. These are temperatures in a range which might be experienced by bees in a colony in Indiana during the winter.

This study also demonstrated that no synergism or antagonism was seen when permethrin and fluvalinate were fed to adult bees together with carbaryl, paraquat or mancozeb. This study did demonstrate that some colonies were more resistant to permethrin and carbaryl than others and that this resistance was related to the race of the queen heading the colony.

Beekeepers, public pesticide applicators and private pesticide applicators were surveyed to examine their knowledge of and attitudes toward the poisoning of honey bee colonies by pesticides. The response rate was not significantly different among the groups. The mean response rate was 75%. Less than 10% of the beekeepers and none of the applicators reported any knowledge of specific incidents in which bees were poisoned by pesticides in 1986. Both the beekeepers and the applicators were concerned about this issue and both groups indicated a willingness to take specific actions to attempt to prevent future poisonings. Each of the three groups showed a poor level of knowledge about pesticides as they relate to bees and about integrated pest management.

Introduction

Insecticides are applied annually to a high percentage of the cultivated acres in Indiana. Some of these insecticides are applied in a manner and at a time as to expose non-target organisms, including the honey bee, to direct sprays or residues (Atkins 1979, E.H.Erickson 1983a). Concern over the impact of pesticides on the beekeeping industry has been expressed by leaders of that industry (Ambrose 1983; Atkins 1980; Crane 1983; E.H.Erickson 1983a, B.J. Erickson 1984a, 1984b; Knol 1983; Stevenson 1978, and others). The public is also becoming more concerned about the impact of pesticides on the environment as evidenced by increased regulation pesticide applicators face arid the removal of products from the marketplace (Adkinsson 1971, Pimentel 1980). As many older, more persistent, chlorinated hydrocarbons were removed from use, farmers turned to newer shorter lived insecticides and often found more applications were necessary to achieve acceptable control.

One class of insecticides that contains many of the newly registered insecticides is the synthetic pyrethroids. As a group, the synthetic pyrethroids are loosely related by their chemistry and mode of action on the target pests. Within the group are a wide range of products that have diverse target pests. Most of these products are characterized by their relatively low mammalian toxicity and their effectiveness against invertebrate pests at low doses (Sine 1988). In Indiana, these products are being utilized in the pest management programs of a growing number of corn, soybean and alfalfa farmers.

Some synthetic pyrethroids are reported to be quite safe to honeybees in some areas of the United States

(Atkins 1979, Johansen 1983, Moffet 1982, Stoner 1984). Early evidence indicates that the toxicity of some synthetic pyrethroids to honeybees may be greater in Midwestern areas than in warmer more arid parts of the country (B.J.Erickson 1983; E.H.Erickson 1983b,1984; Smart 1982). The fact that the toxicity of some synthetic pyrethroids is inversely related to temperature (Georghiou 1964, Morton 1979) and may be important in a contaminated honey bee colony's ability to overwinter in the midwest. It was also evident from early studies that the toxicity of this group to any particular species was very diverse (Atkins 1981, Moffet 1982, Smart 1982). For example, some of these products are highly toxic to mites and are used as miticides (Herbert 1988, Witherell 1988), while others are so safe to mites as to actually increase their population (Flaherty 1981, Flint 1985).

As farmers have become more specialized producers of a declining range of crops, the direct importance of bees as pollinators on the farm has also been declining. While some studies indicated that soybean yields might benefit from insect pollination (Abrams 1978, Erickson 1978, Mason 1979), other crops such as corn, alfalfa grown for hay and wheat require no insect pollination. As these crops captured an increasing proportion of the acreage, the number of nectar producing plants plummeted. The decrease in forested acres, the intensive planting of non-nectar producing plants such as fescue and crown vetch on roadsides and in pastures, the intensive use of herbicides in cropland and the increasing urban demand for land, seriously reduced the nectar resources available to bees. This combined with the low world honey price over the past several years, has driven nearly all commercial beekeepers from many parts of Indiana.

Most of the remaining beekeepers in Indiana are hobbyist or part-time beekeepers who keep bees for pleasure as well as profit. For these beekeepers, the impact of pesticides on their bees is a highly-charged, emotional issue. Much misunderstanding exists between beekeepers and applicators and there is considerable misinformation on both sides. Since discontinuation of the federal governments Beekeeper Indemnification Program, there has been very little effort to evaluate, document, or record reports of pesticide poisonings of honey bee colonies, unless litigation was instituted or threatened (Coleman 1979, Pimentel 1981). It is generally believed, but undocumented, that beekeepers have overestimated the severity of the problem, while pesticide applicatiors have underestimated the extent of the problem.

In Indiana, application of those products which have been classed by the United States Environmental Protection Agency (EPA) as "Restricted Use Pesticides is regulated by the State Chemist's Office. This office has the responsibility for enforcing the laws and regulations relating to pesticides including pesticide applicator testing and certification.

Because many individuals who apply pesticides may never use a restricted product, many private applicators are not required to be trained or show competency in pesticide use. The Indiana Cooperative Extension Service and the Office of the State Chemist have worked together to train pesticides applicators in the safe handling and application of pesticides, as well as safety to the environment. Pesticide applicators are further divided into those who apply pesticides for hire and those who use the products only in conjunction with their own crop production operation. These groups are referred to as Public Pesticide Applicators and Private Pesticide Applicators, respectively.

Among the information which is required for certification is knowledge of the safe handling and use of pesticides, including their toxicity to non-target organisms such as the honey bee. The issue of honey bee poisoning is a complicated one and cannot be covered in depth during applicator training due to time constraints.

The factors which determine the extent to which a given colony of honey bees will be affected by the application of a pesticide to a given field are complex (Atkins 1981, Johansen 1977, Lieberman 1964, Quatitlebaum 1983). The single most important factor is the number of bees from a particular hive that are foraging in the treated area (Nowakawski 1982). This is influenced by many factors such as attractiveness of the crop treated, presence or absence of blooming weeds in the area, distance from the treated field to the colony, strength of the colony, weather, needs of the particular colony, genetic make-up of the colony, etc (Atkins 1977, B.J. Erickson 1983a, Mayer 1983, Mayland 1970, Smirle 1987, Ross 1981, Wailer 1984).

In addition to factors relating to the honey bee colony, factors relating to the pesticide such as the active ingredient, the formulation, the time of application, the method of application, the weather conditions during and following application, etc., will all influence the extent to which a given colony will be affected. There is the additional complicating factor that some pesticide products may cause no observable damage at the time of application, but may cause delayed mortality of the overwintering colony during a period of greater stress.

The synthetic pyrethroid insecticides are of particular concern in this regard due to the inverse relationship between their toxicity and temperature (Yu 1984). Lehner (in press) has shown that the toxicity of permethrin to bees dramatically increases at 20° C over the toxicity at 26° C. Delayed mortality may often not be detected or identified as a result of earlier pesticide exposure. The insecticide stored in the hive may not singularly cause colony mortality, but may act in conjunction with other factors to increase the stress on the hive and cause a decline of the population. This decline may or may not be reversed by the colony as weather and other conditions improve, depending on their reserve strength and size of the initial population.

An understanding by both beekeepers and pesticide applicators of the factors that influence poisoning of colonies of honey bees by pesticides is critical to establishing a situation in which the two groups can operate without conflict. Because this is such an emotional issue, it is often difficult to separate emotion from fact when discussing this subject with either side. Given the complexity of the problem and the limited resources available to try to deal with the situation, a multi-faceted approach to the problem was undertaken. This included laboratory work to examine some the the most critical questions relating to synthetic pyrethroids and bees as well as an examination of the groups of people directly involved; that is beekeepers, farmers and public pesticide applicators in Indiana.

Chapter One: Relative Toxicities and Temperature

This part of the study was designed to answer two basic questions important to further studies. These were: 1) what are the relative effects of four common synthetic pyrethroid insecticides on adult honey bees and 2) what impact does temperature have on each of these products. The synthetic pyrethroids chosen to represent this insecticide class were: fenvalerate (Pydrin 2.4EC), permethrin (Ambush 2E), flucythrinate (Pay-Off 2.5EC) and fluvalinate (Spur 2E). Formulated product at five dosages was fed in 50% sucrose syrup to caged adult bees held in different growth chambers at three temperatures (25° C, 18° C and 12° C). Toxicity was measured by monitoring the number of dead bees daily for five days.

Background and Objectives

Since the introduction of synthetic pyrethroid insecticides, there has been controversy concerning their impact on honeybees. Early reports indicated that these products were safe to bees because they repelled the foraging workers (Atkins 1977, Bos 1983, Moffet 1982, Stoner 1984). Later evidence indicated that the reduced number of foragers observed in treated areas was not due to repellency, but to a disruption of the normal system of communication among foragers or to direct mortality of foraging bees (Rieth 1986, Cox 1987). Because honey bees tend to work the same nectar source until it is depleted, the gradual return of foragers to the treated area could be due to a new set of foraging bees being recruited to the area or due to a loss of repellency as the product degraded.

While this question was still unresolved, reports of serious mortality of bees and the eventual death of colonies of bees exposed to synthetic pyrethroid insecticides began to surface. Laboratory tests indicated that some of these products were toxic to bees when applied topically or ingested, yet field studies did not show the expected damaging effects (Atkins 1981, Johansen 1983, Moffett 1982, Smart 1982, Stevenson 1978). Many of the reports of damage from synthetic pyrethroid insecticides were undocumented because the beekeepers involved did not feel that there was a reasonable chance of reimbursement. Since the USDA Beekeeper Indemnification Program had been terminated, few beekeepers sought restitution for pesticide damage through other legal means (Happ 1971).

The greatest number of reported poisonings came from the midwestern and eastern states. Geographical differences in toxicity to insecticides have been documented in relation to honeybees. As an example, methomyl was found to be safe to honeybees in western states (Atkins, 1979) but highly toxic when used in Wisconsin in sunflower fields leading to massive bee kills (Krause 1983). The differences in toxicity were attributed to the higher humidity of Wisconsin and increased incidences of heavy dew on treated plants which made the pesticide more available to the foraging bees.

In addition to immediate bees kills of greater magnitude, reports by northern beekeepers indicated a greater incidence of winter mortality in colonies exposed to synthetic pyrethroid insecticides during the previous season. If indeed the foraging bees were returning to the hive with nectar or pollen contaminated with sub-lethal doses of synthetic pyrethroid insecticides, perhaps the lower colony temperature in winter was resulting in these doses becoming lethal due to the inverse relationship between the toxicity of synthetic pyrethrbid insecticides and temperature.

In winter, bees control their temperature by forming a cluster. This cluster is roughly a sphere which expands or contracts to regulate the temperature on the surface at about 8° C (45° F) (Owens 1971, Szabo 1985). The center of the cluster may be considerably warmer, depending on the rate of heat loss. In honeybee colonies in the Midwest, the rearing of immatures, or brood, normally begins in January or early February. There is commonly no brood present from October to that time. After the initiation of brood rearing, the temperature at the center of the cluster is maintained at 33° C to 35° C (92° -94° F) (Owens 1971, Szabo 1985). Bees move from the inner areas of the cluster to the outer surface in a slow but constant rotation.

The method of storage of food by honeybees, both honey – the carbohydrate source, and pollen – the protein source, is important in this situation. Nectar is collected from flowers and contains 15-35% sugars. The bees evaporate the excess water from the nectar to increase this sugar concentration to about 80%. During this process, the conversion of most of the 12-carbon sugars to 6-carbon sugars is accomplished by a bee supplied esterase. The resulting product, now correctly called honey, is very stable chemically and biologically. The sugar concentration makes it unsuitable for yeast, fungal or bacterial growth. If pesticides contaminated nectar, they are concentrated in the process of producing honey and end up in a quite stable environment (Barker 1980, Winterlin 1973).

Pollen is also collected from the flowers and stored in the cells. It has been shown to be carried to the hive contaminated with pesticides (Johansen 1972, Mayer 1983, Rhodes 1980, Winterlin 1973). The bees add nectar to this pollen to make it more workable and often seal it under a cap of honey. This situation is also very stable chemically and biologically. Unlike the honey, pollen is not concentrated and is usually consumed only by very young bees and those involved in caring for larvae. Under normal conditions in this highly social colony, most bees spend a period of their early lives caring for larvae. This period coincides with the greatest activity of the hypopharyngeal gland in the head. It is also the period when the most pollen is consumed. In winter when brood rearing is initiated, older bees are often needed to care for larvae and they again begin to consume large amounts of pollen. These older bees are also more likely to have previously been exposed to pesticides than young bees reared in the spring.

In the winter cluster, bees are able to maintain temperature by consuming honey stored from the previous season. If this honey is contaminated with pesticides, it is possible that bees moving from the center where they consumed contaminated honey to the outside of the cluster, could experience a temperature drop from 35° C to 7° C (95° F to 45° F). One of the objectives of this study was to look at the toxicity of the selected products at temperatures within this range.

There were also conflicting reports as to the relative toxicity to bees of the two most commonly available synthetic pyrethroid insecticides of the time, permethrin and fenvalerate (Atkins 1979, B.J.Erickson 1983b, Smart 1982, Stoner 1985). Manufacturers were also seeking registrations for new products for agricultural use. Two of these newer products were chosen because they represented both ends of the spectrum as to bee toxicity. These products were fluvalinate and flucythrinate.

The objectives of this first study were:

To devise an easy method of testing the toxicity of synthetic pyrethroid insecticides to adult honeybees in the laboratory

To determine the relative toxicity of the four selected synthetic pyrethroid insecticides

To determine the effect of three temperature ranges on the toxicity of the selected synthetic pyrethroid insecticides

Materials and Methods

Adult honey bee workers were collected from a single colony headed by an Italian queen purchased from a commercial queen breeder. The queen was introduced into the colony approximately 120 days before the start of the experiments. Adult bees were collected from the honey supers during mid-morning on days when the colonies were in active flight. Collected bees were held without food in one gallon paper Fonda Cups (cardboard ice cream cartons) with screen tops for ventilation until needed.

The bees were anesthetized with CO2 and then counted out directly into one half pint Fonda cups containing a vial of 50% sucrose syrup with the assigned concentration of formulated pesticide. The bees were handled by a leg with larval forceps. Any worker bees appearing damaged or exhibited unusual behavior in any way or any drone bees were discarded.

The cups were filled in numerical order, each cup having been randomly assigned a treatment after numbering. The cups were provided with a seven-dram glass vial of sucrose solution with a perforated plastic cap to serve as a feeder when inverted. The inverted vials were held in place on the side of the cup above the floor by a piece of rubber band stapled on either end to the walls of the cup. The top was replaced by a piece of netting to provide ventilation and to facilitate observation of the bees.

The assigned pesticide treatment was prepared just prior to the introduction of the bees by mixing formulated pesticide with 50% sucrose solution at room temperature with a magnetic stirrer. Dilutions of a stock 1000 PPM solution were used to make concentrations of 100, 33, 10, 3 and 1 PPM. The formulated product was obtained from the manufacturing company without the knowledge of the company as to the nature or purpose of the experiment. Product manufactured for the current year was obtained and the concentration was assumed to be as stated on the label.

Once the cups were filled with the 25 adult bees, the bees were allowed to recover from the anesthesia and recounted. The very occasional bee which did not appear to be behaving normally was replaced. As soon as all the bees had recovered, the cups were moved to one of three environmental chambers that held at 12° C, 18° C or 25° C (± 1,25° C). Since only one temperature could be maintained in any one chamber, the design was a split plot randomized complete block design with randomization restricted by temperature. The three replications were made with each of the three chambers held at each of the temperatures.

Within cup variation was estimated by replication of a selected portion of the treatments in any given replication. These were chosen so that when combined they would represent an additional complete set of treatments. These replicates were randomized within the assigned temperatures.

The bees were held in darkness within the chambers and at 60-70% relative humidity. Mortality was determined at 12 and 24 hours the first day and each 24 hours thereafter for 5 days, for a total of 6 determinations. Bees were considered to be dead if they did not respond to a gentle puff of breath. The CO2 in human breath normally produces a fanning response in worker honeybees.

Preliminary studies had shown that the test insecticides, even at concentrations of 1000 PPM in 50% sucrose solution, did not have any fumigant effect within the growth chambers. The toxicity of the insecticide-tainted sucrose solution did change the bees willingness to consume the solution. The toxicity of the insecticide held in solution for seven days was not different from that of solution made fresh the day of introduction. New solutions were prepared for each replication, however.

Results and Discussion

The relative toxicity of the four synthetic pyrethroid insecticides was found to be constant in nearly all situations after the first observation. Regardless of the temperature, concentration or day of observation, the relative order of toxicity from most toxic to least toxic was: permethrin, flucythrinate, fenvalerate and fluvalinate. The product which showed the greatest actual difference in toxicity between the temperature ranges examined was fenvalerate. It was found that the test procedure was acceptable at 25° C and 18° C. The bees were not observed to cluster at 12° C as they would in the hive and normal behavior was not observed. At 12° C the bees became so inactive that feeding was reduced significantly, resulting in lower mortalities than those observed at 18° C. Even under these conditions, however, the relative order of the toxicities of the test insecticides was unchanged. An accurate test at this temperature would require the use of at least three frame broodless colonies. Normal clustering behavior would be initiated and would have to be controlled so as to not raise the temperature of the bees in the cluster.

Data Handling

The ANOVA, GLM (General Linear Model) and TTEST procedures of SAS (Statistical Analysis System) were used for data analysis as appropriate. The GLM procedure was used when unequal cell sizes were present. The TTEST procedure was used to examine the variance between cups within a treatment. The ANOVA procedure was used for all other tests. Duncan's Multiple Range Test (DMRT) at the 5% level was used to indicate significance of differences between means tested together.

Because the number of dead bees tended to converge on 100% from the higher concentrations of the more toxic products, the mean number of dead bees observed over the five days presented a clearer picture of the relative toxicities. Since the bees were exposed to the test insecticide over the entire time, this mean percentage of dead bees better represented the threat these products present to the bees in an overwintering colony. The mean number of dead bees per observation was used for all calculations unless otherwise noted and results are reported as the mean percentage of dead bees per observation.

Duplicates of one-third of each of the treatments were used to evaluate the within-treatment, between-cup variance. Analysis showed this to be such a minor contribution to total variance (0.96%) that these observations were used as a fourth replicate in further analysis, even though one-third had been run simultaneously with each of the three other replications. Analysis using only the first three replicates did not give different results than using all four.

Overall Results.

Figure 1 shows the mean percentage of test dead bees at each observation period and the standard deviation of this measurement. This represents the mortality for each product at all five concentrations and at all three temperature ranges. The relative order of the toxicity of the products was the same in nearly all of the situations examined. Permethrin was most toxic to the bees followed by flucythrinate, fenvalerate and fluvalinate, in that order. Each of the insecticides was significantly different from the others at the 5% level using Duncans Multiple Range test. Fluvalinate was not significantly different from the control, however.

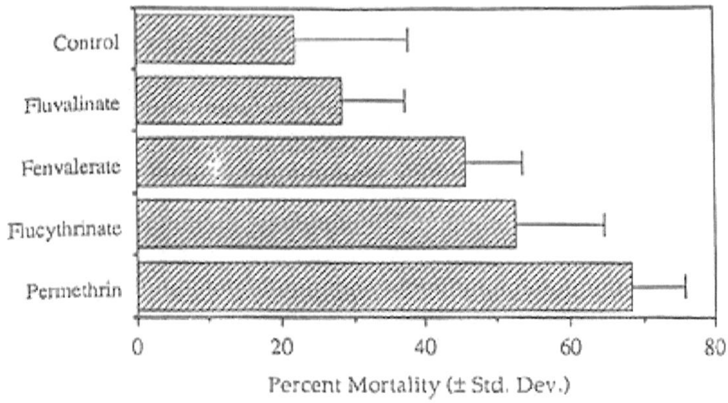

Figure 1 The mean percentage of dead bees per observation by product.

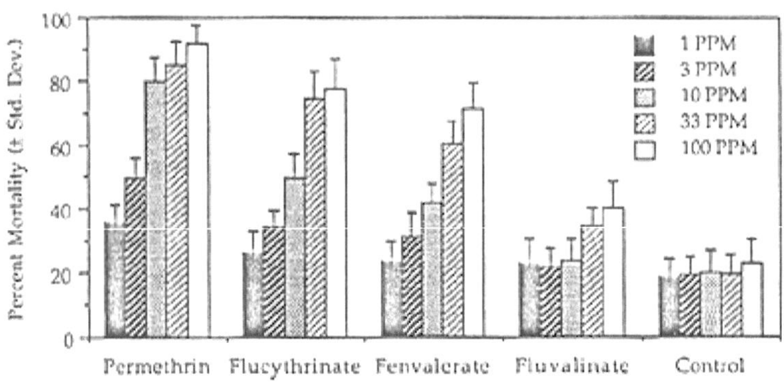

Figure 2 The mean percentage of dead bees per observation for each treatment-concentration combination

Figure 2 shows the mean percentage of dead bees per observation for each treatment by concentration. Cups receiving pure 50% sucrose solution as a control were previously assigned to a concentration for analysis. In an analysis of variance by product, concentration was a significant source of variation at a 0.01 level for all insecticides, including fluvalinate, which was not significantly different from the control in an overall analysis.

Table 1 The mean % of dead bees per observation for each treatment by concentration.					
Mean % of dead bees *					
Treatment	**1PPM**	**3PPM**	**10PPM**	**33PPM**	**100PPM**
Permethrin	36.0a	59.6a	80.1a	84.7a	91.7a
Flucythrinate	26.6b	34.3b	49.5b	74.2b	77.4b
Fnvalerate	24.4b	31.3b	41.9b	60.4c	71.0b
Fluvalinate	23.5b	21.8c	24.0c	34.3d	40.1c
Control	18.9b	19.3c	20.5c	20.1d	23.5d
*Means within a column followed by the same letter are not significantly different at the 5% level by the Duncan's Multiple Range Test					

An analysis of variance by concentration (Table 1) showed significant differences between flucythrinate and fenvalerate only at a concentration of 33 PPM, even though overall means were significantly different. The only concentration at which fluvalinate and the control were significantly different was at 100 PPM. At all concentrations permethrin was significantly more toxic than the other products. Analysis using a pooled estimate of all observations of the value for the control did not change the significance of any differences. The LC50 values calculated using a log-probit analysis were: permethrin – 2.6 PPM, flucythrinate – 8.4 PPM, fenvalerate – 14.7 PPM, and fluvalinate – 799.6 PPM. Each mean represents 6 observations over 5 days for 12 cups, 3 for each of 4 replications, for a total of 120 observations.

Temperature Effects.

The effect of reduced temperature was to increase the toxicity of all of the materials tested over the toxicity observed at 25° C (Figure 3) . The only exception was the mean number of dead bees per observation for permethrin at 12° C, which was lower than for the same product at 25° C. As previously explained, the results at 12° C were affected by the extremely reduced bee activity and food consumption at this temperature. Permethrin caused such a high mortality so quickly that even the relatively high mortality at 12° C was lower than that observed at 25° C or 18° C. The insecticide which showed the greatest increase in the mean percentage of dead bees was fluvalinate, the least toxic of the four synthetic pyrethroid insecticides tested. Fenvalerate showed nearly as large an increase.

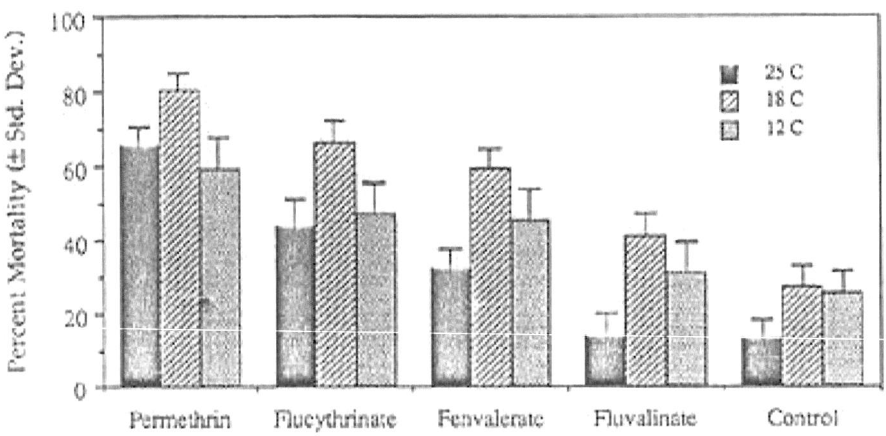

Figure 3

The values for the LC50 of the three most toxic products at 25° and 18° C are shown in Figure 4. This figure points out the increasing numerical differences in toxicity at 18° C compared to 25° C for the less toxic materials. The values for fluvalinate are not shown because of the great difference in magnitude. The values for fluvalinate dropped from an LC50 at 25° C of 800 PPM to 615 PPM at 18° C, however, consistent with the previous observation. This decrease in LC50 for fluvalinate is a 23% drop in LC50 compared to drops in LC5O's of 83.4%, 86.1%and 96.2% for fenvalerate, flucythrinate and permethrin, respectively.

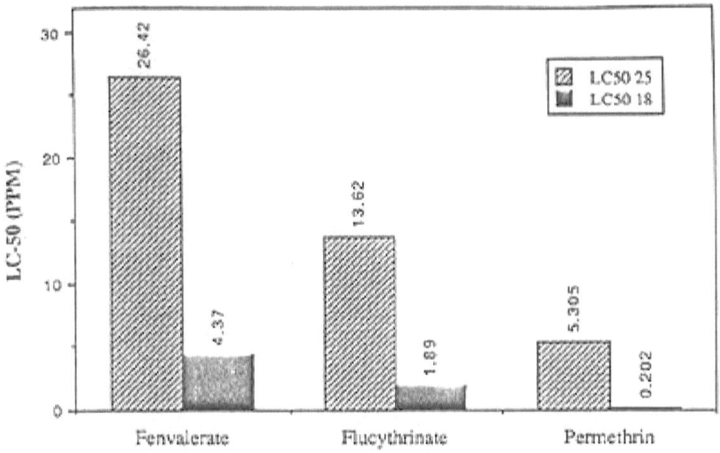

Figure 4

Table 2 The mean % of dead bees per observation for each treatment at three temperatures			
Mean % of dead bees *			
Treatment	**25° C**	**18°C**	**12°C**
Permethrin	65.3a	80.8a	59.2a
Flucythrinate	43.8b	66.3b	47.1b
Fnvalerate	32.3c	59.3c	45.8b
Fluvalinate	13.7d	41.3d	31.2c
Control	13.6d	27.4e	25.7c
*Means within a column followed by the same letter are not significantly different at the 5% level by the Duncan's Multiple Range Test			

Permethrin was significantly more toxic than any of the other treatments at all temperatures (Table 2). Flucythrinate was significantly more toxic than fenvalerate and fluvalinate was significantly more toxic than the control at 18° C only. The relative order of toxicities was the same as had been seen at each concentration and overall for all of the temperatures examined. Each mean represents 6 observations over 5 days of 20 cups, 5 in each of 4 replications, for a total of 120 observations.

Discussion.

Permethrin was found to be the most toxic of the four synthetic pyrethroid insecticides tested at all temperatures and at all concentrations. It was also found to have the largest percentage increase in toxicity with a temperature change from 25° C to 18° C. The extremely low LC50 at 18° C of 0.202 PPM for permethrin certainly suggests that even very small concentrations of permethrin in nectar, collected at temperatures often 25° C or higher, then concentrated by the bees in the process of ripening of nectar to honey, could produce significant mortality in overwintering colonies of bees. If one considers the relative importance of an individual member of a colony of bees in December or January compared to the value of an individual worker in May, June or July, the implications for a significant impact of even the smallest contamination of nectar by permethrin become even more staggering.

Even using optimistic estimates of the rate of degradation of the insecticide while in storage in a solution that is nearly 80% sugars, naturally antibiotic, and sealed within a dark wax cell, the concentration of permethrin in nectar which would result in a 0.202 PPM concentration in the resulting honey, is very small. If we assume a best case scenario of 30% sugar in nectar and a degradation of 80% of the permethrin over the period from collection to consumption, a concentration of only 0.38 PPM permethrin in nectar will result in honey containing the LC50 of permethrin at 18° C.

Flucythrinate was found to be significantly more toxic than fenvalerate overall and at a concentration of 33 PPM, while at other tested concentrations the differences were not significant. Flucythrinate was found to result in a numerically greater mean percentage of dead bees per observation than fenvalerate at all concentrations. The

LC50 of the fenvalerate was approximately double that of flucythrinate at both 25° C and 18° C, however. The percentage of change between the two temperature ranges were approximately equal. Using the same best case scenario as outlined for permethrin, a concentration of 8.8 PPM of fenvalerate or 3.45 PPM of flucythrinate in nectar would result in the LC50 of each being reached at 18° C. These concentrations are in the range of the concentrations of fenvalerate found by Erickson (personal commun.) in nectar being brought back to hives in Wisconsin by bees foraging on sunflowers treated with Pydrin (fenvalerate).

Fluvalinate was significantly less toxic than any of the other synthetic pyrethroid insecticides tested at any concentration or at any temperature It was not found to be significantly different from the control of 50% sucrose except at a concentration of 100 PPM. The LC50 of 615 PPM at 18° C should indicate that even in a worst case scenario considering no degradation of the product after being collected by the bees in nectar and concentrated into honey, the likelihood of a bee receiving a lethal dose is very small. The fact that the product did exhibit some increased toxicity to bees at 18° C should be noted however, especially in light of the recent registration of fluvalinate as a miticide for use in live colonies of bees for control of the Varroa mite, Varroa jacobosoni (Oudemans) (Herbert 1987).

The threat to colonies of honey bees in Indiana from synthetic pyrethroid insecticides as suggested by reports by beekeepers was supported by the results of these tests. The LC50's calculated from the data are certainly not difficult to visualize as a possibility in honey made by colonies foraging on plants treated intentionally or unintentionally with synthetic pyrethroid insecticides. Despite the failure of this methodology to adequately measure the toxicity at temperatures below 18° C, it is reasonable to assume that the same products would show even lower LC50's at the lower extreme of the temperature range a bee might encounter in a hive over the winter in Indiana.

Chapter Two: Pesticide Interactions

As an oversimplification, it may be stated that all pesticides work by interrupting some vital life system in the pest. This same action may also impact a non-target organism such as a honey bee, a bird or the pesticide applicator. Although different pesticides may have different modes of action, the combined effects of seemingly unrelated pesticides can sometimes give surprising results. Because the numerical reality is that no one person, group or research effort could examine all the possible combinations of pesticides on any one organism, this study chose to look at the effect of three common pesticides on bees simultaneously exposed to synthetic pyrethroid insecticides at a sublethal dose.

Background and Objectives

Because honey bee colonies are often exposed to several different pesticides during the year, the question of the interaction of synthetic pyrethroid insecticides and other pesticides becomes important in assessing the total impact on bees. Some of these other pesticides may be other synthetic pyrethroid insecticides, some may be other types of insecticides and some may be fungicides or herbicides which by themselves may not be particularly harmful to honey bees (Moffett 1972, Morton 1974, Stevenson 1978, Stoner 1985). Synergism of the toxicity of insecticides to dipterous pests by herbicides was documented by Lichtenstein (1973) but was not seen by Sonnet (1978) with carbamate and organophosphate insecticides fed in combination with various herbicides to honey bees.

In general, honeybees keep nectar from various sources largely segregated in the combs of the hive. That is, within a given cell which makes up the comb, usually only honey from a single floral source is stored. Further, usually the cells of a given area of the hive are filled at the same time and will represent those honeys made from nectar collected in a rather narrow time range. Most of the honey consumed by the colony over the winter is stored in an area of the hive referred to as the brood nest. This is the area of the hive in which brood is reared through the spring and summer.

In the fall as brood production declines, the bees begin to fill cells no longer needed for brood production with the honey that is being made at the time. This generally leads to honey from various sources being stored in concentric circular areas in the brood nest. Later in the season, the bees will begin to move honey stored in other parts of the hive into the brood nest in preparation for winter. These honeys come from a variety of nectar sources, and may be contaminated with a variety of pesticides. The possibility of pesticide interactions in the honey bee colony is increased by their unique method of storing and utilizing food. In the winter, the cluster may be exposed to any of these products simultaneously.

The importance of interactions between unrelated compounds is well documented. The natural pyrethrins, from which the synthetic pyrethroid insecticides take their name, are isolated from the flowers of certain chrysanthemum plants. They are known for their very quick knockdown of many pest species. They are often formulated in combination with piperonyl butoxide, a compound which by itself has little toxicity to insects. Together, the piperonyl butoxide acts as a synergist increasing the toxicity of the pyrethrins by a factor greater than the additive toxicities would suggest.

The behavior of bees as a unit is highly dependent on chemical cues being passed from individual to individual in the colony. Sharing food among worker bees as they pass in the colony facilitates this form of communication. Because of this food exchange behavior and the pattern of food storage in the brood nest, it is likely that any individual worker may be consuming honey from a number of sources on any one day. In order to simulate this, a study of the effect of exposure to synthetic pyrethroid insecticides simultaneously with one of three types of pesticides was devised.

The fungicide chosen was mancozeb, the active ingredient in the product Dithane M-45, among others. This product is frequently used in vegetable and fruit production (Sine 1988). Bees are attracted to blooming orchards in the spring at a time when they are replenishing honey stores depleted by the winter. Honey produced at this time may be stored for long periods. Because stored honey must be diluted with water prior to consumption by adult bees or for feeding larvae, nectar is preferred as a food over stored honey. If subsequent conditions are good, honey reserves are not needed to sustain the colony through poor nectar flows later in the season. Surpluses that are produced in good nectar years in the spring and stored in empty cells in the brood nest are likely to remain there until winter.

The herbicide chosen was paraquat, the active ingredient in the product Gramoxone 1.5EC. This product was chosen because it is a nonselective contact herbicide used to kill growing vegetation and because it has a higher toxicity to animals than most herbicides (Sine 1988). The areas in which it is applied are more likely to contain vegetation in bloom which is attractive to bees than preplant, preemergence or early postemergence herbicides. It is used throughout the growing season and has seen increased use as the practice of no-till, or reduced tillage farming has become popular in Indiana. In this practice weeds are killed with a herbicide, such as paraquat, in preparation for planting into ground that is not tilled. This may occur early in the season in the case of full season planting or later in the year in preparation for a second or double crop on land from which one crop has already been harvested. In either case, the possibility of some weeds being in bloom and attractive to the bees is high.

The insecticide chosen was Sevin 50W, a formulation of carbaryl, which is a common, general purpose insecticide that is especially toxic to bees. This product is used in nearly all crops, but is of special interest in some parts of Indiana because of the frequency with which it is used in soybeans for control of leaf feeding insects. Soybeans are often treated while in bloom and are often attractive to honeybees as nectar producing plants (Erickson 1979, Kettle 1979, Robacker 1983). Carbaryl is also used in other attractive crops such as sweetcorn, melons, cucumbers and in home gardens and orchards.

Because of the number of pesticide combinations to be examined, only two synthetic pyrethroid insecticides, permethrin and fluvalinate, were used. These products represented the most and least toxic of the insecticides examined in the previous study. The lower temperature range of 12° C was also dropped because of the results of the previous work and to reduce the number of treatments. All of the pesticides were examined at two concentrations, 1 and 10 PPM.

The objective of this study was to determine if there was any synergism or antagonism between the products. This was determined by examining the mortality of bees fed a synthetic pyrethroid insecticide in combination with another pesticide in comparison to the mortality of the two products alone.

Materials and Methods

Bees were collected from the same colony as used in the previous study. The study was conducted one year later, but was headed by the same queen as evidenced by markings placed on her thorax at the time of her introduction to the colony. Bees were collected and handled as previously described. Twenty five bees were again used in cups constructed as before. Each treatment was tested at 18° C and 25° C.

Table 3. – A key to the treatment combinations by number.

Trt. #	Non-Pyrethroid	Conc	Pyrethroid	Conc	Trt. #	Non-Pyrethroid	Conc	Pyrethroid	Conc
1	carbaryl	1	permethrin	1	25	carbaryl	10	Permethrin	1
2	paraquat	1	Permethrin	1	26	paraquat	10	Permethrin	1
3	mancozeb	1	Permethrin	1	27	mancozeb	10	Permethrin	1
4	control	1	Permethrin	1	28	control	10	Permethrin	1
5	carbaryl	1	fluvalinate	1	29	carbaryl	10	Fluvalinate	1
6	paraquat	1	Fluvalinate	1	30	paraquat	10	Fluvalinate	1
7	mancozeb	1	Fluvalinate	1	31	mancozeb	10	Fluvalinate	1
8	control	1	Fluvalinate	1	32	control	10	Fluvalinate	1
9	carbaryl	1	control	1	33	carbaryl	10	control	1
10	paraquat	1	control	1	34	paraquat	10	control	1
11	mancozeb	1	control	1	35	mancozeb	10	control	1
12	control	1	control	1	36	control	10	control	1
13	carbaryl	1	Permethrin	1	37	carbaryl	10	Permethrin	10
14	paraquat	1	Permethrin	1	38	paraquat	10	Permethrin	10
15	mancozeb	1	Permethrin	1	39	mancozeb	10	Permethrin	10
16	control	1	Permethrin	1	40	control	10	Permethrin	10
17	carbaryl	1	Fluvalinate	1	41	carbaryl	10	Fluvalinate	10
18	paraquat	1	Fluvalinate	1	42	paraquat	10	Fluvalinate	10
19	mancozeb	1	Fluvalinate	1	43	mancozeb	10	Fluvalinate	10
20	control	1	Fluvalinate	1	44	control	10	Fluvalinate	10
21	carbaryl	1	control	1	45	carbaryl	10	control	10
22	paraquat	1	control	1	46	paraquat	10	control	10
23	mancozeb	1	control	1	47	mancozeb	10	control	10
24	control	1	control	1	48	Control1	10	control	10

The experimental design was again a split plot randomized complete block design with the restriction on randomization being the two temperature ranges. Treatments (Table 3) consisted of the assigned pesticides mixed in 50% sucrose solution and prepared to give concentrations of 1 or 10 PPM as assigned when mixed with equal volumes of the paired pesticide. All solutions were prepared fresh before the trial with dilutions of 1000 PPM solutions. The study was replicated four times, with each temperature repeated twice in each of two environmental chambers.

Formulated product was used for all solutions assuming the concentration stated on the packaging to be correct. Fresh formulated product was obtained from the manufacturer before the start of the experiment. Tests for possible fumigant action of the products in 50% sucrose indicated no effect from a concentration of 1000 PPM. Concentrations of 1 PPM and 10 PPM were tested for stability in 50% sucrose solution for seven days and had no significant decrease in toxicity.

Bees were checked for mortality at 12 and 24 hours following introduction to the cups and each 24 hours thereafter for 5 days. As previously described, bees not responding to gentle stimuli were considered dead. Two environmental chambers were used for the trial and each chamber was kept at 18° C for two replications and 25° C for the other two replications. The bees were kept in the dark throughout the experiment and at 50% to 70% relative humidity.

Results and Discussion

No evidence of synergism or antagonism was shown in any of the combinations examined. The toxicities of the fungicide and herbicide were not found to be significantly different from the control of 50% sucrose under the conditions examined. Bee mortality for the treatments containing permethrin showed higher mortality at 18° C than at 25° C, as expected from the previous tests.

Data Handling.

The ANOVA and GLM procedures of SAS were used for data analysis. As in the previous study, mortality is indicated by a mean of the percentage of bees dead at each of the six observations made over five days. Means of three or more groups were tested for significance of differences with the Duncan's Multiple Range Test with Type I error rate set at 5%.

Overall Results.

Permethrin and carbaryl were the most toxic (Figure 5) and were not significantly different from each other at the 5% level under the conditions tested. There were also no significant differences between fluvalinate, paraquat, mancozeb or the control of 50% sucrose. A control was grouped with both the synthetic pyrethroid insecticides and the three other pesticides and assigned to both concentrations for completeness of balance in design.

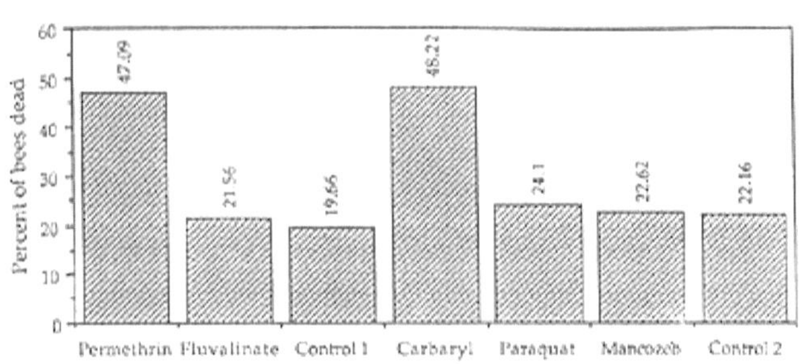

The mean percentage of bees dead per observation for each pesticide tested.

As expected from the previous work, mortality was higher at 18° C than 25° C for all treatments containing permethrin. Figure 6 shows the mean percentage of dead bees per observation for all the treatments by pesticide and temperature. That is, all observations of bees exposed to permethrin, in any combination, at 25° C are represented by the first bar of the graph.

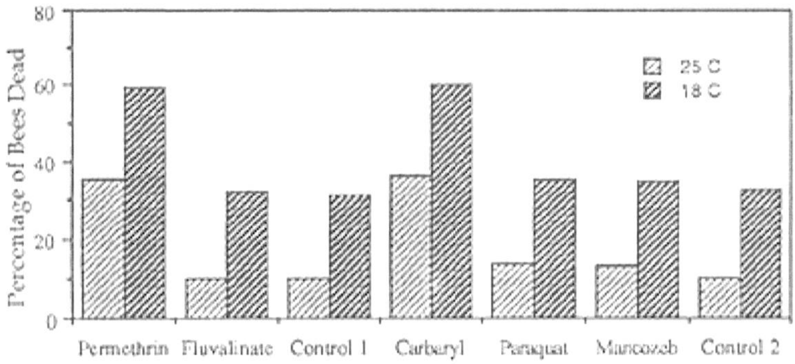

Percentage of bees dead per observation for each pesticide by temperature.

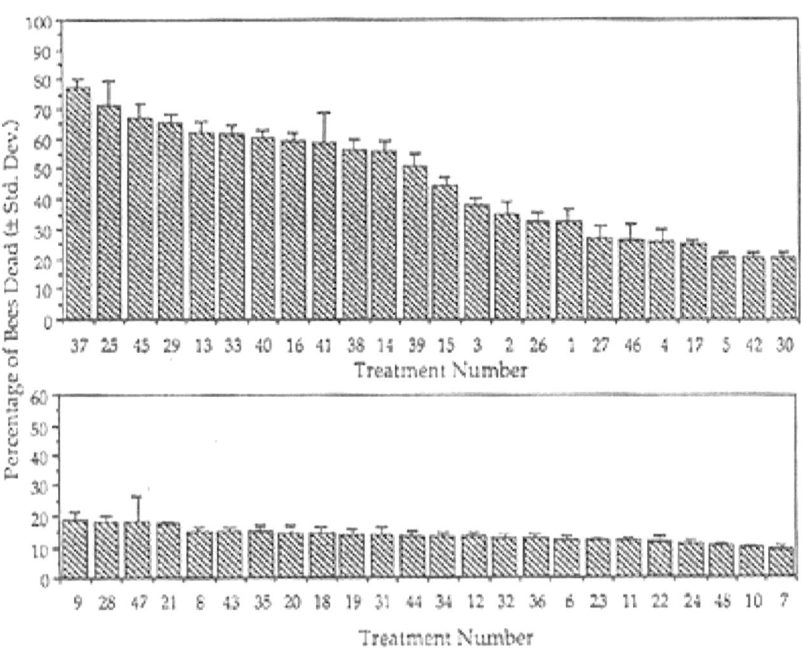

The mean percentage of bees dead per observation by treatment.

Because the value for each pesticide is a mean of all occurrences of that product in combination with all others in a balanced design, much of the increased mortality at lower temperature is due to the combination of that product with premethrin. The mean percentage of dead bees per observation for each of the 48 treatments is given in Figure 7.

Pesticide Interactions.

Comparing the expected mortality of permethrin or fluvalinate with the other pesticides did not indicate that any synergism or interaction was present. The expected mortality was determined by adding the increased mortality over the control for the non-pyrethroid to the increased mortality of the pyrethroid over the control. This calculated, expected value was compared to the observed mortality of the combination minus the control mortality estimated by the control/control mean.

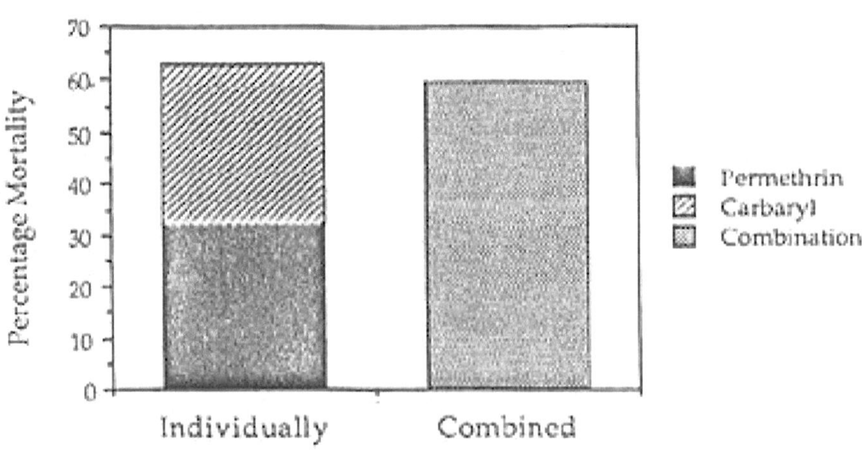

Figure 8. The expected and observed mean mortality from a combination of permethrin and carbaryl.

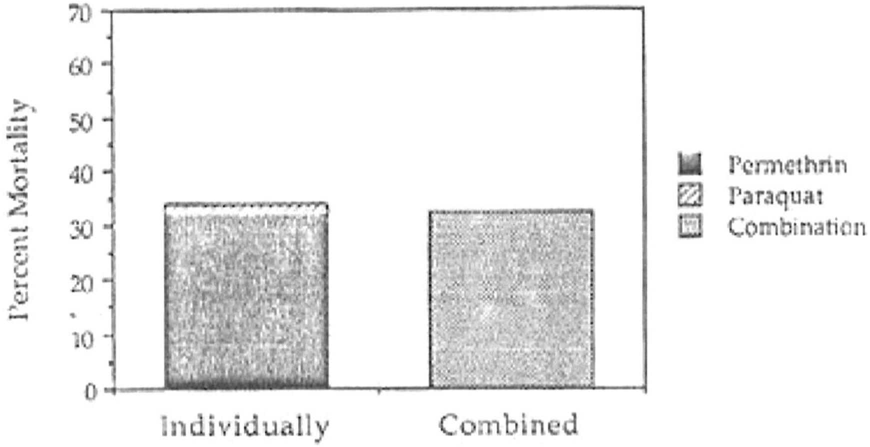

Figure 9. The expected and observed mean mortality from a combination of permethrin and paraquat.

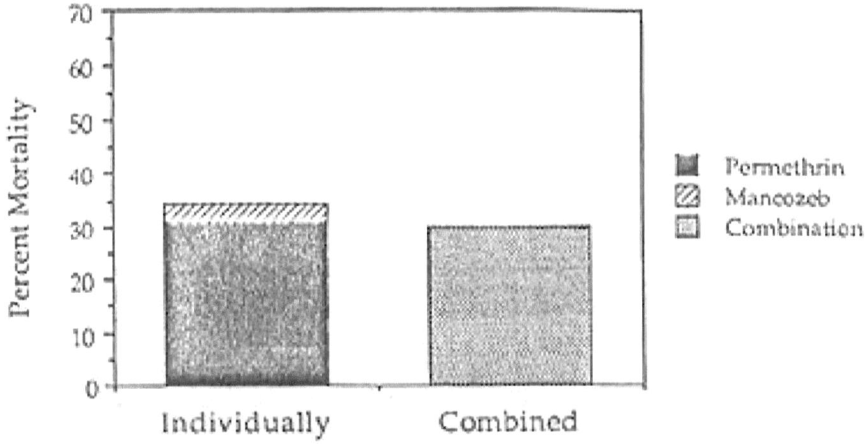

Figure 10. The expected and observed mean mortality from a combination of permethrin and mancozeb.

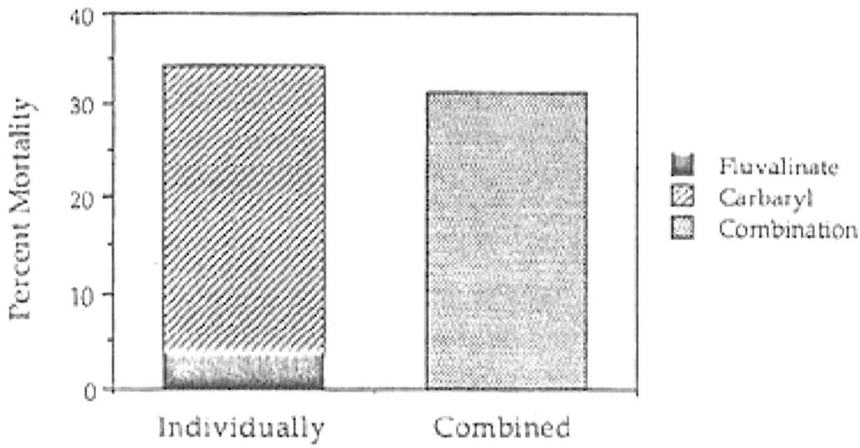

Figure 11. The expected and observed mean mortality from a combination of fluvalinate and carbaryl.

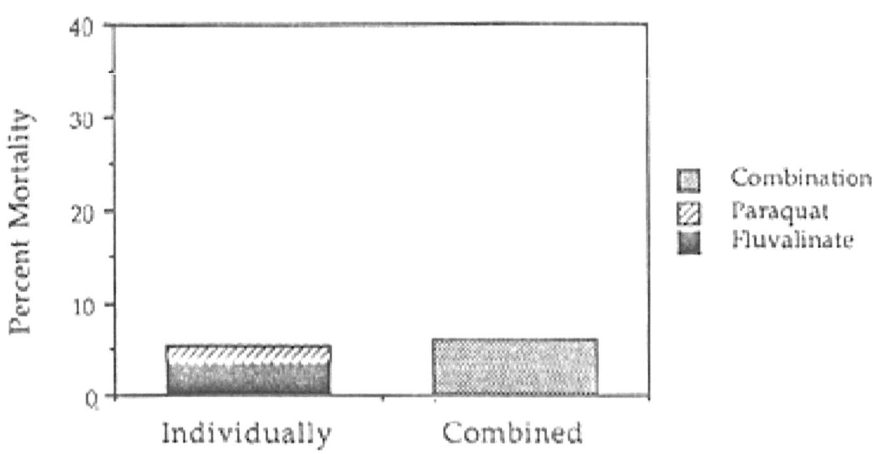

Figure 12. The expected and observed mean mortality from a combination of fluvalinate and paraquat.

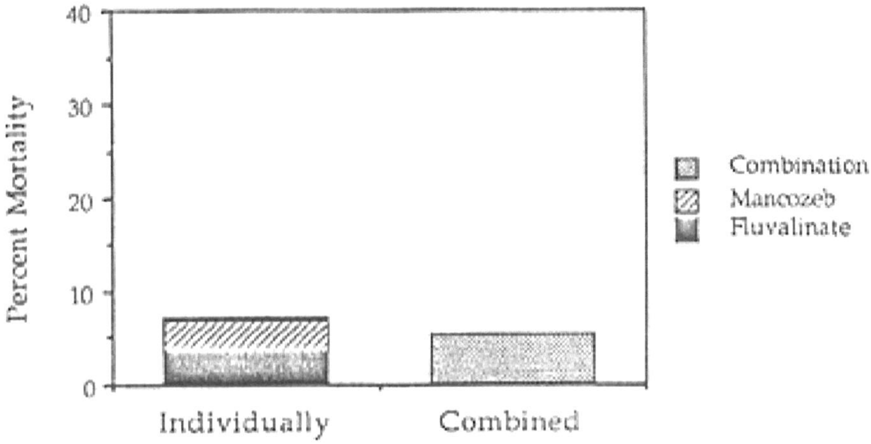

Figure 13. The expected and observed mean mortality from a combination of fluvalinate and mancozeb.

For example, the permethrin/control mean mortality was 43.13%, a 32.17% increase attributable to permethrin over the control/control combination value of 10.96%. The carbaryl/control combination gave a 41.65% mean mortality indicating that 41.65% - 10.96%, or 30.69% was due to carbaryl.

Therefore, the expected mean percent mortality of the permethrin/carbaryl combination was 32.17% + 30.69% or 62.86%. The observed mean percentage of dead bees per observation for the permethrin/carbaryl combination was 70.63%. This value minus the control mortality (70.63% - 10.96%) was 59.67% is the observed combination mortality.

Discussion.

The results of this study indicate that two of the more toxic insecticides to bees, permethrin and carbaryl, have an additive and not a multiplicative toxicity to adult bees. It was also shown that any effects of paraquat and mancozeb are very small in terms of adult mortality and these products do not significantly change the impact of permethrin or fluvalinate on bees. This is especially interesting in light of the perception of many beekeepers and applicators about the impact of herbicides and fungicides on honeybees (see Chapter 4).

Chapter Three: Interhive Variability

Since the 1960's, resistance of insects to a number of insecticides has been documented. Resistance of houseflies to DDT was the first in a long and continuing series of pest species which have become resistant to chemical controls. Only recently has this phenomenon been exploited to select for pesticide resistant beneficial insects and mites. It might be expected that honey bees could also be selected for resistance to pesticides. This study concentrated on examining various populations of honey bees for resistance to synthetic pyrethroid insecticides in all three races of bees commonly sold in North America.

Background and Objectives

The development of resistance in insect populations is influenced by a number of factors. These include the generation time of the species, the amount of insecticide pressure on the population, the mobility of the species, the amount of outbreeding with populations not resistant and whether or not the females mate with more than one male (Georghiou 1980). In a population of leafminers in a greenhouse environment for example, it can be seen how each of these factors favors the development of resistance. A population of honey bees presents a very different picture, however.

Genetic changes in honey bees colonies only occur when the queen is replaced (Collins 1980). This may happen under several conditions, but, it most commonly occurs when the colony swarms or when the old queen is superseded. In managed colonies, the intercession of the beekeeper in replacing the queen must also be considered. The most frequently this might be expected to occur would be once every year or two. This is a very long time considering the development of insecticide resistance.

It is possible that colonies exposed to pesticides will be more likely to supersede their queen. This does provide the opportunity for the selection of a larva that is surviving at a time when the colony is challenged by an insecticide. Neither swarming nor beekeeper replacement of queens is likely to provide any selection for insecticide resistance. Only strong colonies are likely to swarm and it may be assumed that colonies exposed to substantial levels of insecticide would not be strong enough to swarm. Beekeepers do not have a commercial source of pesticide resistant queens available. Queens are not likely to come from colonies which have been selected for any trait other than honey production. Beekeepers raising their own queens may be able to do some indirect selection for pesticide resistance by choosing colonies that perform well in areas of pesticide use.

The mating behavior of queen honey bees is also not well suited for the development of insecticide resistance. The virgin queen leaves the hive at a few days of age and mates with 5 to 10 drones. From this mating she will store all the sperm necessary for egg laying for her entire life. The drones she mates with are likely to be from colonies other than her own, and may come from colonies several miles away. Because drone bees do not contribute to the welfare of a colony, weaker colonies are likely to produce fewer drones than strong colonies. Therefore colonies challenged by insecticides are not likely to produce as many drones as other colonies.

The development of resistance by selection also requires the assumption that some level of natural resistance exists in the population (Graves 1965). In honey bees a number of traits have been selected for at one time or another (Rothenbuhler 1980). Some of these programs have been quite successful, such as breeding for increased pollen collection (Boelter 1984). Other honey bee queen breeders claim to have strains selected for resistance to various honey bee diseases, honey producing ability, overwintering ability, a preferred color or resistance to some pesticides (Tucker 1980). Most of these traits have been selected for from within one race; others from the offspring of selected crosses. Differential resistance to carbaryl was shown between lines of Apis mellifera ligustica Spin, and A. m. scutellata Lepeltier by Danka (1986). The relative impact of race and previous selection pressure on the two lines was not addressed by Danka.

In the United States, the honey bee genetic pool has been closed since 1912 when Congress closed the United States to the importation of live bees. This was done to prevent the introduction of a mite pest which was having devastating effects on the honey bee population of England at the time (Phillips 1925, Rennie 1921). The border was not closed to Canada. Canada had adopted similar laws, but had excluded New Zealand as well as the U.S. and therefore genetic material could come to the U.S. from New Zealand via Canada. Since 1912, there have been some very limited importations under permit from the USDA as well.

At the time the border was closed in 1912, there were four races of bees known to be present in the United

States (Culliney 1983, Severson 1985). These were Apis mellifera ligustica Spin or the Italian, by far the most popular and common; A. m. mellifera L. or the German or Black bee, the first race imported by settlers but an aggressive and unpopular race; A. m. caucasica Gorb. or the Caucasian, a gentle grey bee noted for its ability to overwinter well but not popular for other reasons; and A. m. carnica Pollmann or the Carniolan, also a gentle grey bee but noted for its tendency to swarm. None of these races were genetically pure, with the possible exception of the Italian. The lack of artificial insemination at the time and the sheer number of Italian bees, made keeping pure lines of the other races nearly impossible.

Over time A. m. mellifera became nearly impossible to find in a distinguishable form, although its influence on some strains is still detectable. Renewed interest in the grey races (A. m. carnica and A. m. caucasica) resulted in some breeders selecting strains that showed morphological and behavioral characteristics very close to those of these races as seen in their native areas (Carlisle 1955). Instrumental insemination greatly aided this work because it allowed the use of single drone controlled matings. Today relative pure lines of A. m. ligustica, A. m. caucasica and A. m. carnica are available in the U.S.

A long term selection program for honey bee lines was started in England in 1912 when the native strains were decimated by the Isle of Wright Disease, now believed to be a combination of infection with Nosema apis Zander and infestation with Acarapis woodi Renie (Bullamore 1922). This program was carried out by Brother Adam at Buckfast Abbey. Brother Adam has traveled the world looking for different genetic lines of bees that might contribute to his breeding program and he has imported a number of these races and strains to Buckfast Abbey. His intensive selection and crossing have resulted in a world famous line of bees now known as Buckfast. This line was imported under permit into the United States as eggs which were reared into virgin queens and inseminated with semen also imported under special permit from the USDA. The Buckfast bee is now sold by a single licensed breeder in the U.S. (Sugden 1983). These queens are mated naturally to wild drones, however, so that their progeny are only 50% Buckfast genotype.

The objectives of this study were:

To look for resistance to one or more synthetic pyrethroid insecticides in genetic lines of bees representing all three common races found in North America and a line representing a long term selection program from England.

To determine if any resistance found was related to race of the colony, possible previous selection pressure or simply individual variation between colonies.

Materials and Methods

In the summer of the year previous to the study, colonies of bees were established in a common site near West Lafayette. This was accomplished by removing frames of brood from established colonies and introducing marked queens representing the different lines selected for the study. All of these colonies were established in new hives so that previous pesticide exposure would not be a factor. The old frames removed from other colonies and used to establish the new hives were replaced with new frames as soon as possible.

The queens selected were as follows: two caucasian lines, one each from two breeders; two carniolan lines, one each from two breeders; two Buckfast queens from the sole breeder in the U.S.; two Italians from one breeder and two Italians from hives in a high pesticide use area of southern Indiana which had not been requeened for at least seven years. These ten lines allowed comparisons between races, between queens from the same breeder, and between bees which could be reasonably assumed to have been under selection pressure and those not under such pressure within the same race or from other races.

The seven treatments consisted of a control of 50% sucrose and carbaryl as Sevin 50W, permethrin as Ambush 2EC and fluvalinate as Spur 2.4 EC in 50% sucrose at 1 and 10 PPM each. The bees were handled and mortality measured as previously described. All bees were held in the dark in a single environmental chamber at 25° C. The study was conducted as a randomized complete block design and replicated four times. Each treatment was applied to a single test unit (cup of 25 bees) except for one fourth of the treatments which were duplicated to provide an estimate of between cup variance. The duplicated treatments comprised a fifth complete set of insecticide-bee source combinations which were randomly assigned to a replication.

Results and Discussion

Significant differences were found between the different sources of bees examined. These differences were related to the race of the colony. There was surprising little variation between the different sources of the same race. No significant differences in susceptibility were found in the hives which were assumed to have been exposed to insecticides and those which were assumed to not have been exposed.

Data Handling.

The data were analyzed using the ANOVA, GLM and MEANS procedures of SAS. Mortality is presented as the mean percent of bees dead at each of six observations made over five days. Means were tested for significant differences using Duncan's Multiple Range test with the error rate set at 5%. Two data points of the 350 combinations of 10 hives and 7 treatments for five replications were missing due to the loss of one or more observations used to calculate the mean percentage of bees dead per observation. The significance of differences did not change if the fifth replication was dropped.

Hive to Hive Variation.

There was considerable variation between the hives as to their susceptibility to the products tested (Figure 14). Differences in overall mortality as measured by the mean percentage of bees dead per observation (Table 4) were found to be significant at the 5% level. The ANOVA indicated that variance between duplicate cups within a hive/treatment/replication combination was not a significant component of the total variance (<1%). Analysis by treatment showed that differences between hives were significant for only two treatments, carbaryl and permethrin at 10 PPM (Table 5).

Analysis of Racial Differences.

The influence of race on the hive differences was assessed by grouping the hives into Italian, Caucasian, Carniolan and Buckfast. While this latter group is not a true race, it does represent a different genetic lineage than the other hives and does not fit well into any of the other races, although each of these races was used at some point in the development of the Buckfast bee.

Figure 15 shows the mean percent mortality by race. Each race is represented by two hives except Italian which was represented by four hives. The Italian and Caucasian races were significantly different from the Carniolan and Buckfast races. Further analysis of the four hives grouped as Italian was conducted by splitting the group into

Indiana and California strains. Since the Indiana strain was assumed to have been exposed to the materials in question, while the California strain was selected from breeding stock in Canada in an area of no pesticide use, differences representing selection pressure should have been evident between these two groups. The means and standard deviations of the Indiana and California strains were 13.4%(6.1) and 15.9%(7.4), respectively. The difference was not significant the 5% level.

The racial differences were also examined by treatment. Figure 16 shows the mean percent mortality by treatment. The treatments were: Trt. 1 - carbaryl 1PPM, Trt. 2 - carbaryl 10PPM, Trt. 3 - permethrin, Trt. 4 - permethrin 10PPM, Trt. 5 - fluvalinate 1PPM, Trt. 6 - fluvalinate 10PPM, and Trt. 7 - the control of 50% sucrose only. The Italian race was significantly more sensitive to carbaryl than any other race. The Caucasian race was significantly more sensitive to permethrin than the other races and while not significant, showed an interesting sensitivity to fluvalinate at 10 PPM.

Discussion.

Racial differences in susceptibility to the treatments examined were noted. It was especially interesting to note that the Italian race, the most popular in the United States, was the most susceptible to carbaryl and more susceptible to permethrin than the Carniolan and Buckfast races. The similarity between the Buckfast race and the Carniolan is not surprising since the Carniolan was a major line used by Brother Adam in developing the Buckfast bee. It was also very interesting to note the small amount of variation between the different sources of a race compared to racial differences.

Since the so-named Indiana and California strains of the Italian race were not positively known to have represented lines selected under insecticide pressure and a lack of such pressure respectively, the lack of significance of the difference should not be taken to suggest that selection for increased tolerance to carbaryl or the synthetic pyrethroid insecticides would not be effective. The racial differences would suggest that racial hybrids should be considered in any breeding program whose aim was increased resistance to insecticides.

Appendix E: Miscellaneous Articles

Bee Pathology – pages 171-173, GONCALVES, L. S.; DE JONG, D. (Brazil), MORSE, R. A. (U.S.A.)

Varroa is currently recognized to be the greatest problem for apiculture worldwide. First described in Asiatic bees Apis cerana, in Indonesia, by the Dutch researcher Oudemans in 1904, the mite turned into a problem for Brazilian bees, Apis mellifera, in the 1970s. Transfer of the pest from A. cerana, whose infestation is of little economical importance, to A. mellifera occurred when beekeepers carried honeybees to Asia. Today Varroa is found throughout Asia, in most of Europe, in the Northern region of Africa, in several countries in the Middle East and in the following South American countries: Paraguay, Brazil, Argentina, Uruguay Peru and Bolivia.

In 1971, infested hives were brought to Paraguay from Japan. Soon after, in 1972, an apiculturist from Sao Paulo took some of these Japanese bees to the region of Jundiai, thus initiating infestation in Brazil. However, infestation was first detected in 1978 when the pest had spread considerably and there was no way of eliminating this "new" enemy of Brazilian apiculture. At present, Varroa can be found in 18 Brazilian States, from Rio Grande do Sul to Piaui, so that it is too late to avoid transporting hives from a State to another.

Can we blame the Japanese, or other peoples, or even our own beekeepers for spreading so much these bees without taking precautions against possible consequences such as the introduction of Varroa? In a way, we can. All those who work in apiculture should he aware of the possible consequences of their carelessness, especially those involved in "migrant apiculture" at the international level. A period of quarantine is always recommended to avoid the entry of pests into the country. In 1971, Varroa was little known and the problems of hive mortality, which were already occurring in Russia, were not properly communicated to the Western world. In Eastern Europe also, Varroa was present 4 to 6 years before being detected. Not even West Germany, where apiculture is well organized and controlled, succeeded in containing Varroa.Research on Varroa in Brazil

The Department of Genetics of the University of Sao Paulo, Campus of Ribeirao Preto, started a research program on Varroa in 1979 under our direction and with the financial support of the Foundation for the Advancement of Science in the State of Sao Paulo (FAPESP) and later of the National Research Council (CNPq). In 1980, a collaborative program between USP and the Department of Entomology of Cornell University was started, with the financial support of the National Science Foundation (NSF) and later of the Department of Agriculture, USA (USDA). As part of this program, Dr. De Jong came to Brazil in 1980, where he is currently involved in research.

It was soon observed that Varroa had spread considerably but with low indices of infestation. The mite showed no symptoms to the beekeepers who did not know it, since at the beginning of infestation Varroa is present in very reduced numbers in hives, although it rapidly transfers from hive to hive through new field bees which get lost during the first recognizance flights and through drones, which tend to enter other hives.

What is Varroa?

Varroa is an acarid which attacks honeybees. It is a relatively large brown mite 1-mm long and 1.6-mm wide. Varroa feeds on bee "blood" (hemolymph). Normally, only adult females can be seen. The other female stages (nymphs) and males are found only in bee brood cells and are smaller and white. These also feed on bee "blood" (except for adult males which do not feed on anything). When compared with other types of pests, Varroa, reproduces slowly. Under ideal conditions, the mite leaves on average one to two new female descendants per cycle in worker bee brood. Without taking into account mortality or migration to other hives, an infestation starting with a single female may take approximately 400 days to include 5000 acarids. At the same time, many of these females may be moving to other hives, and therefore it would take even longer for an initial infestation to reach a damaging level. Obviously, if infestation started by introduction of well-infested hives into a beekeeping unit of a migratory type, proliferation becomes more rapid. Also, Varroa multiplies more rapidly at the time of drone rearing, with a single female capable of producing 2 to 4 new females in a drone cell.

Our studies have shown that Varroa reduces the weight and mean life of bees that were infested during development. When infestation is low, however, the beekeeper does not detect this damage. When a single larva is attacked by 5 or more acarids it can still survive, but its body, and the wings in particular, will be visibly damaged. However, very few bees are attacked by so many acarids, and when this happens they are thrown out by the other bees, so that the beekeeper will not see the problem.

Hive mortality caused by Varroa

Millions of hives have already died in Europe because of Varroa. Tens of thousands have died in Argentina. In many regions of the world it is impossible to keep bees without treating them. In Brazil thus far, no hive death caused by Varroa has been reported. even though the mite reached Brazil before being introduced into

Argentina. A completely unexpected phenomenon has occurred, since greater infestation was expected in our climate, which permits Varroa to reproduce throughout the year, in contrast to countries with cold climates, where Varroa does not reproduce in winter.

Fortunately until now we escaped relatively well, since our levels of infestation with the pest are low, whereas they are very high in the rest of the world. However, Varroa still causes serious concern. Even though the level of infestation is low and does not cause death of hives, there is still damage, because all hives in the country are progressively becoming infested. Infestation causes little damage per hive, but if we sum all the hives, great waste occurs in our apiculture.

We can estimate the damage caused by Varroa as follows: our studies have shown that the mean life of bees infested with Varroa during development is reduced by half. Thus, for each two bees affected, we actually lose one. If 2% of worker brood is infested, the population of the hive will be reduced by 1%. We may estimate that his would cause a 1% reduction in honey production. Since national production in 1984 was estimated at about 30,000 tons, the reduction would be approximately 300 tons which, at the cost price of Cr$9,500.00 per kilogram, would cause a loss of about 3 billion cruzeiros (about 300,000 US$). The ideal would be for all beekeepers to be able to treat their bees to save them and to guarantee honey productioin. However, the solution is not so simple, since so far there is no adequate pesticide for the conditions prevailing in Brazil. Either the treatment is too expensive, or is harmful for the bees, or it contaminates the honey. If one opted for treatment with chemical products, such treatment should be applied every year, at a cost that would be higher than the loss caused by Varroa.

The ABC of BEE CULTURE, A.I. Root – 1891 – Pages 62-72

COMB FOUNDATION.

Since the introduction of foundation, within the past few years, many difficult points have been solved completely; such as, how to insure straight combs, how to insure all worker-comb or all drone-comb, as the case may be, and how to furnish the bees with the wax they need without being obliged to secrete it by the consumption of honey. It is so simple a matter to make a practical test of it by hanging a piece in a hive when honey is coming in, that I think I may be excused from describing the way in which the bees use it, at any great length. Neither will it be needful to dwell on the successive steps by which it was discovered, and brought to its present state of perfection. The first mention we have of wax foundations that were accepted by the bees, was published in a German bee-journal as far back as 1857.

Mr. J. Mehring, of Frankinthal, Germany, if I am correct, seems to have been the original inventor. For nearly 20 years the matter seems to have slumbered, although different ones at different times, among whom was our friend Wagner, took it up, made some improvements, and dropped it again. The sheets made in both England and Germany had no side-walls, but simply indentations. Mr. Wagner added shallow side-walls, making it much more like natural comb. Until recently it was all made with a pair of plates; even yet the Given press is preferred by some (see elsewhere); but it did not require much wisdom to decide that such an article, if wanted in large quantities, should be rolled out by machinery. In the latter part of 1875 I talked with a friend of mine who is quite an artist in the way of fine mechanical work and machinery, and told him what I thought was wanted. The result was that he made a machine that would roll out a continuous sheet, with very fair side-walls of wax, and superior to anything ever made. Indeed, so perfect was the workmanship of the rolls, that, even though fifteen years have passed, nothing yet has been constructed which fully equals the foundation from them. Mr. A. Washburn, the mechanic who did the work, made the rolls by stamping – an operation slow, laborious, and consequently expensive. This made the price of these machines from $100 to $125 apiece – a figure beyond the reach of the average bee-keeper, and even of most supply-dealers. In consequence of the call for mills for less money, Mr. Chas. Olm, of Fond du Lac, Wis., invented an automatic machine which cut with a set of knives the embossed surface of the rolls. It was thus made possible for us to manufacture foundation-mills at a price from one-fourth to one-fifth of those first made.

A machine for engraving foundation rolls

As the space here is limited, I can hardly go into minute details showing you how these rolls are made. The following is an engraving of a machine embodying the principles of the original one made by Mr. Olm, but with the added improvements of the foreman of our machine shop, Mr. Washburn.

There are two gravers, as you will notice, held at the proper angles, set in slides operated by a crank and pitman. One of the keen chisels first comes down and makes a cut in the surface of the roll. This first cut raises the edge of the chip, but does not take it out. The other chisel cuts this chip entirely loose, and throws it out. As these knives work back and forth, the carriage holding the roll is spaced automatically until the end of the roll is reached. Here it is again carried back automatically, and after a "click, click," the knives, or gravers, resume their work. This is repeated until the surface of the roll has been indented with the lozenge faces. The side wall is then stamped by a perpendicular punch, likewise fastened into a slide, and operated by a crank and pitman. The machine is run by power, and is almost entirely automatic. The machinist simply operates a set of levers, while the machine responds to his bidding. It can likewise be operated by hand-power whenever occasion demands.

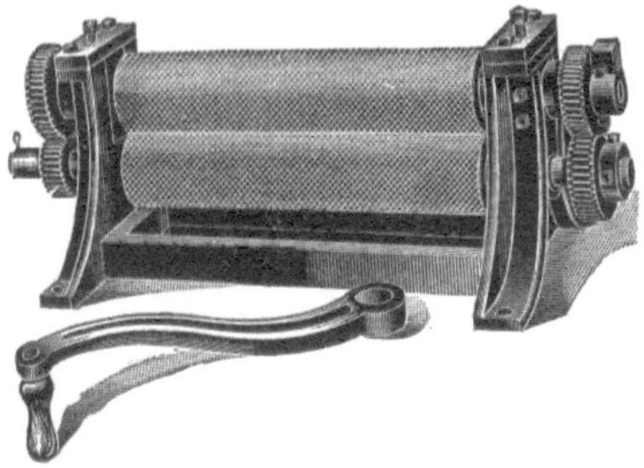

10-inch foundation mill

The cut represents one of the latest improved mills. The wooden-roller attachment will be explained further on. The price of these machines ranges all the way from $15.00 to $40.00. The regular size of a ten-inch machine for the Langstroth frame costs $20.00.

HOW TO REFINE WAX

Under Wax, in the latter part of the work, this subject will be partially treated; but in this place, in order to make a first-class article of foundation, some specific directions will be necessary. Wax cakes are usually of all grades and colors, particularly if your trade is such that you are obliged to make use of the commercial article. The difference in color is due largely to the amount of impurities the wax contains. To cleanse this wax and also reduce it to a uniform color, proceed as follows: Into a receptacle of the proper size (say a wash-boiler, one that your wife will let you have), pour four or five inches of water. Put it on the stove and heat the water, after which put in the wax. When the latter is melted, dip it out and pour into receptacles with sloping sides. The deeper the receptacle the better it will be. The Dadants, who have the reputation of making the finest foundation in the world, use tin cans 10 inches in diameter at the bottom, 12 inches at the top, and 20 inches deep.* If you cannot afford these deep cans, utilize whatever receptacles you can get hold of. Sap-pails or ordinary pails would answer your purpose sufficiently well, perhaps. Having dipped out all the wax from the boiler into the cans, put them in a close room, or, better still, in a cupboard, so that the cooling process may be delayed as long as possible. The longer the cooling the better opportunity is afforded for the impurities to settle to the bottom. When the wax is hard, remove and scrape off the bottom of the cakes, which will be largely foreign settlings and other impurities. If these wax cakes have not, in your judgment, attained the proper color, that is, a bright yellow, repeat the operation once or twice until you are satisfied.

The method already given is essentially the one employed by the Dadants, and I give it here because it is one of the secrets of their success in turning out yellow foundation. If you are making foundation for your own use, it is not necessary to have the wax so thoroughly refined; but as the trade demands yellow foundation you will have to supply what it calls for. We have found, however, that the darker grades of foundation are as readily accepted by the bees as the lighter. As it costs some more to make the yellower foundation, if your customer prefers, let him have the darker for one or two cents per pound less. I might state right here that the wax for thin or surplus foundation should be brighter in color than that intended for the brood-chamber. We make it a practice to save out our yellowest wax for thin foundation.

* Use no receptacles made of galvanized iron – see Wax.

HOW TO MAKE WAX SHEETS.

To be able to do this work successfully, requires not a little skill. Neatness is another important essential. A little carelessness in spilling and dripping wax upon the floor means a great deal of trouble in scrubbing it up afterwards. Indeed, it is well nigh impossible to get a floor clean after particles of wax have become pressed and rubbed into it by great big clumsy feet.

The operation of making wax sheets, in a word, is dipping a thin sheet of wood into a deep vessel of melted wax. A film will cling to the board, which is afterward peeled off. Very simple, isn't it? But I am afraid, my friend, that, before you get through it, you will find it more difficult than you at first imagine. One of the prime essentials for making wax sheets successfully is experience. But with the assistance of a few suggestions, I can save you a great deal of trouble.

To melt wax for dipping, you must be sure not to burn it, otherwise it will be totally spoiled. To insure against this, the receptacle for melting should be inclosed by another larger receptacle containing hot water. This is to be placed upon the stove, and the wax cakes are to be deposited in the inner tank. As the wax cannot get hotter than the boiling-point, there is no danger of burning. But desiring to work as economically as possible, you will feel, perhaps, that you are not able to purchase any more implements than are absolutely necessary. An old wash-boiler, or one that your wife thinks she can spare, can be made to answer nearly as good a purpose. Place it upon the stove and pour in four or five inches of water. Into the water, put the wax cakes. As the latter have a specific gravity lighter than the former, they will float on the water either before or after being melted, and consequently there will be no danger of burning. After putting in a sufficient amount it can be dipped out into the dipping-tank. This is a deep vessel for holding the wax after it is melted. A sufficient quantity should be dipped into this tank so that the dipping-board may be immersed within an inch or so of the upper end.

The dipping-tank should be placed close by the stove, so that the hot wax can be dipped or drawn off readily through a suitable faucet from the melting-tank on the stove. You are now ready for your dipping-boards, which I will presume you have already made. There should be at least two, and more would be an advantage. These boards should be made of the very best straight-grained pine lumber which you can obtain. There are generally only one or two boards in a log which are fit for the purpose, and they are the "heart" boards. These will warp neither one way nor the other, and the grain is not as liable to shale up and catch the wax sheets when being peeled off. They are to be made of a size to suit the frame you are using. If you are using the Langstroth frame, the dipping-boards should be 9 inches wide and about two feet long, or long enough to leave about two inches projecting out of the melted wax for finger room. Before using they should be soaked in brine water for a few hours, the proportion of salt in the water being about a teacupful to two or three pails of water. We have found that the salt serves a double purpose: It acts somewhat as a lubricant in facilitating the removal of the sheets, and as a preventive against the grain rising in the board, and consequently roughening. Before we used the salt, we used to have to sandpaper the boards quite frequently; but we rarely have occasion to do it now.

Besides the melting-tank, dipping-tank, and the dipping-boards, you need a cooling-vat of water, for cooling the wax film adhering to the dipping-boards. An ordinary tub of cold water may answer; but if you propose making very much foundation, you had better make an oblong shallow wooden box, capable of holding water. This cooling-vat should be close at hand.

Two can work to the best advantage – one to dip, and the other to peel off the sheets. In order to make the dipping a success, the wax must be neither too hot nor too cold. We find that we get the best results when it is at about the temperature of 165° or 170° F (74° -77° C)). It is too cold if there is a small film, or little spots of cooling wax on top of the melted liquid from which you are dipping. If too cold, it will leave little ripples on the sheets, and the surface of the sheets will be wavy and the thickness irregular. If the wax is too hot, the sheets will crack in peeling off. It is very important, as you will find by experience, to do the dipping when the wax is at the right temperature. Properly made sheets will work better in the rolls than when they have been subjected to either extreme of temperature. If they begin at any time to stick to the plate, rub a rag, moistened in a weak solution of lye, such as is made from an ash-leach, on both surfaces of the board, and you will probably have no more trouble. If this fails, then the sides of the boards have become roughened, and, of course, nothing will do then but to sandpaper them down again after they are dry.

We make five kinds of foundation; viz., heavy brood, from 4 to 5 ft. per lb.; medium brood, 5 to 6 ft. per lb.; light brood, 7 to 8 ft.; thin surplus, about 10 ft. to the lb.; and extra thin surplus, from 11 to 12 ft. To make sheets for the first named, five dippings will be required; for the second, three; for the third, two; and for the last, one short quick dip.

After each successive dip into the tank, before immersing again a low all the ripples to run off till the board is smooth. Immerse quickly, and draw out as quickly. The number of dippings will have to be varied, however, according to circumstances. The adjustment of the mill, the temperature of the wax, and the quickness of the plunge of the dipping-board, all have their influence. It may be an advantage to reverse the dipping-board, i.e., dipping the other end. After the boards are dipped they should be placed immediately into the vat of cool water, which we before described. After the boards are cold, scrape the edges with a knife. Peel up a corner of the sheet, and pull it off. As you proceed in your work, the wax in the dipping-tank will become cool, and the water* in the cooling-vat will become warm. Of course, both must be restored to their proper temperature. To bring the wax in the dipping-tank to the right point, pour in a dipperful from the melting-tank on the stove. Add another dipperful, if necessary. To cool the water in the cooling-vat, draw off a portion of it and add cold water.

I have thus given minute details in regard to making wax sheets, because beginners usually fail on this feature of the work more than in any other.

*Use soft water whenever you can in foundation making.

ROLLING THE WAX SHEETS.

I will presume that you have carried out faithfully the foregoing instructions, and that you have already purchased a foundation-machine. Procure a box or small table about three feet high, and upon this screw down the machine. You will also need two other small tables, one in the rear of the machine and the other in front. The latter is to hold the piles of sheets after they have been embossed on the rolls. The former is to hold a shallow vat for holding the sheets – the latter immersed in three or four inches of water. This vat should be made of tin, long enough to accommodate the length of the sheets, and of suitable width. We find that, when the sheets are taken from lukewarm briny water (110° F or 43° C)), they work much better; indeed, we now regard this tempering of the sheets quite a necessity. In order that you may get a proper idea of the arrangement as above given, I submit the engraving on next page, taken from a photograph, as the two helpers were making foundation.

Rolling out foundation

At the left of lady No. 1 is the oblong shallow vat containing the sheets immersed in tepid water. For the sake of economy of space, and general convenience, we have a couple of tables made exactly right for the purpose. The engraving will make their manner of construction self-evident. We use a similar table for holding the piles of wax sheets after being run through the rolls.

Before proceeding with the operation of rolling, see that the room is properly warmed, say about 80° F (27° C). It has been found by experience that this temperature is best. This is rather too warm to work with comfort; but in making fine quality of foundation, comfort is not to be looked after. Next, you need some sort of lubricant. Various mixtures have been advocated, such as soap made into a lather; a weak solution of lye, obtained from an ordinary ash-leach; a saturated solution of salt and water; a solution of slippery-elm bark; and ordinary starch paste, such as woman use for wall-paper. After testing most thoroughly all of the different ones mentioned, we have decided in favor of the paste, with the addition of a tablespoonful of salt to the pint, as being by far the best. I believe the Dadants use the soap lather; but for some reason or other we have not been able to make it answer as well as the starch paste.

Your enthusiasm may prompt you to run a dry sheet through the rolls, just to "see how it will work." Just as sure as you do, you will find your ardor greatly diminished, for the wax will cling to both rolls, and can be removed only by a method to be described further on. Having prepared your starch paste (and we suppose every woman knows how that is made), add about a tablespoonful of salt to a pint of paste. This should, of course, be added in the preparation of the paste, in order to be quite thoroughly mixed throughout. When cold, fill the tin tray under the roll. Dip your hand into the paste, and rub it over the rolls until they are thoroughly lubricated. If possible they should be warmed to about 95° F (35° C)) in order to work best. Place the mill near the stove for a little while before you expect to use it.

Referring to the engraving again, No. 1 is to feed the sheets and turn the crank. We will suppose that you assume the position of No. 1 while an assistant acts as No. 2. If the end of the sheet is too thick, cut it off with a knife.* Feed the sheet into the mill and turn the crank about half a revolution. Now raise the wooden roller until it is level with the upper metallic roll. The office of this wooden roller is to keep the sheet, after it has passed through the mill, from coming in contact with the lower roll before it should. It also causes the sheet to be fed evenly. As soon as the sheet is run through an inch or so, the end will stick on one of the rolls and must be picked out with a blunt hickory bodkin. A shawl-pin made blunt would be better, but you must be careful not to let it scratch the surface of the rolls. You will find that the first three or four sheets will give you more trouble than those succeeding; and, likewise, that a new mill will give more trouble at first than after you have used it some. After you have loosened the end of the sheet in the manner indicated, No. 2 is to grasp it with the grippers, made as shown in the accompanying engraving. The manner of using them is shown above in the right hand of No. 2.

Referring to the large engraving again, No. 1 rolls out the sheet, and watches carefully to see that no foreign particles adhere, either to the upper or under side of the sheet, such as would damage the surface of the rolls. No. 1 receives the sheet and deposits it on the table at her right.

*The sheets as they leave the dipping-boards are, as a general thing, a little ragged, and sometimes a little thickened at the ends. Instead of trimming each sheet individually before passing it through the mill, take a pile of them and trim all at once, evenly and squarely, with a large butcher-knife, as will be explained presently. Put this pile into the vat of water, and you are ready to roll.

HOW TO ADJUST THE MILL FOR LIGHT AND HEAVY FOUNDATION.

In adjusting the mill from thin to thick foundation, give the adjusting top bolts each an equal turn somewhere about one quarter of a turn up. If the sheets roll bowing on one edge, the rolls are screwed down too much on one side. If you are running on heavy foundation, and desire to turn the mill down to medium, an eighth of a turn will probably be entirely sufficient. Be careful not to screw down the mill too much, or you will bruise the surface of the lozenge faces. If the bottom of the cell is thick on one side, with a screw-driver loosen the screw in the cam one-eighth of a turn, and follow up with the one on the opposite side of the cam which you will find on one end of the top roll. Be sure to oil often.

CAUTION

I have already incidentally remarked in one or two places in regard to the danger of running pieces of metal through the mills. To prevent the occurrence of such accidents, be sure that all nails and pins are kept out of the room. We used to box our wax in the same room where we rolled out the wax sheets. By some means, the nails would get on to the tables by the piles of wax sheets, and we had trouble later. A nail is an innocent looking thing when lying on a table, to be sure; but let some one heedlessly lay a pile of wax sheets on top, and that nail will be sure to imbed itself in the sheet above it. As it will be pretty apt to elude scrutiny, it will be passed through the mill, clinging to the sheet, and the consequence is a big nail-mark on the surface of each roll. After having invested twenty-five or thirty dollars in a foundation-mill, and damaging it, you will find, as Josh Billings says, that "egsperiens keeps a gude skule, but the tuishen is ruther hi." Only one little nail, that's all! We have also had the rolls injured by the bodkin, or little implement used for lifting up the sheets from the rolls. It would be laid carelessly in front of the mill, and, in some strange way, would get imbedded into the sheet, only to repeat the mischief. We now have them suspended by a rubber cord from the ceiling, in such a way as to hang four or five inches above the rolls. When it is necessary to use it, the bodkin can be drawn down. After usage it is let go, when it will draw up out of the way, where it cannot get entangled in the sheets.

HOW TO CLEAN THE FOUNDATION-ROLLS.

Now, after you have been using your comb-mill for a day or so, the rolls will become clogged, or dirty, from small particles of wax collecting in the interstices. The most expeditious way we have found for removing all such particles is to turn a jet of steam upon the rolls for five or ten minutes, or until the rolls feel hot to the hand. While the steam is blowing, the rolls should be turned backward and forward. The action of the steam is to melt the particles of wax, and then blow them off. Next scour with a brush and boiling soapsuds. Where it is not convenient to use steam, a stream of boiling water from a tea-kettle will answer nearly as well as the steam, though it does not do its work as rapidly.

If you do not succeed in making nice foundation, clean the rolls as I have just directed, and you will be surprised at the difference in results. Unless you do keep your rolls clean you will probably become disgusted with the whole business.

MAKING FOUNDATION IN LARGE QUANTITIES.

The foregoing directions in regard to making wax sheets, and passing them through the mill, apply to those who either desire to make foundation for their own use, or to supply a moderate trade which they may have. Where the article is to be made by the ton, the wax should be melted by steam, by means of a series of coiled pipes, or by heating water surrounding the vat of wax. Either plan is very simple; and where large quantities are to be melted, it is by far the best. Steam is not only a great convenience in melting the wax and cleaning the foundation-rolls, but it may be made a very useful servant in turning the rolls themselves. Very recently, comb-foundation machines have been built, to be operated by steam-power. The following engraving illustrates one of these machines.

A power foundation mill

For some time it was a problem as to how these mills could be started instantly and stopped instantly, and yet in no way inconvenience or endanger the operator while manipulating the wax sheets. The problem was successfully solved by means of friction-rollers. The treadle B communicates, as you will notice, with a light iron rod. This operates another lever, A, which in turn operates a friction-pulley. Pressure upon the treadle brings the friction-pulley in contact with the lower pulley, C. The mill can be instantly started or stopped. Before we adopted power attachment, our employees complained a good deal in consequence of the tiresome work of turning the crank on the hand-mills, and we found it necessary to employ a good strong man. Since the adoption of these power-mills, the services of the latter have been entirely dispensed with; and only one woman (rarely two) operates the machine easily alone. Reversal of motion is accomplished, what little there is of it, by hand. The large balance-wheel can be turned backward or forward. When ready to roll, power is applied. The general directions which have been given for the hand-mills will apply to the power-mills.

TRIMMING AND SQUARING THE SHEETS.

As the sheets are taken from the rolls, lay them squarely upon each other until you have a pile 2 or 3 inches high. Now lay on them a board cut the exact size you wish the foundation to be, and with a sharp, thin-bladed butcher or other knife, cut through the whole, all around the board. To prevent the knife from sticking, dip it occasionally in the starch, such as is used in rolling the sheets. To have the knife work nicely, you should have a coarse whetstone nearby, with which to keep the edge keen. As the board is liable to shrink, warp, and get the edges whittled off, where a great number of sheets of a particular size is wanted, we have frames, made sharp on their edges and lined with tin. The tin is folded, and put on so that the knife-edge does not strike it, if the blade is held in the proper position.

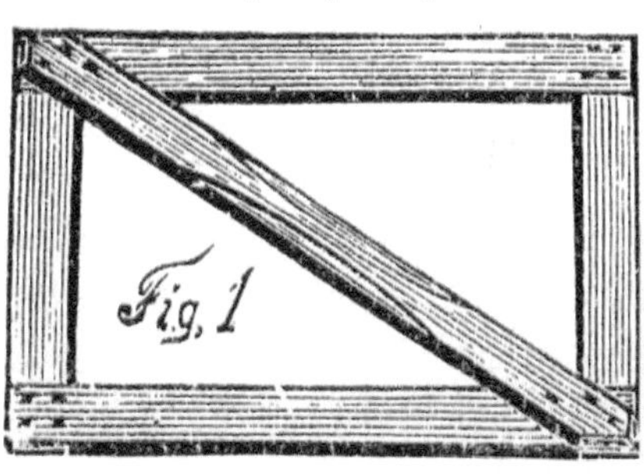

Frames for cutting sheets for brood-frames

To cut the sheets we have frames made as follows:

The diagonal piece in figure 1 serves as a brace to keep it true and square, and also for a handle to lift it by. The frame is placed over the sheet so as to cut to the best advantage, and the knife is run around it.

Figure 1 is for cutting sheets 12 by 18, and figure 2 for the L. frame, 8 by 16-1/2 in. For the wired frames shown on page 65, the sheets are to be cut 8-5/8 X 17-1/8.

For cutting a great number of small pieces, such as starters for sections, a pair of frames like those shown in the engravings below are very convenient.

Fig. 3.

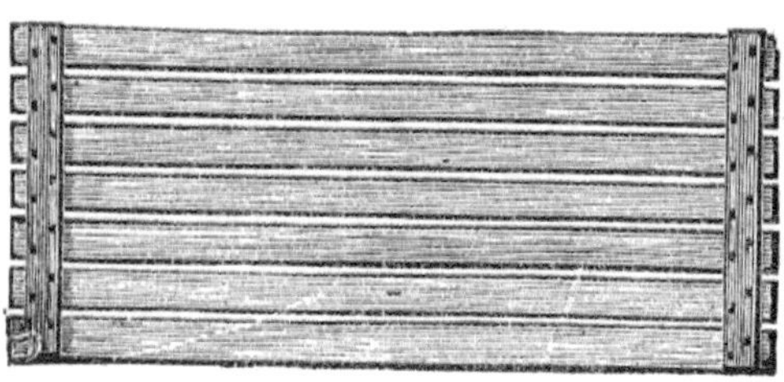

Fig. 4.

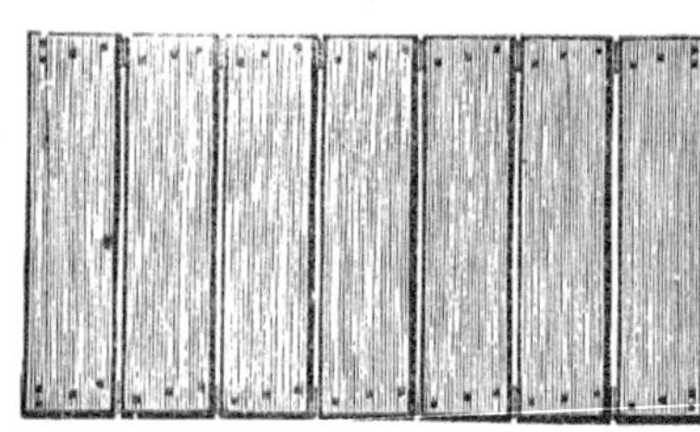

Machine for cutting starters

Fig. 3 is composed of seven 1/4-inch strips, 1-3/4 inches wide, by about 20 inches long. The spaces are just wide enough to allow the knife to run between them. Fig. 4 is composed of the same number of boards, but they are 3-5/8 wide, by about 16 long. You will observe that this allows one frame to be placed over the other, each fitting in between the cleats of the other. To use the machine, place a sheet (or sheets) of foundation, say 12 by 18, on Fig. 3, and lay Fig. 4 over it. Run the knife through all the spaces, and then turn the whole machine over. Now run it through as before, and your sheet is cut into oblong pieces, just such as we put in the 4-1/4 section boxes when we ship them in hives complete. We should, perhaps, use pieces somewhat larger, were it not that there would be greater danger of their breaking out with the rough handling they get when the hives are sent by freight. The pieces, as made with the above frames, are 1-3/4 by 3-5/8 inches.* If much work is to be done with these frames, they had better be covered with tin, like frames before mentioned.

*Nearly all our prominent honey-producers, however, are strongly in favor of having the thin foundation entirely fill the sections; and for the one-pound sections, they are cut 3-3/4 X 3-3/4, made of foundation with the base about as thin as natural comb. To make starters this size the slats in both Fig. 3 and Fig. 4 should be 3-3/4 in. wide.

FOUNDATION FOR COMB HONEY.

The only trouble with it for comb honey is that, under some circumstances occurring very rarely I believe, the bees will build on to the foundation, without thinning the center at all, as they usually do. I believe this is more apt to occur when a good yield of honey comes during rather cool weather, the bees being unable to get the wax warm enough to work readily. The remedy for this will be in making the base of the cells of the foundation exceedingly thin, and the small 6-inch machines seem best for this purpose. We have made machines for making the foundation four, four and a half, and five cells to the inch. The latter is intended to be used in brood-rearing, unless, per-chance one may desire to rear drones. In that case, four cells to the inch should be used. As the queens are not as apt to deposit eggs in drone-cells, it was once thought that drone foundation would be more desirable in the surplus-apartment. But notwithstanding this, more recently a decided preference has been shown for thin worker foundation (five cells to the inch).

In order to get nice thin foundation, the rolls should be screwed down as closely as they may be (according to directions already given), so as to get the base of the cells nearly if not quite as thin as the natural base. If it is made a little too thick, the base is very easily detected in the comb honey, and has been called, not inappropriately, "fishbone."

Flat-bottom foundation has been made, which some think is the best surplus foundation. It is nothing but a sheet of wax, embossed with hexagonal cells inclosing a flat base. While it makes very nice comb honey, yet the testimony of many of those who have tried it is to the effect that it is not readily accepted by the bees, and consequently valuable time is lost. We do know this much, that they remodel and rebuild the cells before drawing them out. Notwithstanding this, there are two or three large honey-producers in the State of New York who consider it the best surplus foundation Mr. P. H. Elwood, of Starkville, N. Y., an extensive bee-keeper of large experience, among the number. There are other New York bee-keepers who think as he does.

SAGGING OF THE FOUNDATION, AND HOW TO PREVENT IT.

Many devices have been tried to prevent the sagging of the foundation, and consequently slight elongation of the cells, in the upper part of the comb. With the L. frames, this is so slight that it occasions no serious trouble with the greater part of the wax of commerce; but with deeper frames, or with some specimens of natural wax, the sagging is sufficient to allow the bees to raise drones in the upper cells. Paper has been tried, and succeeds beautifully, while the bees are getting honey; but during dearth, when they have nothing to do, they are liable at any time to tear the nice combs all to bits, to get out the paper, which I have supposed they imagine to be the web of the moth-worm. In our apiary I have beautiful combs built on thin wood; but as the bottom of the cell is flat, they are compelled to use wax to fill out the interstices, and the value of this surplus wax, it seems to me, throws the wood base entirely out of the question. I do not like the foundation with wire rolled in it, on account of the greater expense, and because we cannot fasten it in the frames as securely as we can where the wires are first sewed through the frames.

Aside from the avoidance of drone-cells, we want combs that will not break out of the frames in shipping, handling, or extracting, in either hot or cold weather; we also want frames that will not sag in the middle, no matter how heavily they may be filled with honey.

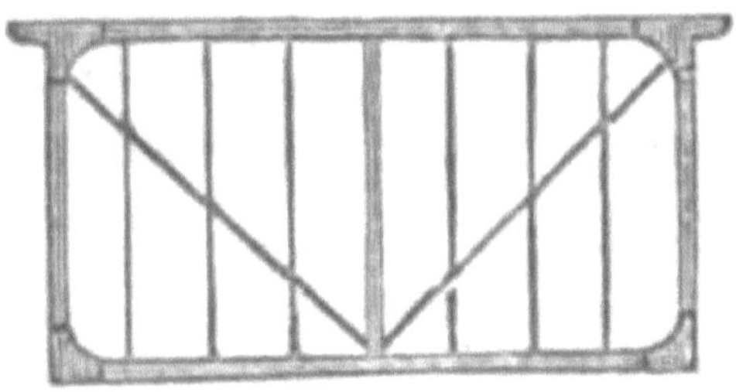

L. Frame wired ready for use

For several years we wired all our combs as shown in the accompanying engraving. The top and bottom bars were pierced at regular distances, through which the wire was threaded back and forth. If a thin top-bar that is, one not more than 1/4 inch is used, a folded tin bar will be necessary.

Latterly we have employed the method shown below, and it is what we call the Keeney plan. Perpendicular wiring is apt to bow up the bottom-bar if the wires are drawn tight, and to pull the top-bar down if it is not thicker than 1/4. True, we can avoid that by the use of folded tin bars, but bees seldom build over them nicely. The Keeney method of wiring takes less wire and less time, and it brings the entire strain upon the four corners of the frame the point where there is the greatest strength. No piercing of top-bars or bottom-bars is necessary. A 1-1/4-inch wire nail is driven through the end-bars 3/4 of an inch from the top and bottom bars. They are then bent into the form of a hook by means of round-nosed pliers. To do this rapidly, string a lot of frames over a narrow board, so that the end-bars will lie in contact side by side, and then support the two projecting ends of the boards. With a straight-edge and pencil draw a line 3/4 inch from the top-bars, and then a line 3/4 inch from the bottom-bars. This gives you the location for each wire nail as regards the top and bottom bars. Before taking the frames off the board, drive the nails in. Then slide them off en masse, and afterward bend the points, as shown in the accompanying engraving. Cut your wire 69 inches long. Twist a loop in one end; catch the wire over hook No. 1, and pass successively to hooks 2, 3, 4, and back to 1; then draw. Next pass the wire under the wire at 5, catch over the hook at 2, draw the wire taut, and fasten by twisting.

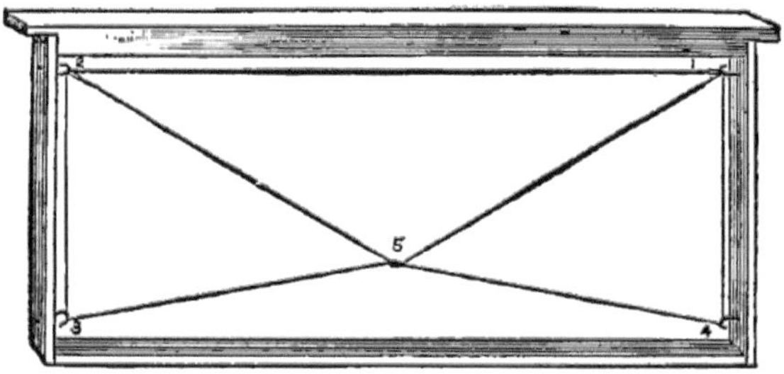

Keeney's method of wiring, improved

To get your wire the right length, wind it over a long board 5 or 6 inches wide, and rounded at the end to a feather edge. The length of this board should be just half the length of the wire you use; namely, for the L. frame, 34-1/2 inches. After you have wound the whole coil of wire on this board from end to end, take an old pair of shears and cut all the strands in two, right where they bend over the end; and to keep them from flying all over when cut, slip a couple of rubber bands over each end of the board. Now, when you are ready to wire, just simply pull the wire out from one end.

This method of wiring is very expeditions and satisfactory for the ordinary bee-keeper. It is not as substantial as the perpendicular-wiring plan, but enough so for practical purposes. The two perpendicular wires, 2 and 3, 1 and 4, hold the ends of the foundation from flopping out of position. The horizontal wires, 1 and 2, hold the top, also, permanent.

The wire used is No. 30, tinned iron wire. After the wires are in and drawn up tight, the foundation is cut so as to fill the frame, and the wires are then imbedded into the wax by means of one of the various devices for that purpose. During this operation the foundation is supported on a level board cut so as to just slip inside the frame, and come up against the wires. The board is to be kept wet with a damp cloth, to prevent the wax sticking to it.

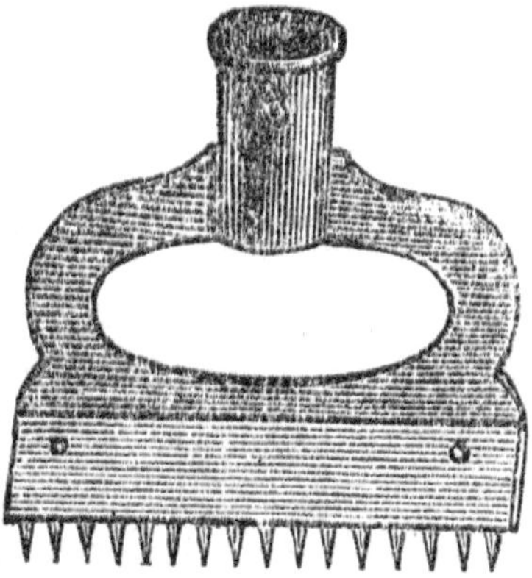

Carpet-stretcher

A common carpet stretcher, like the cut above, is fitted with a short handle, and then the wax is warmed up so as to be quite soft. The wires are imbedded by laying the points along the wire, and pressing down while the foundation is supported by a board in the manner already given. By the use of the carpet-stretcher, the bees finish out the cells as perfectly as if nothing of the kind had ever touched them.

In putting in foundation on the Keeney plan, slip the top edge up in the groove where the comb-guide would go if the frame were not wired. Then imbed the wires in the foundation.

Easterday's foundation-fastener

Still later, the implement figured in the cut below has found favor, and our girls now consider it quicker and easier to use than any other thing heretofore tried. You see, the points strike one at a time, therefore no very great pressure is needed; and yet by rocking the implement the work is done very rapidly.

This press has found considerable favor with a few. With a pair of dies just the size of the inside of the frame, plain sheets of wax are made into foundation, and the wires imbedded into it at one and the same operation. The objections to it are, the price is much more than the price of rolls; that it makes sheets of only one size; that the wire used for it must be considerably finer than No. 30. No. 36, I believe, is generally used, and this we find too frail for our use, shipping bees, etc. As yet, I believe it does not put foundation into wired frames so that they will bear shipment, while that put in by hand can be shipped safely anywhere during warm weather. Neither is it adapted to making sheets of foundation that entirely fill the frames; and I should always want the sheets to come clear up to the wood on all sides.

Given foundation press

FASTENING STARTERS IN SECTION BOXES.

For this purpose the foundation is made in narrow strips, as has been before explained. For the one-pound section we have dipping-boards 3-3/4 inches wide; and after being rolled, they are then cut up into pieces that nearly fill the sections, or as much less as the taste or purse of the bee-keeper demands. The pieces are fastened only to the top-bar of the section, and this is done by either of the accompanying machines shown.

STARTERS FOR SECTION BOXES.

Many bee-keepers want the starter to fill the section as nearly as possible, leaving a space of only 1/4 or 3/8 inch at the sides and bottom. Even with so large a starter as this, the bees sometimes fail to fasten the comb at the sides and bottom. It is especially desirable to have it fastened at the bottom, to prevent breaking out in shipping; but even if long enough to touch the bottom, the bees do not always finish it down. Perhaps a safer way is to fasten a starter at the bottom, 3/4 inch wide or deep; then fasten at the top a 3-1/4 inches deep. This makes a sure thing of having the comb fastened to the bottom-bar. Such starters properly fastened with a Clark fastener have been safely hauled on the trot to an out-apiary. If cut 3-7/8 instead of 3-1/4, the swing, and the consequent liability to fall out, would be much greater. The idea is, to rub or press a thin edge of the wax into the dry wood of the section. The motion of the machine spreads the wax down, and mashes it into the wood, as it were. Below is the Parker machine, which is used quite largely; in fact, many thousands of them have been sold. It does very nice work; but where thousands of starters are to be put in, it becomes a little tiresome on the hands.

The one next illustrated is what is called Clark's starter machine. Instead of rubbing the foundation it presses it into the wood. Pressure is exerted entirely by the foot. This not only gives more power, but it leaves both hands free to pick up the sections, adjust the foundation, and, after fastening, remove them.

Parker machine for fastening starters in sections

To operate, screw it down to a bench or table, so that the treadle just clears the floor. Make a little paddle, say 8 or 10 inches long, 1/4 inch thick, and 1 to 1-1/2 inches wide. Nail upon one side of it a piece of felt, or two or three thicknesses of old soft cloth, equal to the length of the presser-tongue, then whittle off the handle end, saturate the cushioned part well with salt water, renewing it if it should get dry. To moisten the tongue, lay your paddle under it, press with the feet just as when fastening in a starter, and then throw the paddle in your lap till needed again. This takes less time, and is more thorough, than to use the brush. You may need to moisten the tongue for each starter, or you may need it only after fastening several starters. It is a good plan to have a little tin dish of salt water in which the tongue may be so set as to keep in soak overnight, so as to be in good trim for next day's work. With one hand pick up a section, and with the other put the foundation in position, directly under the tongue. Bring the latter down with the feet, and let the feet come back with a rebound, and the whole performance is quickly and easily done. If the presser-tongue is so sharp at the edge that it cuts off the foundation, round it off a little with sand-paper. For the first few trials, the wax may stick to the tongue rather than to the section. Scrape the former off smooth with a knife; wet it thoroughly with water or paste. The foundation, before insertion, should be warmed up to a temperature of about 110°. If the sheets are put in the direct rays of the sun, shining through a window, they will be soft enough. Some prefer to put the foundation in piles of perhaps 50, and then heat only one edge by means of a hot brick or a body of water in some kind of vessel kept heated by a lamp. Foundation must be tolerably soft or it will not stick firmly to the sections. This is the machine that is recommended and used by Dr. Miller, referred to elsewhere in this work.

Clark's starter machine

American Bee Journal June 1923 (Vol 63 No 6) starting on page 299

IS THE HONEYBEE NATIVE OF AMERICA?

A Discourse Intended to Commemorate the Discovery of America by Christopher Columbus.

By Jeremy Belknap.

Delivered at the request of the Historical Society of Massachusetts on the 23rd of October, 1792

Dissertation No. 3, on the question whether the honeybee is a native of America.

Mr. Jefferson, in his notes on Virginia, has said that "The honeybee is not a native of our continent. The Indians concur with us in the tradition that it was brought from Europe, but when and by whom we know not. The bees have generally extended themselves into the country, a little in advance of the white settlers. The Indians called them the white man's fly; and consider their approach as indicating the approach of the settlement of the whites." He allows that "in Brazil there is a species of honeybee without a sting, but that is very different from the one we have, which perfectly resembles that of Europe." The facts adduced by the respectable author are true; but they will not warrant his conclusion that "the honeybee, meaning the one resembling that of Europe, is not a native of our continent."

There is one circumstance in the history of Columbus which proves that bees were known in the islands of the West Indies, at the time of his discovery. When on his first return to Europe he was in danger of perishing at sea, he wrote an account of his discovery on parchment, which he inclosed in a cake of wax, and put into a tight cask, committing the whole to the sea, in hope of it's being driven on shore or taken up. This was procured in the island of Hispaniola, which he had visited, and it was one of the first fruits of his discovery.

The indefatigable Purchas gives us an account of the revenues of the Empire of Mexico, before the arrival of the Spaniards, as described in its annals; which were pictures drawn on cotton cloth. Among other articles he exhibits the figures of covered pots with two handles, which are said to be pots of "bees' honey." Of these pots, two hundred are depicted in one tribute-roll, and one hundred in several others.

This account is confirmed by the late history of Mexico, written by the Abbe Clavigero, a native of Vera Cruz who from a residence of thirty-six years in Mexico, and a minute inquiry into the natural history and antiquities of his country must be supposed to be well informed, and competent to give a just account. He tells us that a part of every useful production of nature or art was paid in tribute to the kings of Mexico, and among other articles of revenue he reckons "600 cups of honey" paid annually by the inhabitants of the southern part of the empire. He also says, "that though they extracted a great quantity of wax from the honeycomb, they either did not know how or were not at pains to make lights of it."

In his enumeration of the insects of Mexico, he reckons six different kinds of bees which make honey, four of which have no stings, and one of the other two which have stings, one "agrees with the common bee of Europe, not only in size, shape and color, but also disposition and manners, and in qualities of its honey and wax."

In the account given by Purchas, of the travels of Ferdinado de Soto, in Florida, it is observed that when he came to Chiaha, which by the description was one of the upper branches of the Mobile (now in the State of Georgia) he found among the provisions of the natives "a pot full of honey of bees." This was A.D. 1540, when there were no Europeans settled on the continent of America, but in Mexico and Peru.

From these authorities it is evident that honeybees were known in Mexico and the islands, before the arrival of the Europeans; and that they had extended as far northward as Florida, a country so denominated from the numberless flowers, which grow there in the wild luxuriance and afford a plenty of food for this useful tribe of insects. The inference is, that bees were not imported by the Spaniards; for however fond they might be of honey as an article of food, or of wax to make tapers for common use, or for the illumination of their churches, yet as bees were known to be in the country there could be no need of importing them. The report of honey and wax being found in the islands, in Mexico, and in Florida, had reached Europe and had been published there long before any emigrations were made to the northward; therefore, if these had been considered as articles of subsistence or of commerce, the sanguine spirit of the first adventurers would have rather led them to think of finding them in America, than of transporting bees from Europe to make them.

As to the circumstance of the bees "extending themselves a little in advance of white settlers," it cannot be considered as a conclusive argument in favor or their having been first brought from Europe. It is well known that where land is cultivated bees find a greater plenty of food than in the forest. The blossoms of fruit trees, of grasses and grain, particularly clover and buckwheat, afford them a rich and plentiful repast, and they are seen in vast numbers in our fields and orchards at the season of those blossoms. They therefore delight in the neighborhood of "the white settlers", and are able to increase in numbers, as well as to augment their quantity of stores, by availing themselves of the labors of man. May it not be from this circumstance that the Indians have given them the name of "the white man's fly"; and that they "consider their approach (or frequent appearance) as indicating the approach of the settlement of the whites?"

The first European settlement in Virginia was made about seventy years after the expedition of De Soto, in Florida, and the first settlement in New England was ten years posterior to that of Virginia. The large intermediate country was uncultivated for a long time afterward. The southern bees, therefore, could have no inducement to extend themselves very far into the northward for many years after the settlements were begun, and within that time bees were imported from Europe.

That honey and wax were not known to the Indians of New England is evident from this, that they had no words in their language for them. When Mr. Eliot translated the Bible into the Indian language, wherever these terms occurred he used the English words, though sometimes with Indian termination.

Joffelyn, who visited New England first in 1638, and afterward in 1663, and wrote an account of his voyage with some sketches of natural history in 1673, speaks of the honeybee in these words: "the honeybees are carried over by the English, and thrive there exceedingly."

There is a tradition in New England that the person who first brought a hive of bees into the country was rewarded with a grant of land; but the person's name, or the place where the land lay or by whom the grant was made, I have not been able to learn.

It appears then that the honeybee is a native of America, and that its productions were found by the first European visitors as far northward as Florida and Georgia. It is also true that bees were imported from Europe into New England, and probably into Virginia; but whether if this importation had not taken place, the bees of the southern parts would not have extended themselves northerly, or whether those which we now have are not a mixture of native and imported bees, cannot be determined. It is however certain that they have multiplied exceedingly, and that they are frequently found in New England, in a wild state, in the trunks of hollow trees, as far northward as cultivation and settlements have extended, which is nearly to the 45th degree of latitude.

I have made an inquiry of several persons from Canada, but have not learned that bees were known during their residence in that country. It is, however, not improbable that as cultivation extends, the bees may find their way to the northward of the lakes and rivers of Canada, even though none should be transported thither by the inhabitants.

WAS THERE A NATIVE HONEY_BEE?

By Frank C. Pellett.

We are indebted to our good friend Harold L. Kelly of Washington D.C. for the above article written by Mr. Belknap in 1792. Mr. Kelly found this among some old writings on bees in the Library of Congress. Written, as it was, so long ago, it is of much historical interest, although not altogether convincing to one who is familiar with the bees of the tropics.

The first statement to the effect that Columbus enclosed an account of his discovery of a cake of wax is not conclusive for the reason that the wax could easily have been obtained from the stingless bees.

The Mexican bees are said to be of six different kinds, of which four have no stings and the other two have stings. The stingless bees, of course, could not be mistaken for the honey-bee. One of the stinging kind was very probably the honey-making wasp, which was described by the writer in this Journal in January, 1921.

Since the description which was supposed to refer to the honey-bee was written long after the Spaniards had settled in Mexico it does not follow that it was a native species to which the writer referred.

The reference to the "pot full of honie of bees" found in what is now Georgia is more convincing, but even this might easily refer to the product of the bumblebee, which produces a small amount of honey. The natives were often attracted by food in small quantity and travelers finding them eating honey might mention the fact without calling attention to the source.

The fact that the honeybee extended its range so rapidly and in advance of settlement indicates that it was an introduced species. This is not absolute, however, since a change in conditions often affects the spread of a native species. The breaking up of the prairies caused the Colorado potato beetle, a native insect, to change its food plant from buffalo burr to the potato, and then to spread all over the continent. This change however, came with the advance of settlement and not ahead of it. It is possible, of course, that there was a native honeybee confined to a limited area which remained in its original habitat and that it was the European species which did in fact spread over the country following its introduction.

There is a persistent opinion that the honeybee was native to America although no proof of the fact has yet been brought forward. We are much interested in establishing the fact or definitely proving to the contrary and appreciate such information as the above which Mr. Kelly has found. Through his kindness we are able also to republish, for our readers, notes concerning bees in Mexico to which Mr. Belknap refers.

EXCERPT FROM THE HISTORY OF MEXICO

By Abbe D. Francesco Saverio Clavigero (1731-1787)

Translated from the original Italian in 1806 by Chas Cullen, Esq.

Excerpt from Book 1, of Volume 1.

Bees

There are at least six different kinds of bees. The first is the same as the common bee of Europe, with which it agrees, not only in size, shape and color, but also in its disposition and manners, and in the qualities of its honey and wax.

The second species which differs from the first only in having no sting, is the bee of Yucatan and Chiapa, which makes the fine, clear honey of Estabentun, of an aromatic flavor, superior to that of all other kinds of honey with which we are acquainted. The honey is taken from them six times a year, that is once in every other month; but the best is that which is got in November, being made from a white flower like Jessamine, which blooms in September, called in that country Estabentun, from which the honey has derived its name. The honey of Estabentun is in high estimation with the English and French, who touch at the ports of Yucatan; and I have known the

French of Buarico to buy it sometimes for the purpose of sending it as a present to the king.

The third species resembles in its form, the winged ants, but is smaller than the common bee, and without a sting. This insect, which is peculiar to warm and temperate climates, forms nests, in size and shape resembling sugar loaves, and even sometimes greatly exceeding these in size, from trees, and particularly from the oak. The populousness of these hives is much greater than those of the common bee. The nymphs of this bee, which are eatable, are white and round, like a pearl. The honey is of a grayish color, but of a fine flavor.

The fourth species is a yellow bee, smaller than the common one, but like it, furnished with a sting. Its honey is not equal to those already mentioned.

The fifth is a small bee furnished with a sting which constructs its hives of an orbicular form. In subterranean cavities; and the honey is sour and somewhat bitter.

The Tlalpiprolli, which is the sixth species, is black and yellow of the size of the common bee, but has no sting.

Wasp

The Xicotli or Xicote, is a thick black wasp, with a yellow belly, which makes a very sweet honey, in holes made by it in walls. It is provided with a strong sting, which gives a very painful wound. The cuicalmiahautl has likewise a sting, but whether it makes honey or not, we do not know.

Other References

At one time all of these were available for free online. An internet search on the title and authors should be sufficient to find them if they are still available.

Reproduction of Varroa destructor in South African honey bees: does cell space influence Varroa male Survivorship, Stephen J. MARTIN, Per KRYGER

The influence of brood comb cell size on the reproductive behavior of the ectoparasitic mite Varroa destructor in Africanized honey bee colonies Giancarlo A. Piccirillo1 and D. De Jong

Survival of a Commercial Beekeeper in Norway, Hans Otto Johnsen

Trial of HoneySuperCell® Small Cell Combs, Randy Oliver

Effects of Comb Cell Diameter on Parasitic Mite Infestations in Honey Bee Colonies, Eric Erickson Jr.

Life span of worker honeybees reared in colonies kept on small-cell combs, Krzysztof Olszewski, Grzegorz Borsuk, Jerzy Paleolog, Aneta Strache

A model of the mite parasite, Varroa destructor, on honeybees (Apis mellifera) to investigate parameters important to mite population growth, D Wilkinson, G.C Smith

Analysis of small cell test designs, Erik Österlund

About the Authors

Ed and Dee are commercial beekeepers in AZ. Dee and Ed work side by side in all phases of their operation. In addition to the conventional activities of beekeeping, they mill their own woodenware and wax foundation, select and maintain a stock of slightly smaller bees highly adapted to their area, and produce their own queens.

Their non-chemical 'back to basics' approach to beekeeping leads them to spend much of their spare time in libraries where they search for obscure bits of information which, when assembled in logical order, yield insights into old problems such as bee kills due to the use of pesticides, and new problems like parasitic mites. Such has been their pursuit of an understanding of the importance of comb cell diameter

The Lusbys found that comb cell diameter differs among the various sources of foundation manufactured in the United States and around the world. Following publication of this discovery in 1990, they undertook to resolve the question of natural cell diameter and its impact on colony vigor. Having identified, optimal cell diameter for their geographic area (Southern Arizona), they have nearly completed converting their operation to a 'natural system' incorporating smaller cells. They have widely reported to beekeepers that their use of optimal natural cell diameter has significantly reduced disease and mite infestation in their colonies while simultaneously increasing brood viability and colony productivity. Convinced, a number of beekeepers have embraced the Lusby's management strategies.– American Bee Journal

Note: Ed died Wed, 25th of October 2006

www.ingramcontent.com/pod-product-compliance
Ingram Content Group UK Ltd.
Pitfield, Milton Keynes, MK11 3LW, UK
UKHW041851031026
15587UKWH00005B/1153/J

9 781614 761044